W9-AUP-231
Firetech Consulting
720-981-3413

# Handbook for Electrical Safety in the Workplace

# Handbook for Electrical Safety in the Workplace

Edited by

Ray A. Jones Kenneth G. Mastrullo Jane G. Jones

With the complete text of the 2004 edition of NFPA 70E, *Standard for Electrical Safety in the Workplace*

NFPA® National Fire Protection Association
Quincy, Massachusetts

Product Manager: Charles Durang
Project Editor: Irene Herlihy
Copy Editor: Joyce Grandy
Text Processing: Maureen White
Composition: Omegatype Typography, Inc.
Art Coordinator: Cheryl Langway
Illustrations: Rolin Graphics and George Nichols, with contributions by J. Philip Simmons
Interior Design: Cheryl Langway
Cover Design: Cameron, Inc.
Manufacturing Manager: Ellen Glisker
Printer: R.R. Donnelley/Willard

Copyright © 2004
National Fire Protection Association, Inc.
One Batterymarch Park
Quincy, Massachusetts 02169-7471

All rights reserved. No part of the material protected by this copyright notice may be reproduced or utilized in any form without acknowledgment of the copyright owner nor may it be used in any form for resale without written permission from the copyright owner.

**Notice Concerning Liability:** Publication of this handbook is for the purpose of circulating information and opinion among those concerned for fire and electrical safety and related subjects. While every effort has been made to achieve a work of high quality, neither the NFPA nor the contributors to this handbook guarantee the accuracy or completeness of or assume any liability in connection with the information and opinions contained in this handbook. The NFPA and the contributors shall in no event be liable for any personal injury, property, or other damages of any nature whatsoever, whether special, indirect, consequential, or compensatory, directly or indirectly resulting from the publication, use of, or reliance upon this handbook.

This handbook is published with the understanding that the NFPA and the contributors to this handbook are supplying information and opinion but are not attempting to render engineering or other professional services. If such services are required, the assistance of an appropriate professional should be sought.

**Notice Concerning Code Interpretations:** This 2004 edition of *Handbook for Electrical Safety in the Workplace* is based on the 2004 edition of NFPA 70E, *Standard for Electrical Safety in the Workplace.* All NFPA codes, standards, recommended practices, and guides are developed in accordance with the published procedures of the NFPA by technical committees comprised of volunteers drawn from a broad array of relevant interests. The handbook contains the complete text of NFPA 70E and any applicable Formal Interpretations issued by the Association. These documents are accompanied by explanatory commentary and other supplementary materials.

The commentary and supplementary materials in this handbook are not a part of the Standard and do not constitute Formal Interpretations of the NFPA (which can be obtained only through requests processed by the responsible technical committees in accordance with the published procedures of the NFPA). The commentary and supplementary materials, therefore, solely reflect the personal opinions of the editor or other contributors and do not necessarily represent the official position of the NFPA or its technical committees.

The following are registered trademarks of the National Fire Protection Association:

*National Electrical Code®* and *NEC®*

NFPA No.: 70EHB04
ISBN (book): 0-87765-581-2
ISBN (CD): 0-87765-693-2
Library of Congress Card Control No.: 2004112802

Printed in the United States of America
05 06 07 08 5 4 3

# Dedication

Working on a document as special as NFPA 70E, *Standard for Electrical Safety in the Workplace,* is a heavy responsibility. The individuals who serve on the technical committee attend long meetings, often have to "hang in" when the debates don't go their way, and always come away with tired brains and bodies. Most of the men and women, members and alternates, who serve on the committee get to know one another as they go through a lot together and wrestle with important safety issues. In time, the group becomes a true brother/sisterhood. To a person, members and alternates of the technical committee feel passionate about the issues involved, and they are especially passionate about preventing electrical incidents—saving limbs and lives of workers.

In this business, we often talk about "near misses"—incidents or injuries that almost occurred but didn't quite happen. We can be grateful that those involved in the near misses are not recorded as injury statistics, which means that some folks escaped serious consequences. The labor of the individuals on the NFPA 70E Technical Committee enables many workers to walk away unhurt. Those workers may never even realize the danger they could have been in, thanks to the procedures put in place as required by NFPA 70E. Therefore, no statistics appear on how many lives were saved. All those volunteers who have served on the NFPA 70E Technical Committee may not know the exact numbers or statistics, but they do know the importance of their work. It is to those people that this handbook is dedicated. We appreciate your efforts, and we salute you.

Ray A. Jones

Ken Mastrullo

Jane G. Jones

# Contents

# Preface

The *Handbook for Electrical Safety in the Workplace* contains the text of NFPA 70E, *Standard for Electrical Safety in the Workplace*, 2004 edition, in its entirety. Explanatory commentary immediately follows the text of the standard and is set in blue type for easy identification. The commentary in the handbook is offered only as guidance to the user and is therefore written in nonmandatory language. The commentary was written by the editors to help the user understand how to apply the requirements of NFPA 70E. It is not a substitute for the actual wording of the standard or the text of the codes and standards that are incorporated by reference. Illustrations and other graphics are included where they were deemed helpful.

## NFPA 70E

The 2004 edition of NFPA 70E, *Standard for Electrical Safety in the Workplace,* was prepared by the Technical Committee on Electrical Safety in the Workplace and acted on by the National Fire Protection Association at its November Meeting held November 17–19, 2003, in Reno, Nevada. This seventh edition was issued by the Standards Council on January 14, 2004, with an effective date of February 11, 2004, and supersedes all previous editions. It was approved as an American National Standard on February 11, 2004.

The seventh edition of NFPA 70E reflects several significant changes to the document. The major changes emphasize safe work practices and also enhance the clarity and usability of the document. The name of the document was changed to NFPA 70E, *Standard for Electrical Safety in the Workplace.* The entire document was reformatted to comply with the *NEC® Style Manual,* thus providing a unique designation for each requirement. The existing parts were renamed as chapters and were reorganized to emphasize safety-related work practices by moving that section to the front of the document.

Chapter 1, which addresses safety-related work practices, also was reorganized to emphasize working on live parts as the last alternative work practice. The document also incorporates an energized electrical work permit and related requirements. Several definitions were modified or added to enhance the usability of the document, and Chapter 4 was updated to correlate with the 2002 edition of the *National Electrical Code®(NEC®).*

## Chapter 4 of NFPA 70E

Essential to the proper use of Chapter 4 is the understanding that it is not intended to be applied as a design, installation, modification, or construction standard for an electrical installation or system. Its content has been intentionally limited in comparison to the content of the *NEC* to apply to an electrical installation or system as part of an employee's workplace. This standard is compatible with corresponding provisions of the *NEC,* but it is not intended to, nor can it be, used in lieu of the *NEC.*

Chapter 4 of NFPA 70E was intended to serve a very specific need of OSHA. Omission of any requirements presently in the *NEC* does not affect the *NEC,* nor should these omitted requirements be considered as unimportant. They are essential to the *NEC* and its intended application, which is to be used by those who design, install, and inspect electrical installations. NFPA 70E, on the other hand, is intended to be used by employers, employees, and OSHA.

# NFPA 70E,
# *Standard for Electrical Safety in the Workplace,*
# 2004 Edition, and Commentary

# ARTICLE 90

# Introduction

The scope of NFPA 70E, *Standard for Electrical Safety in the Workplace,* is similar to the scope of the 2002 edition of the *National Electrical Code*® (NFPA 70®, *NEC*®, or *Code*). NFPA 70E is intended to be in harmony with the *NEC.* At the request of OSHA, NFPA chartered the NFPA 70E Technical Committee and assigned to it the responsibility of defining safe work practices for use in general industry.

A subtle but important difference between the *NEC* scope and the NFPA 70E scope is that the *NEC* applies to *installations,* whereas NFPA 70E applies to *workplaces.* To ensure consistency between requirements contained in the *NEC* and NFPA 70E, all *NEC* panels and the NFPA 70E Technical Committee report to the Association through the same Technical Correlating Committee.

This concept is essential to provide consistency between installation rules and safety-related work practice requirements for workers. *NEC* requirements that are associated with personal safety are extracted and placed in NFPA 70E. The content of NFPA 70E was the basis for requirements contained in OSHA 29 CFR 1910, Subpart S. The requirements defined in NFPA 70E are consistent with 29 CFR 1926, Subpart K.

### 90.1 Scope

**(A) Covered.** This standard addresses those electrical safety requirements for employee workplaces that are necessary for the practical safeguarding of employees in their pursuit of gainful employment. This standard covers the installation of electric conductors, electric equipment, signaling and communications conductors and equipment, and raceways for the following:

(1) Public and private premises, including buildings, structures, mobile homes, recreational vehicles, and floating buildings
(2) Yards, lots, parking lots, carnivals, and industrial substations

FPN: For additional information concerning such installations in an industrial or multibuilding complex, see ANSI C2–2002, *National Electrical Safety Code.*

The basic intent of NFPA 70E is to define requirements for work practices that workers should use to avoid injury from a release of electrical energy when performing a work task(s). Defined requirements are based on exposure to an electrical hazard and apply to all types of employers or facilities, including in-house employees, contractor employees, general industrial workplaces, and construction workplaces.

(3) Installations of conductors and equipment that connect to the supply of electricity

The requirements of this standard apply to all electrical conductors that are installed at a workplace, including those that connect with the utility supply.

(4) Installations used by the electric utility, such as office buildings, warehouses, garages, machine shops, and recreational buildings, that are not an integral part of a generating plant, substation, or control center

Workplaces that are owned and operated by an electrical utility but are not associated with generating, transmitting, or distributing electricity are covered by NFPA 70E.

**(B) Not Covered.** This standard does not cover the following:

(1) Installations in ships, watercraft other than floating buildings, railway rolling stock, aircraft, or automotive vehicles other than mobile homes and recreational vehicles
(2) Installations underground in mines and self-propelled mobile surface mining machinery and its attendant electrical trailing cable
(3) Installations of railways for generation, transformation, transmission, or distribution of power used exclusively for operation of rolling stock or installations used exclusively for signaling and communications purposes
(4) Installations of communications equipment under the exclusive control of communications utilities located outdoors or in building spaces used exclusively for such installations
(5) Installations under the exclusive control of an electric utility where such installations:
   a. Consist of service drops or service laterals, and associated metering, or
   b. Are located in legally established easements, rights-of-way, or by other agreements either designated by or recognized by public service commissions, utility commissions, or other regulatory agencies having jurisdiction for such installations, or
   c. Are on property owned or leased by the electric utility for the purpose of communications, metering, generation, control, transformation, transmission, or distribution of electric energy.

The *National Electrical Code* defines requirements for installations. NFPA 70E, on the other hand, applies to workplaces, not to installations. An *installation* involves equipment, devices, and materials, but a *workplace* involves a worker interacting with equipment, devices, and materials.

### 90.2 Organization

This standard is divided into the following four chapters and thirteen annexes:

(1) Chapter 1, Safety-Related Work Practices
(2) Chapter 2, Safety-Related Maintenance Requirements
(3) Chapter 3, Safety Requirements for Special Equipment
(4) Chapter 4, Installation Safety Requirements
(5) Annex A, Referenced Publications
(6) Annex B, Informational Publications
(7) Annex C, Limits of Approach
(8) Annex D, Sample Calculation of Flash Protection Boundary
(9) Annex E, Electrical Safety Program
(10) Annex F, Hazard/Risk Evaluation Procedure
(11) Annex G, Sample Lockout/Tagout Procedure
(12) Annex H, Simplified, Two-Category, Flame-Resistant (FR) Clothing System
(13) Annex I, Job Briefing and Planning Checklist
(14) Annex J, Energized Electrical Work Permit

(15) Annex K, General Categories of Electrical Hazards
(16) Annex L, Typical Application of Safeguards in the Cell Line Working Zone
(17) Annex M, Cross-Reference Tables

The 2004 edition of NFPA 70E has been reorganized to enhance its user-friendliness. The 2000 edition consisted of four parts, each containing chapters, a format that created redundant numbering. The Standards Council approved a proposal to reformat the standard to conform to the *NEC Style Manual.* This format provides two benefits. First, the document embraces the premise that many of the users of NFPA 70E also are familiar with the *NEC.* Second, it creates a unique designation for every topic in the document.

The 70E Technical Committee voted to change the document title to *Standard for Electrical Safety in the Workplace* to align user expectations about the use of the standard with a comprehensive approach to worker safety. A cross-reference table, in Annex M, has been developed to help the user familiar with the 2000 edition locate provisions within the 2004 edition. The table is also available on the NFPA website (http://www.NFPA.org).

Committee members considered several proposals to reorganize the material to enhance usability and comprehension. They agreed to move Part II from the 2000 edition, Safety-Related Work Practices, to Chapter 1, for two reasons: First, the main focus of the document is to provide safety-related work practices for workers, and, second, most questions about the document focus on safety-related work practices. Part III, Safety-Related Maintenance Practices, was moved to Chapter 2, and Part IV, Special Equipment, was relocated to Chapter 3. Part I, Safety-Related Installation Practices, was relocated to Chapter 4. The committee agreed to retain the *NEC* installation requirements in the body of the standard to provide a basis for OSHA enforcement of installation requirements, which are important for worker safety. The annexes now follow Chapter 4. Metric units of measurement, followed by inch-pound units in parentheses, are now provided to comply with the *NFPA Manual of Style.* This addition also should increase the usability of the document in the international arena.

## REFERENCES CITED IN COMMENTARY

*Manual of Style for NFPA Technical Committee Documents,* 2003 edition, National Fire Protection Association, Quincy, MA [available online at www.nfpa.org].

*National Electrical Code (NEC®) Style Manual,* 2003 edition, National Fire Protection Association, Quincy, MA [available online at www.nfpa.org].

OSHA 29 CFR 1910, Subpart S, U.S. Government Printing Office, Washington, DC.

OSHA 29 CFR 1926, Subpart K, U.S. Government Printing Office, Washington, DC.

CHAPTER 1

# Safety-Related Work Practices

Consensus standards historically have defined safety requirements in terms of equipment, facilities, and installations. Experience, however, suggests that unsafe equipment and unsafe conditions associated with equipment and facilities, including maintenance, account for roughly one-third of all electrical incidents and injuries. The remaining two-thirds of electrical incidents and injuries can be attributed to unsafe acts—how people interact with equipment. The primary focus of NFPA 70E is related to defining requirements that impact the actions of workers when they are exposed to electrical hazards, especially equipment that is inadequately maintained or otherwise in an abnormal condition.

This edition of NFPA 70E attempts to consider the most important aspects of working with equipment in an abnormal condition by addressing those aspects first. The objective of NFPA 70E is to *minimize* exposure to electrical hazards and resulting injuries. Safety-related work practices are about *people* and how to help them work as safely as possible.

## ARTICLE 100 Definitions

Interviews with people who have been involved in a workplace injury or in a close call most often suggest that those people have misunderstood a word or term. In some instances, communication between the supervisor and the worker was incomplete. Frequently, disagreements and arguments occur when both parties agree essentially but are unable to communicate. Disagreements or arguments frequently result in increased exposure to hazards for the persons involved. In some instances, incomplete communication is caused by different definitions and understandings of the same or similar words or terms. Experience shows that definitions vary from one geographic area to another as well as among people of different heritage. The defined words and terms in this section can help to build a common vocabulary with common definitions to achieve more complete communication between employers and employees.

The definitions in Article 100 are derived from two sources. Commonly defined terms and installation-related terms are extracted from the 2002 edition of the *National Electrical Code®* (NFPA 70, *NEC®*, or *Code*). *All NEC references are from the 2002 edition.* Unique safety-related terms have been defined by the technical committee over the seven editions of NFPA 70E. Understanding these definitions is essential to understanding the requirements in 70E.

**Scope.** This article contains only those definitions essential to the proper application of this standard. It is not intended to include commonly defined general terms or commonly defined technical terms from related codes and standards. In general, only those terms that are used in two or more articles are defined in Article 100. Other definitions are included in the article in which they are used but may be referenced in Article 100.

Part I of this article contains definitions intended to apply wherever the terms are used throughout this standard. Part II contains definitions applicable only to the parts of articles specifically covering installations and equipment operating at over 600 volts, nominal.

The definitions in this article shall apply wherever the terms are used throughout this standard.

Commonly defined general terms, including those defined in general English language dictionaries, are not defined in this standard unless they are used in a unique or restricted manner. Some commonly defined technical terms, such as volt (abbreviated V) and ampere (abbreviated A), are found in the *Authoritative Dictionary of IEEE Standards Terms.*

Some definitions that are not listed in Article 100 are defined in their appropriate article. These articles follow the common format according to the *NEC Style Manual* and list the section number as XXX.2, Definition(s). For example, the definitions applicable to electrolytic cells can be found in the *NEC*, 310.2, Definitions.

## I. General

**Accessible (as applied to equipment).** Admitting close approach; not guarded by locked doors, elevation, or other effective means.

Exhibit 100.1 illustrates a few examples of equipment considered to be *accessible (as applied to equipment).* The main rule for switches and circuit breakers used as switches is shown as (a) and is according to 404.8(A) of the *NEC.* In illustration (b), the busway installation is according to *NEC* 368.12. The exceptions to the main rule are illustrated in (c), the installation of busway switches according to *NEC* 404.8(A), Exception No. 1. Illustration (d) shows a switch installed adjacent to a motor according to *NEC* 404.8(A), Exception No. 2. Illustration (e) shows a hookstick-operated isolating switch installed according to *NEC* 404.8(A), Exception No. 3.

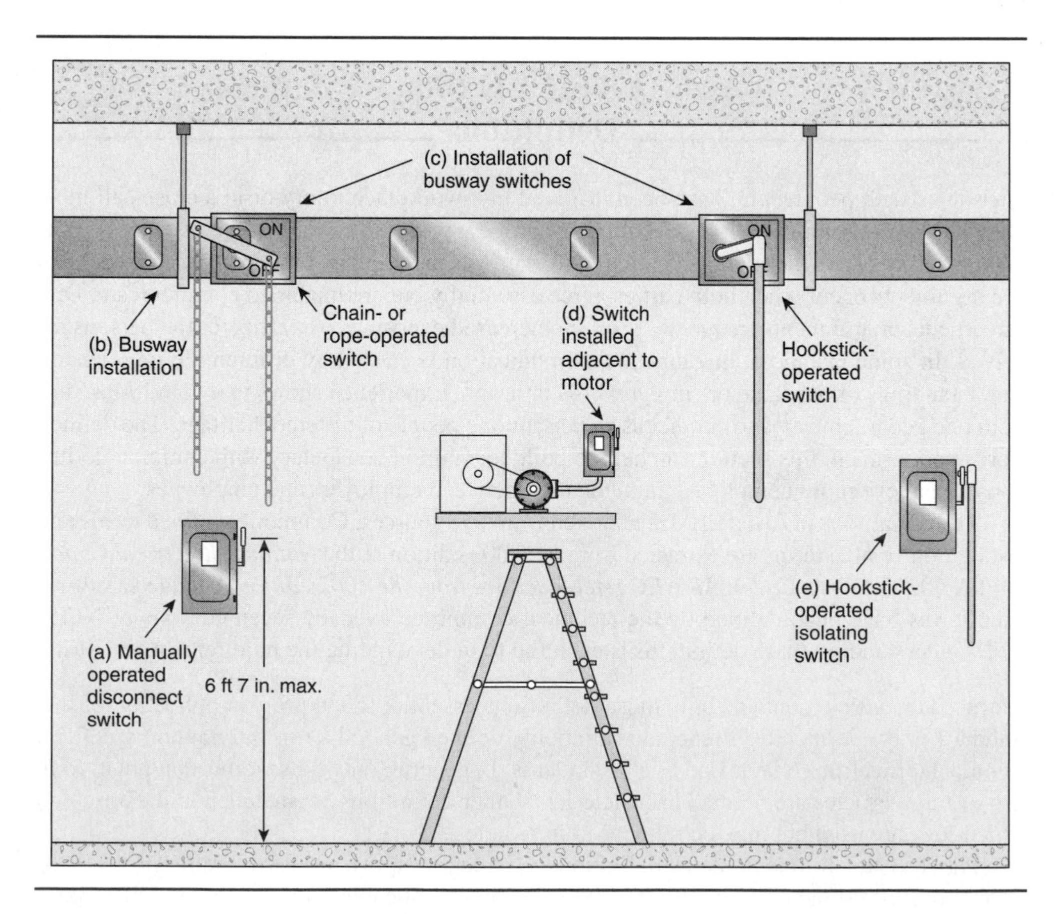


***EXHIBIT 100.1.*** *Examples of busway and of switches considered accessible, even if located above 6 ft, 7 in.*

**Accessible (as applied to wiring methods).** Capable of being removed or exposed without damaging the building structure or finish or not permanently closed in by the structure or finish of the building.

Wiring methods located behind removable panels designed to allow access are not considered permanently enclosed and are considered exposed as applied to wiring methods. See 300.4(C) of the *NEC* regarding cables located in spaces behind accessible panels.

Exhibit 100.2 illustrates examples of wiring methods and equipment that are considered to be *accessible (as applied to wiring methods).*

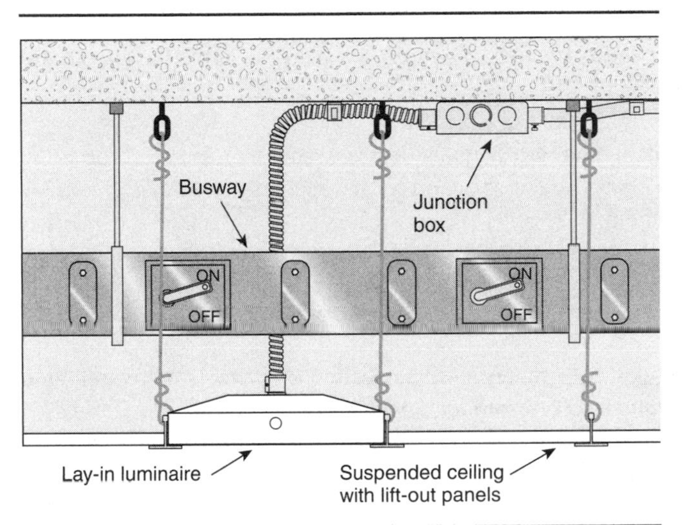


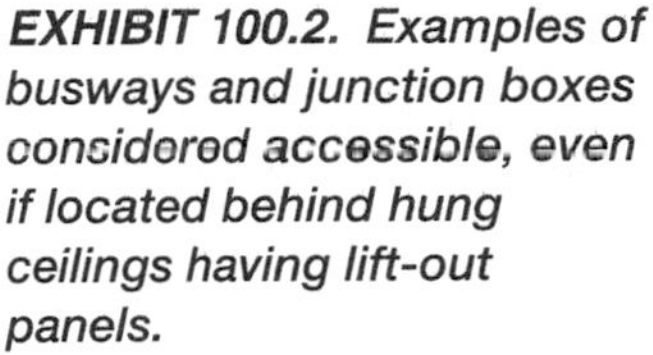
***EXHIBIT 100.2.*** ***Examples of busways and junction boxes considered accessible, even if located behind hung ceilings having lift-out panels.***

**Accessible, Readily (Readily Accessible).** Capable of being reached quickly for operation, renewal, or inspections without requiring those to whom ready access is requisite to climb over or remove obstacles or to resort to portable ladders, and so forth.

The definition of *readily accessible* does not preclude the use of a locked door for service equipment or rooms containing service equipment, provided those for whom ready access is necessary have a key (or lock combination). For example, 230.70(A)(1) and 230.205(A) of the *NEC* require service-disconnecting means to be readily accessible. *NEC* 225.32 requires that feeder disconnecting means for separate buildings be readily accessible. A commonly used, permitted practice is to locate the disconnecting means in the electrical equipment room of an office building or large apartment building and to keep the door to that room locked, to prevent access by unauthorized persons.

**Ampacity.** The current, in amperes, that a conductor can carry continuously under the conditions of use without exceeding its temperature rating.

The definition of the term *ampacity* states that the maximum current a conductor carries varies continuously with the conditions of use as well as with the temperature rating of the conductor insulation. For example, ambient temperature is a condition of use. A conductor with insulation rated at 60°C, installed near a furnace where the ambient temperature is continuously maintained at 60°C, has no current-carrying capacity. Any current flowing through the conductor will raise its temperature above the 60°C insulation rating. Therefore, the ampacity of this conductor, regardless of its size, is zero. See the Correction Factors section for temperature at the bottom of *NEC* Table 310.16 through Table 310.20, or see Annex B of the *NEC.*

Another condition of use is the number of conductors in a raceway or cable. [See 310.15(B)(2) of the *NEC.*] Increasing the number of conductors in a restricted space limits the ability of the conductor to dissipate heat.

**Appliance.** Utilization equipment, generally other than industrial, that is normally built in standardized sizes or types and is installed or connected as a unit to perform one or more functions such as clothes washing, air conditioning, food mixing, deep frying, and so forth.

**Approved.** Acceptable to the authority having jurisdiction.

See the definition of *authority having jurisdiction* in 110.2 of the *NEC* for a better understanding of the approval process and of the definition of *approved.* Understanding *NEC* terms such as *listed, labeled,* and *identified (as applied to equipment)* also can assist the user in understanding the approval process.

**Arc Rating.** The maximum incident energy resistance demonstrated by a material (or a layered system of materials) prior to breakopen or at the onset of a second-degree skin burn. Arc rating is normally expressed in cal/cm$^2$.

FPN: *Breakopen* is a material response evidenced by the formation of one or more holes in the innermost layer of flame-resistant material that would allow flame to pass through the material.

The definition of *arc rating* provides consistency in the selection of protective apparel, as it correlates with ASTM F 1506, *Standard Performance Specification for Flame Resistant Textile Materials for Wearing Apparel for Use by Electrical Workers Exposed to Momentary Electric Arc and Related Thermal Hazards,* for protective apparel and ASTM F 1891, *Standard Specification for Arc and Flame-Resistant Rainwear,* for protective raingear. The definition is consistent with the terminology in the applicable ASTM standards and in the selection of personal protective equipment (PPE).

**Armored Cable.** A fabricated assembly of insulated conductors in a metallic enclosure.

**Attachment Plug (Plug Cap) (Plug).** A device that, by insertion in a receptacle, establishes a connection between the conductors of the attached flexible cord and the conductors connected permanently to the receptacle.

Standard *attachment plugs* are available with built-in options, such as switching, fuses, or even ground-fault circuit interrupter protection. Attachment plug contact blades have specific shapes, sizes, and configurations so that a receptacle or cord connector will not accept an attachment plug of a different voltage or current rating than that for which the device is intended. Configuration charts from NEMA Standard WD 6, *Wiring; Devices—Dimensional Requirement,* for general-purpose nonlocking and specific-purpose locking plugs and receptacles are shown in Exhibits 420.2 and 420.3, respectively. (See Chapter 4.)

**Automatic.** Self-acting, operating by its own mechanism when actuated by some impersonal influence, as, for example, a change in current, pressure, temperature, or mechanical configuration.

**Bare Hand Work.** A technique of performing work on live parts, after the employee has been raised to the potential of the live part.

The expertise necessary to perform *bare hand work* can be acquired only by specialized training. The technique is not readily applicable to work tasks other than working on bare overhead transmission lines. The technique sometimes requires the worker to wear special conductive mesh over his or her clothing to avoid touch potential conditions across his or her body. The "bare hand work" method is not applicable to facilities constructed under the guidance of the *NEC.*

**Barricade.** A physical obstruction such as tapes, cones, or A-frame-type wood or metal structures intended to provide a warning about and to limit access to a hazardous area.

**Barrier.** A physical obstruction that is intended to prevent contact with equipment or live parts or to prevent unauthorized access to a work area.

**Bathroom.** An area including a basin with one or more of the following: a toilet, a tub, or a shower.

**Bonding (Bonded).** The permanent joining of metallic parts to form an electrically conductive path that ensures electrical continuity and the capacity to conduct safely any current likely to be imposed.

The purpose of *bonding* is to establish an effective path for fault current that, in turn, facilitates the operation of the overcurrent protective device. This is explained in 250.4(A)(3) and (A)(4) and 250.4(B)(3) and (B)(4) in the *NEC.* Specific bonding requirements are found in Part V of Article 250 of the *NEC* and in other sections of that *Code* as referenced in *NEC* 250.3.

**Bonding Jumper.** A reliable conductor to ensure the required electrical conductivity between metal parts required to be electrically connected.

Both concentric- and eccentric-type knockouts can impair the electrical conductivity between the metal parts and can actually introduce unnecessary impedance into the grounding path. Installing *bonding jumper(s)* is one method often used between metal raceways and metal parts to ensure electrical conductivity. Bonding jumpers can be found at service equipment [*NEC* 250.92(B)], bonding for over 250 volts (*NEC* 250.97), and expansion fittings in metal raceways (*NEC* 250.98). Exhibit 100.3 illustrates the difference between concentric- and eccentric-type knockouts as well as one method of applying bonding jumpers at these types of knockouts.

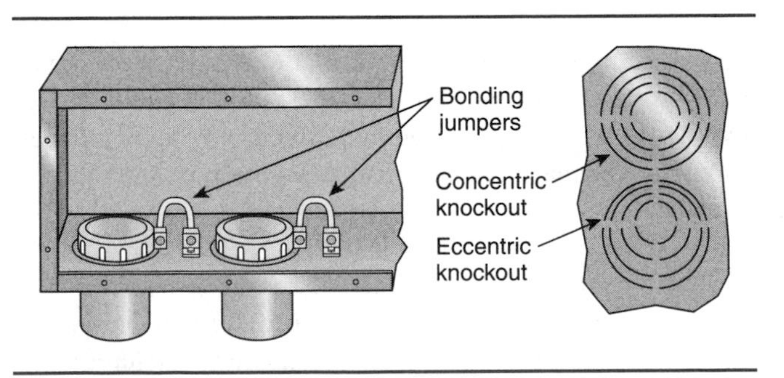


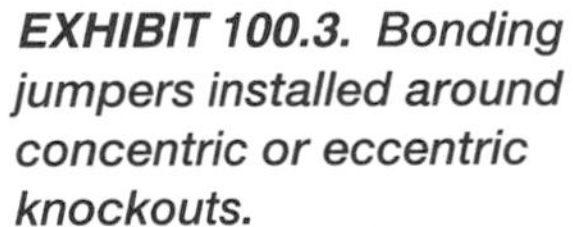

***EXHIBIT 100.3.*** ***Bonding jumpers installed around concentric or eccentric knockouts.***

**Branch Circuit.** The circuit conductors between the final overcurrent device protecting the circuit and the outlet(s).

Exhibit 100.4 illustrates the difference between *branch circuits* and feeders. Conductors between the overcurrent devices in panelboards and duplex receptacles are branch-circuit conductors. Conductors between service equipment or the source of separately derived systems and the panelboards are feeders.

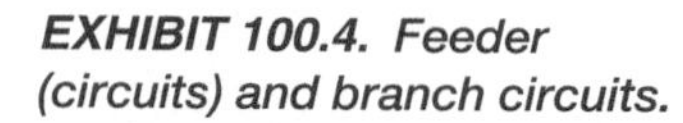

***EXHIBIT 100.4. Feeder (circuits) and branch circuits.***

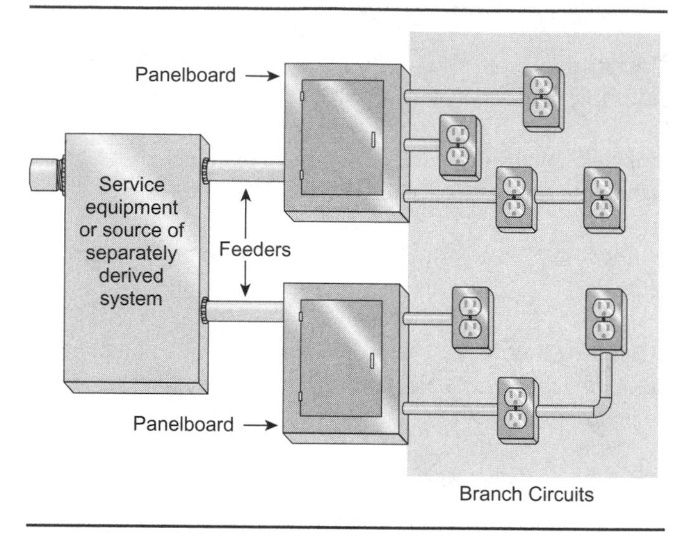


**Building.** A structure that stands alone or that is cut off from adjoining structures by fire walls with all openings therein protected by approved fire doors.

A *building* is generally considered to be a roofed or walled structure that can be used for supporting or sheltering any use or occupancy. However, it might also be a separate structure such as a pole, billboard sign, or water tower.

Definitions of the terms *fire walls* and *fire doors* are the responsibility of building codes. Generically, a fire wall may be defined as a wall that separates buildings or subdivides a building to prevent the spread of fire and that has a fire resistance rating and structural stability. Fire doors (and fire windows) are used to protect openings in walls, floors, and ceilings against the spread of fire and smoke within, into, or out of buildings.

**Cabinet.** An enclosure that is designed for either surface mounting or flush mounting and is provided with a frame, mat, or trim in which a swinging door or doors are or can be hung.

Both cabinets and cutout boxes are covered in Article 312 of the *NEC. Cabinets* are designed for surface or flush mounting with a trim to which a swinging door(s) is hung. Cutout boxes are designed for surface mounting with a swinging door(s) secured directly to the box. Panelboards are electrical assemblies designed to be placed in a cabinet or cutout box. (See the definitions of *cutout box* and *panelboard.*)

**Cablebus.** An assembly of insulated conductors with fittings and conductor terminations in a completely enclosed, ventilated protective metal housing. Cablebus is ordinarily assembled at the point of installation from the components furnished or specified by the manufacturer in accordance with instructions for the specific job. This assembly is designed to carry fault current and to withstand the magnetic forces of such current.

**Circuit Breaker.** A device designed to open and close a circuit by nonautomatic means and to open the circuit automatically on a predetermined overcurrent without damage to itself when properly applied within its rating.

**Class I Locations.** Class I locations are those in which flammable gases or vapors are or may be present in the air in quantities sufficient to produce explosive or ignitible mixtures. Class I locations shall include those specified in Division 1 or Division 2.

*Class I, Division 1.* A Class I, Division 1 location is a location:

(1) In which ignitible concentrations of flammable gases or vapors can exist under normal operating conditions, or

(2) In which ignitible concentrations of such gases or vapors may exist frequently because of repair or maintenance operations or because of leakage, or

(3) In which breakdown or faulty operation of equipment or processes might release ignitible concentrations of flammable gases or vapors and might also cause simultaneous failure of electrical equipment in such a way as to directly cause the electrical equipment to become a source of ignition.

FPN No. 1: This classification usually includes the following locations:

(1) Where volatile flammable liquids or liquefied flammable gases are transferred from one container to another
(2) Interiors of spray booths and areas in the vicinity of spraying and painting operations where volatile flammable solvents are used
(3) Locations containing open tanks or vats of volatile flammable liquids
(4) Drying rooms or compartments for the evaporation of flammable solvents)
(5) Locations containing fat- and oil-extraction equipment using volatile flammable solvents
(6) Portions of cleaning and dyeing plants where flammable liquids are used
(7) Gas generator rooms and other portions of gas manufacturing plants where flammable gas may escape
(8) Inadequately ventilated pump rooms for flammable gas or for volatile flammable liquids
(9) The interiors of refrigerators and freezers in which volatile flammable materials are stored in open, lightly stoppered, or easily ruptured containers
(10) All other locations where ignitible concentrations of flammable vapors or gases are likely to occur in the course of normal operations

FPN No. 2: In some Division 1 locations, ignitible concentrations of flammable gases or vapors could be present continuously or for long periods of time. Examples include the following:

(1) The inside of inadequately vented enclosures containing instruments normally venting flammable gases or vapors to the interior of the enclosure
(2) The inside of vented tanks containing volatile flammable liquids
(3) The area between the inner and outer roof sections of a floating roof tank containing volatile flammable fluids
(4) Inadequately ventilated areas within spraying or coating operations using volatile flammable fluids
(5) The interior of an exhaust duct that is used to vent ignitible concentrations of gases or vapors

Experience has demonstrated the prudence of avoiding the installation of instrumentation or other electric equipment in these particular areas altogether or, where it cannot be avoided because it is essential to the process and other locations are not feasible, using electric equipment or instrumentation approved for the specific application or consisting of intrinsically safe systems.

Fine print note No. 2 following the definition of *Class I locations* describes locations that are defined in 500.5(B)(1) of the *NEC* as Class I, Division 1 locations. These locations are Class I, Zone 0 locations in accordance with *NEC* 505.5(B)(1). Where classified as a Zone 0 location, only intrinsically safe equipment and wiring suitable for Zone 0 locations is permitted to be used.

*Class I, Division 2.* A Class I, Division 2 location is a location:

(1) In which volatile flammable liquids or flammable gases are handled, processed, or used, but in which the liquids, vapors, or gases will normally be confined within closed containers or closed systems from which they can escape only in case of accidental rupture or breakdown of such containers or systems or in case of abnormal operation of equipment, or
(2) In which ignitible concentrations of gases or vapors are normally prevented by positive mechanical ventilation, and which might become hazardous through failure or abnormal operation of the ventilating equipment, or
(3) That is adjacent to a Class I, Division 1 location, and to which ignitible concentrations of gases or vapors might occasionally be communicated unless such communication is prevented by adequate positive-pressure ventilation from a source of clean air and effective safeguards against ventilation failure are provided.

FPN No. 1: This classification usually includes locations where volatile flammable liquids or flammable gases or vapors are used but that, in the judgment of the authority having jurisdiction, would become hazardous only in case of an accident or of some unusual operating condition. The quantity of flammable material that might escape in case of accident, the adequacy of ventilating equipment, the total area involved, and the record of the industry or business with respect to explosions or fires are all factors that merit consideration in determining the classification and extent of each location.

FPN No. 2: Piping without valves, checks, meters, and similar devices would not ordinarily introduce a hazardous condition even though used for flammable liquids or gases. Depending on factors such as the quantity and size of the containers and ventilation, locations used for the storage of flammable liquids or liquefied or compressed gases in sealed containers may be considered either hazardous (classified) or unclassified locations. See NFPA 30-2000, *Flammable and Combustible Liquids Code,* and NFPA 58-2001, *Liquefied Petroleum Gas Code.*

**Class I, Zone 0, 1, and 2 Locations.** Class I, Zone 0, 1, and 2 locations are those in which flammable gases or vapors are or may be present in the air in quantities sufficient to produce explosive or ignitible mixtures. Class I, Zone 0, 1, and 2 locations shall include those specified as follows:

*Class I, Zone 0.* A Class I, Zone 0 location is a location in which

(1) Ignitible concentrations of flammable gases or vapors are present continuously, or
(2) Ignitible concentrations of flammable gases or vapors are present for long periods of time.

FPN No. 1: As a guide in determining when flammable gases or vapors are present continuously or for long periods of time, refer to ANSI/API RP 505-1997, *Recommended Practice for Classification of Locations for Electrical Installations of Petroleum Facilities Classified as Class I, Zone 0, Zone 1, or Zone 2*; ISA 12.24.01-1998, *Recommended Practice for Classification of Locations for Electrical Installations Classified as Class I, Zone 0, Zone 1, or Zone 2;* IEC 60079-10-1995, *Electrical Apparatus for Explosive Gas Atmospheres, Classifications of Hazardous Areas;* and *Area Classification Code for Petroleum Installations, Model Code, Part 15,* Institute of Petroleum.

FPN No. 2: This classification includes locations inside vented tanks or vessels that contain volatile flammable liquids; inside inadequately vented spraying or coating enclosures, where volatile flammable solvents are used; between the inner and outer roof sections of a floating roof tank containing volatile flammable liquids; inside open vessels, tanks, and pits containing volatile flammable liquids; the interior of an exhaust duct that is used to vent ignitible concentrations of gases or vapors; and inside inadequately ventilated enclosures that contain normally venting instruments utilizing or analyzing flammable fluids and venting to the inside of the enclosures.

FPN No. 3: It is not good practice to install electrical equipment in Zone 0 locations except when the equipment is essential to the process or when other locations are not feasible (see FPN No. 2). If it is necessary to install electrical systems in a Zone 0 location, it is good practice to install intrinsically safe systems.

*Class I, Zone 1.* A Class I, Zone 1 location is a location

(1) In which ignitible concentrations of flammable gases or vapors are likely to exist under normal operating conditions; or
(2) In which ignitible concentrations of flammable gases or vapors may exist frequently because of repair or maintenance operations or because of leakage; or
(3) In which equipment is operated or processes are carried on, of such a nature that equipment breakdown or faulty operations could result in the release of ignitible concentrations of flammable gases or vapors and also cause simultaneous failure of electrical equipment in a mode to cause the electrical equipment to become a source of ignition; or
(4) That is adjacent to a Class I, Zone 0 location from which ignitible concentrations of vapors could be communicated, unless communication is prevented by adequate positive pressure ventilation from a source of clean air and effective safeguards against ventilation failure are provided.

FPN No. 1: Normal operation is considered the situation when plant equipment is operating within its design parameters. Minor releases of flammable material could be part of normal operations. Minor releases include the releases from mechanical packings on pumps. Failures that involve repair or shutdown (such as the breakdown of pump seals and flange gaskets, and spillage caused by accidents) are not considered normal operation.

FPN No. 2: This classification usually includes locations where volatile flammable liquids or liquefied flammable gases are transferred from one container to another. In areas in the vicinity of spraying and painting operations where flammable solvents are used; adequately ventilated drying rooms or compartments for evaporation of flammable solvents; adequately ventilated locations containing fat and oil extraction equipment using volatile flammable solvents; portions of cleaning and dyeing plants where volatile flammable liquids are used; adequately ventilated gas generator rooms and other portions of gas manufacturing plants where flammable gas could escape; inadequately ventilated pump rooms for flammable gas or for volatile flammable liquids; the interiors of refrigerators or freezers in which volatile flammable materials are stored in the open, lightly stoppered, or in easily ruptured containers; and other locations where ignitible concentrations of flammable vapors or gases are likely to occur in the course of normal operation but not classified Zone 0.

*Class I, Zone 2.* A Class I, Zone 2 location is a location

(1) In which ignitible concentrations of flammable gases or vapors are not likely to occur in normal operation and, if they do occur, will exist only for a short period; or
(2) In which volatile flammable liquids, flammable gases, or flammable vapors are handled, processed, or used but in which the liquids, gases, or vapors normally are confined within closed containers of closed systems from which they can escape, only as a result of accidental rupture or breakdown of the containers or system, or as a result of the abnormal operation of the equipment with which the liquids or gases are handled, processed, or used; or
(3) In which ignitible concentrations of flammable gases or vapors normally are prevented by positive mechanical ventilation but which may become hazardous as a result of failure or abnormal operation of the ventilating equipment; or
(4) That is adjacent to a Class I, Zone 1 location, from which ignitible concentrations of flammable gases or vapors could be communicated, unless such communication is prevented

by adequate positive-pressure ventilation from a source of clean air and effective safeguards against ventilation failure are provided.

FPN: The Zone 2 classification usually includes locations where volatile flammable liquids or flammable gases or vapors are used but which would become hazardous only in case of an accident or of some unusual operating condition.

**Class II Locations.** Class II locations are those that are hazardous because of the presence of combustible dust. Class II locations shall include those in Division 1 and Division 2.

*Class II, Division 1.* A Class II, Division 1 location is a location

(1) In which combustible dust is in the air under normal operating conditions in quantities sufficient to produce explosive or ignitible mixtures, or
(2) Where mechanical failure or abnormal operation of machinery or equipment might cause such explosive or ignitible mixtures to be produced, and might also provide a source of ignition through simultaneous failure of electric equipment, through operation of protection devices, or from other causes, or
(3) In which combustible dusts of an electrically conductive nature may be present in hazardous quantities.

FPN: Combustible dusts that are electrically nonconductive include dusts produced in the handling and processing of grain and grain products, pulverized sugar and cocoa, dried egg and milk powders, pulverized spices, starch and pastes, potato and wood-flour, oil meal from beans and seed, dried hay, and other organic materials that could produce combustible dusts when processed or handled. Only Group E dusts are considered to be electrically conductive for classification purposes. Dusts containing magnesium or aluminum are particularly hazardous, and the use of extreme precaution is necessary to avoid ignition and explosion.

*Class II, Division 2.* A Class II, Division 2 location is a location

(1) Where combustible dust is not normally in the air in quantities sufficient to produce explosive or ignitible mixtures, and dust accumulations are normally insufficient to interfere with the normal operation of electrical equipment or other apparatus, but combustible dust may be in suspension in the air as a result of infrequent malfunctioning of handling or processing equipment, and
(2) Where combustible dust accumulations on, in, or in the vicinity of the electrical equipment may be sufficient to interfere with the safe dissipation of heat from electrical equipment or may be ignitible by abnormal operation or failure of electrical equipment.

FPN No. 1: The quantity of combustible dust that may be present and the adequacy of dust removal systems are factors that merit consideration in determining the classification and may result in an unclassified area.

FPN No. 2: Where products such as seed are handled in a manner that produces low quantities of dust, the amount of dust deposited could not warrant classification.

**Class III Locations.** Class III locations are those that are hazardous because of the presence of easily ignitible fibers or flyings, but in which such fibers or flyings are not likely to be in suspension in the air in quantities sufficient to produce ignitible mixtures. Class III locations shall include Division 1 and Division 2.

*Class III, Division 1.* A Class III, Division 1 location is a location in which easily ignitible fibers or materials producing combustible flyings are handled, manufactured, or used.

FPN No. 1: Such locations usually include some parts of rayon, cotton, and other textile mills; combustible fiber manufacturing and processing plants; cotton gins and cotton-seed mills; flax-processing plants; clothing manufacturing plants; woodworking plants; and establishments and industries involving similar hazardous processes or conditions.

FPN No. 2: Easily ignitible fibers or flyings include rayon, cotton (including cotton linters and cotton waste), sisal or henequen, istle, jute, hemp, tow, cocoa fiber, oakum, baled waste kapok, Spanish moss, excelsior, and other materials of similar nature.

*Class III, Division 2.* A Class III, Division 2 location is a location in which easily ignitible fibers are stored or handled other than in the process of manufacture.

**Concealed.** Rendered inaccessible by the structure or finish of the building. Wires in concealed raceways are considered concealed, even though they may become accessible by withdrawing them.

Raceways and cables supported or located within hollow frames or permanently closed in by the finish of buildings are considered to be *concealed.* Open-type work—such as raceways and cables in exposed areas; in unfinished basements; in accessible underfloor areas or attics; attached to the surface of finished areas; or behind, above, or below panels designed to allow access and that may be removed without damage to the building structure or finish—is not considered to be concealed. [See the definition of *exposed (as applied to wiring methods).*]

**Conductive.** Suitable for carrying electric current.

**Conductor, Bare.** A conductor having no covering or electrical insulation whatsoever.

**Conductor, Covered.** A conductor encased within material of composition or thickness that is not recognized by this standard as electrical insulation.

Typical *covered conductors* are the green-covered equipment grounding conductors contained within a nonmetallic-sheathed cable or the uninsulated grounded system conductors within the overall exterior jacket of a Type-SE cable. Covered conductors should always be treated as bare conductors for working clearances because they are really uninsulated conductors.

**Conductor, Insulated.** A conductor encased within material of composition and thickness that is recognized by this standard as electrical insulation.

For the covering on a conductor to be considered insulation, the conductor with the covering material generally is required to pass minimum testing required by a product standard. One such product standard is UL 83, *Thermoplastic-Insulated Wires and Cables.* To meet the requirements of UL 83, specimens of finished single-conductor wires must pass specified tests that measure (1) resistance to flame propagation, (2) dielectric strength, even while immersed, and (3) resistance to abrasion, cracking, crushing, and impact. Only wires and cables that meet the minimum fire, electrical, and physical properties required by the applicable standards are permitted to be marked with the letter designations found in Table 310.13 and Table 310.61 of the *NEC.* See *NEC* 310.13 for the exact requirements of insulated conductor construction and applications.

**Conduit Body.** A separate portion of a conduit or tubing system that provides access through a removable cover(s) to the interior of the system at a junction of two or more sections of the system or at a terminal point of the system.

FPN: Boxes such as FS and FD or larger cast or sheet metal boxes are not classified as conduit bodies.

*Conduit bodies* are portions of a raceway system with removable covers to allow access to the interior of the system. They include the short-radius type as well as capped elbows and service-entrance elbows.

Some conduit bodies are referred to in the trade as "condulets" and include the LB, LL, LR, C, T, and X designs. (See 300.15 and Article 314 of the *NEC* for rules on the usage of conduit bodies.)

Types FS- and FD-boxes are not classified as conduit bodies; they are listed with boxes in *NEC* Table 314.16(A).

**Controller.** A device or group of devices that serves to govern, in some predetermined manner, the electric power delivered to the apparatus to which it is connected.

A *controller* may be a remote-controlled magnetic contactor, switch, circuit breaker, or device that is normally used to start and stop motors and other apparatus and, in the case of motors, is required to be capable of interrupting the stalled-rotor current of the motor. Stop-and-start stations and similar control circuit components that do not open the power conductors to the motor are not considered controllers.

**Cooking Unit, Counter-Mounted.** A cooking appliance designed for mounting in or on a counter and consisting of one or more heating elements, internal wiring, and built-in or mountable controls.

**Cutout Box.** An enclosure designed for surface mounting that has swinging doors or covers secured directly to and telescoping with the walls of the box proper.

**Dead Front.** Without live parts exposed to a person on the operating side of the equipment.

**Deenergized.** Free from any electrical connection to a source of potential difference and from electrical charge; not having a potential different from that of the earth.

The term *deenergized* describes an operating condition of electrical equipment. The term should be used for no other purpose. *Deenergized* does not describe a safe condition.

**Device.** A unit of an electrical system that is intended to carry but not utilize electric energy.

Components (such as switches, circuit breakers, fuseholders, receptacles, attachment plugs, and lampholders) that distribute or control but do not consume electricity are considered to be *devices.*

**Dielectric Heating.** Heating of a nominally insulating material due to its own dielectric losses when the material is placed in a varying electric field.

**Disconnecting Means.** A device, or group of devices, or other means by which the conductors of a circuit can be disconnected from their source of supply.

For *disconnecting means* for service equipment, see Part VI of *NEC* Article 230; for fuses, see Part IV of *NEC* Article 240; for circuit breakers, see Part VII of *NEC* Article 240; for appliances, see Part III of *NEC* Article 422; for space-heating equipment, see Part III of *NEC* Article 424; for motors and controllers, see Part IX of *NEC* Article 430; and for air-conditioning and refrigerating equipment, see Part II of *NEC* Article 440.

**Effective Ground-Fault Current Path.** An intentionally constructed, permanent, low-impedance electrically conductive path designed and intended to carry current under ground-fault conditions from the point of a ground-fault on a wiring system to the electrical supply source.

OSHA requested that the definition of an *effective ground-fault current path* be defined in this document. An effective grounding path is essential for the proper operation of overcurrent protective devices. The key issue is that the ground-fault current path be *effective.*

**Electric Sign.** A fixed, stationary, or portable self-contained, electrically illuminated utilization equipment with words or symbols designed to convey information or attract attention.

**Electrical Hazard.** A dangerous condition such that contact or equipment failure can result in electric shock, arc flash burn, thermal burn, or blast.

FPN: Class 2 power supplies, listed low voltage lighting systems, and similar sources are examples of circuits or systems that are not considered an electrical hazard.

Fire, shock, and electrocution have been considered to be *electrical hazards* for many years. Since the publication of NFPA 70E, 1995 edition, arc flash has been considered to be an electrical hazard. The arc flash hazard currently is defined to consider only the thermal aspects of an arcing fault. Other hazards include flying parts and pieces and the pressure wave that is generated in an arcing fault. Other electrical hazards also might be associated with an arcing fault.

**Electrical Safety.** Recognizing hazards associated with the use of electrical energy and taking precautions so that hazards do not cause injury or death.

*Electrical safety* can be achieved only after all of the electrical hazards have been identified, a comprehensive plan to mitigate the hazards has been formulated, and protective schemes and training for both qualified and unqualified persons have been provided.

**Electrical Single-Line Diagram.** A diagram that shows, by means of single lines and graphic symbols, the course of an electric circuit or system of circuits and the component devices or parts used in the circuit or system.

**Electrically Safe Work Condition.** A state in which the conductor or circuit part to be worked on or near has been disconnected from energized parts, locked/tagged in accordance with established standards, tested to ensure the absence of voltage, and grounded if determined necessary.

After an *electrically safe work condition* has been established, electrical hazards no longer exist, and protective clothing is no longer needed. See Article 120.

**Enclosed.** Surrounded by a case, housing, fence, or wall(s) that prevents persons from accidentally contacting energized parts.

**Enclosure.** The case or housing of apparatus, or the fence or walls surrounding an installation to prevent personnel from accidentally contacting energized parts, or to protect the equipment from physical damage.

Commentary Table 1.1 summarizes the intended uses of the various types of *enclosures* for nonhazardous locations. Enclosures that comply with the requirements for more than one type of enclosure may be marked with multiple designations. Enclosures marked with a type may also be marked as follows:

- Type 1: "Indoor Use Only"
- Types 3, 3S, 4, 4X, 6, or 6P: "Raintight"
- Type 3R: "Rainproof"
- Type 4, 4X, 6, or 6P: "Watertight"
- Type 4X or 6P: "Corrosion Resistant"
- Type 2, 5, 12, 12K, or 13: "Driptight"
- Type 3, 3S, 5, 12K, or 13: "Dusttight"

***COMMENTARY TABLE 1.1*** *Environmental Protections for Nonhazardous Locations, by Type of Enclosure*

| *Enclosure Type Number* | *Environmental Condition(s)*[1] |
|---|---|
| 1 | Indoor use |
| 2 | Indoor use, limited amounts of falling water |
| 3R | Outdoor use, undamaged by the formation of ice on the enclosure[2] |
| 3 | Same as 3R plus windblown dust |
| 3S | Same as 3R plus windblown dust; external mechanisms remain operable while ice laden |
| 4 | Outdoor use, splashing water, windblown dust, hose-directed water, undamaged by the formation of ice on the enclosure[2] |
| 4X | Same as 4 plus resists corrosion |
| 5 | Indoor use to provide a degree of protection against settling airborne dust, falling dirt, and dripping noncorrosive liquids |
| 6 | Same as 3R plus entry of water during temporary submersion at a limited depth |
| 6P | Same as 3R plus entry of water during prolonged submersion at a limited depth |
| 12, 12K | Indoor use, dust, dripping noncorrosive liquids |
| 13 | Indoor use, dust, spraying water, oil, and noncorrosive coolants |

[1]All enclosure types provide a degree of protection against ordinary corrosion and against accidental contact with the enclosed equipment when doors or covers are closed and in place. All types of enclosures provide protection against a limited amount of falling dirt.

[2]All outdoor-type enclosures provide a degree of protection against rain, snow, and sleet. Outdoor enclosures are also suitable for use indoors if they meet the environmental conditions present.

*Source:* Underwriters Laboratories, *General Information Directory,* 2001 edition.

For equipment designated *raintight,* testing designed to simulate exposure to a beating rain will not result in entrance of water. For equipment designated *rainproof,* testing designed to simulate exposure to a beating rain will not interfere with the operation of the apparatus or result in wetting of live parts and wiring within the enclosure. *Watertight* equipment is so constructed that water does not enter the enclosure when subjected to a stream of water. *Corrosion-resistant* equipment is constructed so that it provides a degree of protection against exposure to corrosive agents such as salt spray. *Driptight* equipment is constructed so that falling moisture or dirt does not enter the enclosure. *Dusttight* equipment is constructed so that circulating or airborne dust does not enter the enclosure.

**Energized.** Electrically connected to or having a source of voltage.

The definition of *energized* was editorially revised for the 2002 *Code* by substituting the term *voltage* for *potential difference.* For a more thorough understanding of *energized,* also see the definitions of *exposed (as applied to live parts)* and *live parts.*

**Equipment.** A general term including material, fittings, devices, appliances, luminaires (fixtures), apparatus, and the like used as a part of, or in connection with, an electrical installation.

**Explosionproof Apparatus.** Apparatus enclosed in a case that is capable of withstanding an explosion of a specified gas or vapor that may occur within it and of preventing the ignition of a specified gas or vapor surrounding the enclosure by sparks, flashes, or explosion of the gas or vapor within, and that operates at such an external temperature that a surrounding flammable atmosphere will not be ignited thereby.

**Exposed (as applied to live parts).** Capable of being inadvertently touched or approached nearer than a safe distance by a person. It is applied to parts that are not suitably guarded, isolated, or insulated.

For a more thorough understanding of *exposed (as applied to live parts),* also see the definitions of *energized* and *live parts.* Requirements for guarding of live parts may be found in 110.27 of the *NEC.*

**Exposed (as applied to wiring methods).** On or attached to the surface or behind panels designed to allow access.

See Exhibit 100.2, where wiring methods located behind a suspended ceiling with lift-out panels are considered to be *exposed (as applied to wiring methods).*

**Exposed.** For the purposes of Article 450, the word *exposed* means that the circuit is in such a position that, in case of failure of supports or insulation, contact with another circuit may result.

**Externally Operable.** Capable of being operated without exposing the operator to contact with live parts.

**Feeder.** All circuit conductors between the service equipment, the source of a separately derived system, or other power supply source and the final branch-circuit overcurrent device.

See the commentary following the definition of *branch circuit,* including Exhibit 100.4, which illustrates the difference between branch circuits and *feeders.*

**Fitting.** An accessory such as a locknut, bushing, or other part of a wiring system that is intended primarily to perform a mechanical rather than an electrical function.

Examples of *fittings* are such items as condulets, conduit couplings, EMT connectors and couplings, and threadless connectors.

**Flame-Resistant (FR).** The property of a material whereby combustion is prevented, terminated, or inhibited following the application of a flaming or non-flaming source of ignition, with or without subsequent removal of the ignition source.

FPN: Flame resistance can be an inherent property of a material, or it can be imparted by a specific treatment applied to the material.

Both the terms *flame-resistant* and *flame-retardant* have been used to describe clothing characteristics. To establish consistency with other standards, the committee approved the term *flame-resistant.* This definition is modified from NFPA 2112, *Standard on Flame-Resistant Garments for Protection of Industrial Personnel Against Flash Fire,* and is very similar to ASTM F 1891, *Standard Specification for Arc and Flame Resistant Rainwear.*

**Flash Hazard.** A dangerous condition associated with the release of energy caused by an electric arc.

Arc flash incidents involving workers who are not properly protected results in more than 2000 workers being admitted to burn centers each year. The best protection from the arc *flash hazard* is to establish an electrically safe work condition. If that is not feasible, the alternative is to conduct a flash hazard analysis according to a published safety procedure, then wear appropriate FR clothing.

**Flash Hazard Analysis.** A study investigating a worker's potential exposure to arc-flash energy, conducted for the purpose of injury prevention and the determination of safe work practices and the appropriate levels of PPE.

An arc occurs when an insulating medium such as air is breached by a conducting component. An arc flash, defined as the energy released during an arcing fault, occurs when current flows through a medium that is not intended to conduct electrical current. Because the arc current is not intended, the arc current releases energy that also is not intended, thus exposing a worker to unexpected hazards.

A *flash hazard analysis* must consider the possibility that an arcing fault might occur. If an arcing fault occurs, workers are exposed to unexpected conditions. The flash hazard analysis must determine the degree of the hazards associated with the unintended conditions.

In an arc flash event, electrical energy is converted to other forms. The other forms of energy include heat, pressure, IR, UV, visible light, and energy at other electromagnetic frequencies. As currently practiced, however, the flash hazard analysis considers only exposure to thermal energy. The analysis must determine the amount of thermal energy to enable a worker to select appropriate PPE. (See Article 130.3.)

**Flash Protection Boundary.** An approach limit at a distance from exposed live parts within which a person could receive a second degree burn if an electrical arc flash were to occur.

The *Flash Protection Boundary* is the first issue to be defined in a flash hazard analysis. The Flash Protection Boundary defines the point at which FR protection is necessary to avoid a second-degree burn. Any or all body parts of a worker are required to be protected. If a worker's hand and arm are within the Flash Protection Boundary, the hand and arm must be protected from the thermal hazard. If a worker's head is within the Flash Protection Boundary, the worker's head must be protected from the thermal hazard. (See Article 130.7.)

**Flash Suit.** A complete FR clothing and equipment system that covers the entire body, except for the hands and feet. This includes pants, jacket, and bee-keeper-type hood fitted with a face shield.

The personal protective equipment industry has experienced a dramatic evolution of protective schemes and equipment for workers. During this time, the term *flash suit* has not been used consistently throughout the industry. This definition clarifies the specific components that comprise a flash suit and points out that a face shield may not be used when a flash suit is required.

**Fuse.** An overcurrent protective device with a circuit-opening fusible part that is heated and severed by the passage of overcurrent through it.

FPN: A fuse comprises all the parts that form a unit capable of performing the prescribed functions. It may or may not be the complete device necessary to connect it into an electrical circuit.

**Ground.** A conducting connection, whether intentional or accidental, between an electrical circuit or equipment and the earth or to some conducting body that serves in place of the earth.

**Grounded.** Connected to earth or to some conducting body that serves in place of the earth.

**Grounded Conductor.** A system or circuit conductor that is intentionally grounded.

**Grounded, Effectively.** Intentionally connected to earth through a ground connection or connections of sufficiently low impedance and having sufficient current-carrying capacity to prevent the buildup of voltages that may result in undue hazards to connected equipment or to persons.

**Grounding Conductor.** A conductor used to connect equipment or the grounded circuit of a wiring system to a grounding electrode or electrodes.

**Grounding Conductor, Equipment.** The conductor used to connect the non–current-carrying metal parts of equipment, raceways, and other enclosures to the system grounded conductor, the grounding electrode conductor, or both, at the service equipment or at the source of a separately derived system.

See 250.118 of the *NEC* for types of equipment grounding conductors. Proper sizing of the *equipment grounding conductor* is found in *NEC* 250.122 and its associated *NEC* Table 250.122.

**Grounding Electrode Conductor.** The conductor used to connect the grounding electrode(s) to the equipment grounding conductor, to the grounded conductor, or to both, at each service, at each building or structure where supplied from a common service, or at the source of a separately derived system.

The grounding electrode conductor is covered extensively in Article 250, Part III, of the *NEC*. The *grounding electrode conductor* is required to be copper, aluminum, or copper-clad aluminum. It is used to connect the equipment grounding conductor or the grounded conductor (at the service or at the separately derived system) to the grounding electrode or electrodes for either grounded or ungrounded systems. The grounding electrode conductor is sized according to the requirements of 250.66 and Table 250.66 of the *NEC*.

**Ground Fault.** An unintentional, electrically conducting connection between an ungrounded conductor of an electrical circuit and the normally non–current-carrying conductors, metallic enclosures, metallic raceways, metallic equipment, or earth.

**Ground-Fault Circuit Interrupter.** A device intended for the protection of personnel that functions to deenergize a circuit or portion thereof within an established period of time when a current to ground exceeds the values established for a Class A device.

See the commentary following 410.4 for specific information regarding the requirements for GFCIs. Commentary Table 410.1 provides a list of applicable cross-references for *ground-fault circuit interrupters (GFCIs)*. (See Chapter 4.)

FPN: Class A ground-fault circuit-interrupter trips when the current to ground has a value in the range of 4 mA to 6 mA. For further information, see UL 943, *Standard for Ground-Fault Circuit Interrupters*.

The definition of *ground-fault circuit interrupter* was revised for the 2002 *Code* with the addition of an FPN that describes how personal protection can be achieved.

**Ground-Fault Current Path.** An electrically conductive path from the point of a ground fault on a wiring system through normally non–current-carrying conductors, equipment, or the earth to the electrical supply source.

FPN: Examples of ground-fault current paths could consist of any combination of equipment grounding conductors, metallic raceways, metallic cable sheaths, electrical equipment, and any other electrically conductive material such as metal water and gas piping, steel framing members, stucco mesh, metal ducting, reinforcing steel, shields of communications cables, and the earth itself.

A person must be exposed to a potential difference of 50 volts or more for a shock hazard to exist. The purpose of establishing an effective *ground-fault current path* is to eliminate the possibility that a potential difference of 50 volts or more can exist.

**Guarded.** Covered, shielded, fenced, enclosed, or otherwise protected by means of suitable covers, casings, barriers, rails, screens, mats, or platforms to remove the likelihood of approach or contact by persons or objects to a point of danger.

**Health Care Facilities.** Buildings or portions of buildings in which medical, dental, psychiatric, nursing, obstetrical, or surgical care is provided. Health care facilities include, but are not limited to, hospitals, nursing homes, limited care facilities, clinics, medical and dental offices, and ambulatory care centers, whether permanent or movable.

**Heating Equipment.** For the purposes of Article 430, the term includes any equipment used for heating purposes whose heat is generated by induction or dielectric methods.

**Hoistway.** Any shaftway, hatchway, well hole, or other vertical opening or space in which an elevator or dumbwaiter is designed to operate.

See Article 620 of the *NEC* for the installation of electrical equipment and wiring methods in *hoistways.*

**Identified (as applied to equipment).** Recognizable as suitable for the specific purpose, function, use, environment, application, and so forth, where described in a particular code or standard requirement.

FPN: Examples of ways to determine suitability of equipment for a specific purpose, environment, or application include investigations by a qualified testing laboratory (listing and labeling), an inspection agency, or other organizations concerned with product evaluation.

**Incident Energy.** The amount of energy impressed on a surface, a certain distance from the source, generated during an electrical arc event. One of the units used to measure incident energy is calories per centimeter squared (cal/cm$^2$).

The definition of *incident energy* was added to provide clarity for the user. Incident energy could be expressed in several different terms, such as calories per square centimeter, joules per square centimeter, or calories per square inch. However, incident energy must be expressed in the same terms in which the PPE is thermally rated, which usually is calories per square centimeter.

**Induction Heating.** The heating of a nominally conductive material due to its own $I^2R$ losses when the material is placed in a varying electromagnetic field.

**Insulated.** Separated from other conducting surfaces by a dielectric (including air space) offering a high resistance to the passage of current.

FPN: When an object is said to be insulated, it is understood to be insulated for the conditions to which it is normally subject. Otherwise, it is, within the purpose of these rules, uninsulated.

**Irrigation Machine.** An electrically driven or controlled machine, with one or more motors, not hand portable, and used primarily to transport and distribute water for agricultural purposes.

**Isolated (as applied to location).** Not readily accessible to persons unless special means for access are used.

See the definition of *accessible, readily.*

**Labeled.** Equipment or materials to which has been attached a label, symbol, or other identifying mark of an organization that is acceptable to the authority having jurisdiction and concerned with product evaluation, that maintains periodic inspection of production of labeled equipment or materials, and by whose labeling the manufacturer indicates compliance with appropriate standards or performance in a specified manner.

Equipment and conductors required or permitted by the *NEC* are acceptable only if they have Fbeen approved for a specific environment or application by the authority having jurisdiction, as stated in 110.2 of the *NEC.* See *NEC* 90.7 regarding the examination of equipment for safety. Listing or labeling by a qualified testing laboratory provides a basis for approval.

**Lighting Outlet.** An outlet intended for the direct connection of a lampholder, a luminaire (lighting fixture), or a pendant cord terminating in a lampholder.

**Limited Approach Boundary.** An approach limit at a distance from an exposed live part within which a shock hazard exists.

The term *Limited Approach Boundary* is used to identify an imaginary distance beyond which special considerations are necessary to protect the worker. *Limited Approach Boundary* is related to mitigating exposure to shock only.

**Listed.** Equipment, materials, or services included in a list published by an organization that is acceptable to the authority having jurisdiction and concerned with evaluation of products or services, that maintains periodic inspection of production of listed equipment or materials or periodic evaluation of services, and whose listing states that the equipment, material, or services either meets appropriate designated standards or has been tested and found suitable for a specified purpose.

FPN: The means for identifying listed equipment may vary for each organization concerned with product evaluation, some of which do not recognize equipment as listed unless it is also labeled. Use of the system employed by the listing organization allows the authority having jurisdiction to identify a listed product.

The definition of *listed* was revised for the 2002 *NEC.* This slightly reworded definition is now in line with the definition of *listed* found in the NFPA Regulations Governing Committee Projects. Reviewing other *NEC*-defined terms such as *approved, authority having jurisdiction, labeled,* and *identified (as applied to equipment)* will help the user understand the approval process.

**Live Parts.** Energized conductive components.

The definition of *live parts* was changed for the 2002 *NEC.* The condition of a hazard has been removed from this new definition. Live parts are associated with all voltage and energy levels.

**Location, Damp.** Locations protected from weather and not subject to saturation with water or other liquids but subject to moderate degrees of moisture. Examples of such locations include partially protected locations under canopies, marquees, roofed open porches, and like locations, and interior locations subject to moderate degrees of moisture, such as some basements, some barns, and some cold-storage warehouses.

**Location, Dry.** A location not normally subject to dampness or wetness. A location classified as dry may be temporarily subject to dampness and wetness, as in the case of a building under construction.

**Location, Wet.** Installations under ground or in concrete slabs or masonry in direct contact with the earth; in locations subject to saturation with water or other liquids, such as vehicle washing areas; and in unprotected locations exposed to weather.

It is intended that the inside of a raceway in a wet location or a raceway installed under ground be considered a *wet location.* Therefore, any conductors contained therein would be required to be suitable for wet locations.

See *NEC* 300.6(C) for some examples of wet locations and *NEC* 410.4(A) for information on luminaires installed in wet locations. See *patient care area* in *NEC* 517.2 for a definition of wet locations in a patient care area.

**Medium Voltage Cable.** A single or multiconductor solid dielectric insulated cable rated 2001 volts or higher.

**Metal-Clad Cable.** A factory assembly of one or more insulated circuit conductors with or without optical fiber members enclosed in an armor of interlocking metal tape, or a smooth or corrugated metallic sheath.

**Metal Wireways.** Sheet metal troughs with hinged or removable covers for housing and protecting electric wires and cable and in which conductors are laid in place after the wireway has been installed as a complete system.

**Mineral-Insulated Metal-Sheathed Cable.** A factory assembly of one or more conductors insulated with a highly compressed refractory mineral insulation and enclosed in a liquidtight and gastight continuous copper or alloy steel sheath.

**Mobile X-Ray.** X-ray equipment mounted on a permanent base with wheels, casters, or a combination of both to facilitate moving the equipment while completely assembled.

**Motor Control Center.** An assembly of one or more enclosed sections having a common power bus and principally containing motor control units.

**Nonmetallic-Sheathed Cable.** A factory assembly of two or more insulated conductors having an outer sheath of nonmetallic material.

**Nonmetallic Wireways.** Flame-retardant, nonmetallic troughs with removable covers for housing and protecting electric wires and cables in which conductors are laid in place after the wireway has been installed as a complete system.

**Open Wiring on Insulators.** An exposed wiring method using cleats, knobs, tubes, and flexible tubing for the protection and support of single insulated conductors run in or on buildings.

**Outlet.** A point on the wiring system at which current is taken to supply utilization equipment.

An *outlet* could be a lighting outlet or a receptacle outlet.

**Outline Lighting.** An arrangement of incandescent lamps or electric discharge lighting to outline or call attention to certain features such as the shape of a building or the decoration of a window.

See Article 600 of the *NEC* for details on *outline lighting.*

**Oven, Wall-Mounted.** An oven for cooking purposes and consisting of one or more heating elements, internal wiring, and built-in or separately mountable controls.

**Overcurrent.** Any current in excess of the rated current of equipment or the ampacity of a conductor. It may result from overload, short circuit, or ground fault.

FPN: A current in excess of rating may be accommodated by certain equipment and conductors for a given set of conditions. Therefore, the rules for overcurrent protection are specific for particular situations.

**Overload.** Operation of equipment in excess of normal, full-load rating, or of a conductor in excess of rated ampacity that, when it persists for a sufficient length of time, would cause damage or dangerous overheating. A fault, such as a short circuit or ground fault, is not an overload.

**Panelboard.** A single panel or group of panel units designed for assembly in the form of a single panel, including buses and automatic overcurrent devices, and equipped with or without switches for the control of light, heat, or power circuits; designed to be placed in a cabinet or cutout box placed in or against a wall, partition, or other support; and accessible only from the front.

See Article 408 of the *NEC* for details on *panelboards.*

**Power and Control Tray Cable.** A factory assembly of two or more insulated conductors, with or without associated bare or covered grounding conductors under a nonmetallic jacket, for installation in cable trays, in raceways, or where supported by a messenger wire.

**Power-Limited Tray Cable.** Type PLTC nonmetallic-sheathed cable is a factory assembly of two or more insulated conductors under a nonmetallic jacket.

**Premises Wiring (System).** That interior and exterior wiring, including power, lighting, control, and signal circuit wiring together with all their associated hardware, fittings, and wiring devices, both permanently and temporarily installed, that extends from the service point or source of power, such as a battery, a solar photovoltaic system, or a generator, transformer, or converter windings, to the outlet(s). Such wiring does not include wiring internal to appliances, luminaires (fixtures), motors, controllers, motor control centers, and similar equipment.

**Prohibited Approach Boundary.** An approach limit at a distance from an exposed live part within which work is considered the same as making contact with the live part.

*Prohibited Approach Boundary* is related only to shock.

**Qualified Person.** One who has skills and knowledge related to the construction and operation of the electrical equipment and installations and has received safety training on the hazards involved.

The definition of the term *qualified person* has been modified in the 2004 edition of NFPA 70E to include safety training. No longer is it adequate for an electrical worker to be "familiar with the hazards involved"; the electrical worker of today also must have safety training on the hazards involved.

**Raceway.** An enclosed channel of metal or nonmetallic materials designed expressly for holding wires, cables, or busbars, with additional functions as permitted in this standard. Raceways include, but are not limited to, rigid metal conduit, rigid nonmetallic conduit, intermediate metal conduit, liquidtight flexible conduit, flexible metallic tubing, flexible metal conduit, electrical metallic tubing, electrical nonmetallic tubing, underfloor raceways, cellular concrete floor raceways, cellular metal floor raceways, surface raceways, wireways, and busways.

Cable trays (see Article 392 of the *NEC*) are not considered *raceways* in the *NEC*.

**Receptacle.** A receptacle is a contact device installed at the outlet for the connection of an attachment plug. A single receptacle is a single contact device with no other contact device on the same yoke. A multiple receptacle is two or more contact devices on the same yoke.

Exhibit 100.5 shows one single and two multiple *receptacles*.

***EXHIBIT 100.5. Receptacles.***

**Receptacle Outlet.** An outlet where one or more receptacles are installed.

**Restricted Approach Boundary.** An approach limit at a distance from an exposed live part within which there is an increased risk of shock, due to electrical arc over combined with inadvertent movement, for personnel working in close proximity to the live part.

The *Restricted Approach Boundary* is related only to shock.

**Separately Derived System.** A premises wiring system whose power is derived from a battery, from a solar photovoltaic system, or from a generator, transformer, or converter windings, and that has no direct electrical connection, including a solidly connected grounded circuit conductor, to supply conductors originating in another system.

**Service.** The conductors and equipment for delivering electric energy from the serving utility to the wiring system of the premises served.

The definition of *service* was modified for the 1999 *NEC* to state that electrical energy to a service can be supplied only by the serving utility. If electrical energy is supplied by other than the serving utility, the supplied conductors and equipment are considered to be feeders, not a service.

**Service Cable.** Service conductors made up in the form of a cable.

**Service Conductors.** The conductors from the service point to the service disconnecting means.

The definition of *service conductors* was revised in the 1999 *NEC* to be more precise. The phrase "or other source of power" was deleted. Service conductors originate at the service point (where the serving utility ends) and end at the service disconnect. Due to the associated 1999 *NEC* change in the definition of *service,* service conductors may originate only from the serving utility.

*Service conductors* is a broad term and may include service drops, service laterals, and service-entrance conductors, but this term specifically excludes any wiring on the supply side (serving utility side) of the service point.

If the utility has specified that the service point is at the utility pole, the service conductors from an overhead distribution system originate at the utility pole and terminate at the service disconnecting means. If the utility has specified that the service point is at the utility manhole, the service conductors from an underground distribution system originate at the utility manhole and terminate at the service disconnecting means. Where utility-owned primary conductors are extended to outdoor pad-mounted transformers on private property, the service conductors originate at the secondary connections of the transformers only if the utility has specified that the service point is at the secondary connections.

See Article 230, Part VIII, of the *NEC* for service conductors exceeding 600 volts, nominal.

**Service Drop.** The overhead service conductors from the last pole or other aerial support to and including the splices, if any, connecting to the service-entrance conductors at the building or other structure.

In Exhibit 100.6, the overhead *service-drop* conductors run from the utility pole and connect to the service-entrance conductors at the service point. Conductors on the utility side of the service point are not covered by the *NEC.* Instead, the utility specifies the location of the service point, which varies from utility to utility, as well as from occupancy to occupancy.

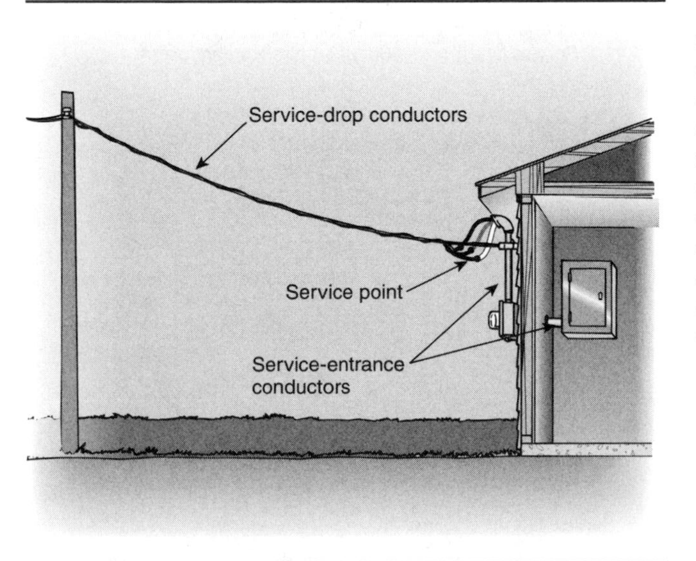


***EXHIBIT 100.6.** Overhead system showing a service drop from a utility pole to attachment on a house and service-entrance conductors from point of attachment (spliced to service-drop conductors), down the sides of the house, through the meter socket, and terminating in the service department.*

**Service-Entrance Cable.** A single conductor or multiconductor assembly provided with or without an overall covering, primarily used for services, and of the following types:

*Type SE.* Service-entrance cable having a flame-retardant, moisture-resistant covering.

*Type USE.* Service-entrance cable, identified for underground use, having a moisture-resistant covering, but not required to have a flame-retardant covering.

**Service-Entrance Conductors, Overhead System.** The service conductors between the terminals of the service equipment and a point usually outside the building, clear of building walls, where joined by tap or splice to the service drop.

See Exhibit 100.6 for an illustration of *service-entrance conductors* in an overhead system.

**Service-Entrance Conductors, Underground System.** The service conductors between the terminals of the service equipment and the point of connection to the service lateral.

FPN: Where service equipment is located outside the building walls, there may be no service-entrance conductors, or they may be entirely outside the building.

See Exhibit 100.7 for an illustration of *service-entrance conductors* in an *underground system.* As illustrated, the underground service laterals may be run from poles (top) or from trans-

***EXHIBIT 100.7.*** *Underground systems showing service laterals run from a pole and from a transformer.*

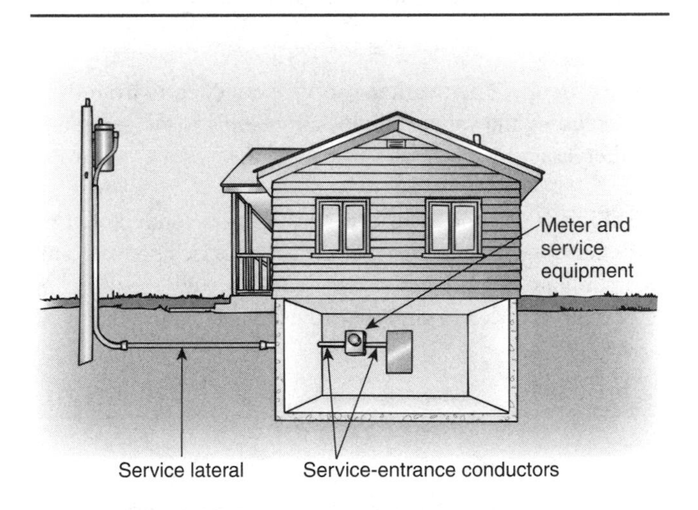


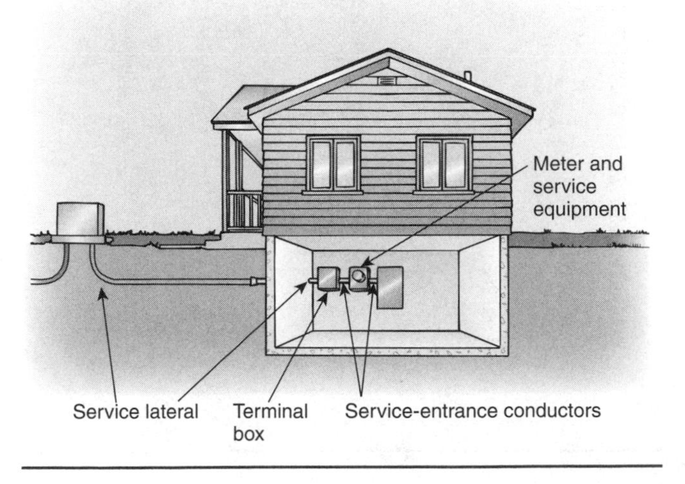

formers (bottom) and with or without terminal boxes, provided they begin at the service point. Conductors on the utility side of the service point are not covered by the *NEC*. The utility specifies the location of the service point, which vary from utility to utility, as well as from occupancy to occupancy.

**Service Equipment.** The necessary equipment, usually consisting of a circuit breaker(s) or switch(es) and fuse(s), and their accessories, connected to the load end of service conductors to a building or other structure, or an otherwise designated area, and intended to constitute the main control and cutoff of the supply.

**Service Point.** The point of connection between the facilities of the serving utility and the premises wiring.

The *service point* is the point of demarcation between the serving utility and the premises wiring. The service point is the point on the wiring system where the serving utility ends and the premises wiring begins. The serving utility generally specifies the location of the service point.

Because the location of the service point is generally determined by the utility, the service-drop conductors and the service-lateral conductors may or may not be part of the service as covered by the *NEC*. Only conductors physically located on the premises wiring side of the service point are covered by the *NEC*. Conductors located on the utility side of the service point not covered in the definition of service conductors therefore are not covered by the *NEC*.

Generally, based on the definitions of the terms *service point* and *service conductors,* any conductor on the serving utility side of the service point is not covered by the *NEC*. For example, a typical suburban residence has an overhead service drop from the utility pole to the house. If the utility specifies that the service point is at the point of attachment of the service drop to the house, the service-drop conductors are not considered service conductors because the service drop is not on the premises wiring side of the service point. Alternatively, if the service point is specified as "at the pole" by the utility, the service-drop conductors are considered service conductors, and the *NEC* would apply to the service drop.

Exact locations for a service point may vary from utility to utility, as well as from occupancy to occupancy.

**Shock Hazard.** A dangerous condition associated with the possible release of energy caused by contact or approach to live parts.

**Show Window.** Any window used or designed to be used for the display of goods or advertising material, whether it is fully or partly enclosed or entirely open at the rear and whether or not it has a platform raised higher than the street floor level.

See 220.3(B)(7), 220.12(A) of the *NEC* for *show window* lighting load requirements.

**Signaling Circuit.** Any electric circuit that energizes signaling equipment.

**Special Permission.** The written consent of the authority having jurisdiction.

The authority having jurisdiction for enforcement of the *NEC* is responsible for making interpretations and granting *special permission* contemplated in a number of the rules, as stated in 90.4 of the *NEC*. For specific examples of special permission, see 110.26(A)(1)(b), 230.2(B), and 426.14 of the *NEC*.

**Step Potential.** A ground potential gradient difference that can cause current flow from foot to foot through the body.

**Switch, Isolating.** A switch intended for isolating an electric circuit from the source of power. It has no interrupting rating, and it is intended to be operated only after the circuit has been opened by some other means.

**Switch, Motor Circuit.** A switch rated in horsepower that is capable of interrupting the maximum operating overload current of a motor of the same horsepower rating as the switch at the rated voltage.

**Switchboard.** A large single panel, frame, or assembly of panels on which are mounted on the face, back, or both, switches, overcurrent and other protective devices, buses, and usually instruments. Switchboards are generally accessible from the rear as well as from the front and are not intended to be installed in cabinets.

Busbars on a *switchboard* are required to be so arranged as to avoid inductive overheating. Service busbars are required to be isolated by barriers from the remainder of the switchboard. Most modern switchboards are totally enclosed to minimize the probability of spreading fire to adjacent combustible materials and to guard live parts. See Article 408 of the *NEC* for more information regarding switchboards.

**Touch Potential.** A ground potential gradient difference that can cause current flow from hand to hand or hand to foot through the body.

**Unqualified Person.** A person who is not a qualified person.

Workers who might be exposed to an electrical hazard as a work task is performed must be trained to recognize that a hazard exists and how to avoid that hazard. Any person who has not received specific training is an *unqualified person.* A worker who has been trained to perform a task might be qualified to perform that task and still be unqualified to perform any other task. The characteristics of being qualified and unqualified are task dependent.

**Utilization Equipment.** Equipment that utilizes electric energy for electronic, electromechanical, chemical, heating, lighting, or similar purposes.

**Ventilated.** Provided with a means to permit circulation of air sufficient to remove an excess of heat, fumes, or vapors.

**Volatile Flammable Liquid.** A flammable liquid having a flash point below 38°C (100°F), or a flammable liquid whose temperature is above its flash point, or a Class II combustible liquid that has a vapor pressure not exceeding 276 kPa (40 psia) at 38°C (100°F) and whose temperature is above its flash point.

The flash point of a liquid is defined as the minimum temperature at which it gives off sufficient vapor to form an ignitible mixture with the air near the surface of the liquid or within the vessel used to contain the liquid. An ignitible mixture is defined as a mixture within the explosive or flammable range (between upper and lower limits) that is capable of the propagation of flame away from the source of ignition when ignited. Some emission of vapors takes place below the flash point but not in sufficient quantities to form an ignitible mixture.

**Voltage (of a Circuit).** The greatest root-mean-square (rms) (effective) difference of potential between any two conductors of the circuit concerned.

FPN: Some systems, such as 3-phase 4-wire, single-phase 3-wire, and 3-wire direct-current, may have various circuits of various voltages.

Common 3-phase, 4-wire wye systems are 480/277 volts and 120/208 volts. The *voltage* of the circuit is the highest voltage between any two conductors (i.e., 480 volts or 208 volts). The voltage of the circuit of a 2-wire feeder or branch circuit (single phase and the grounded conductor) derived from these systems would be the voltage between the two conductors at the lower voltage (i.e., 277 volts or 120 volts). The same applies to dc or single-phase, 3-wire systems where there are two voltages.

**Voltage, Nominal.** A nominal value assigned to a circuit or system for the purpose of conveniently designating its voltage class (e.g., 120/240 volts, 480Y/277 volts, 600 volts). The actual voltage at which a circuit operates can vary from the nominal within a range that permits satisfactory operation of equipment.

FPN: See ANSI C84.1-1995, *Electric Power Systems and Equipment—Voltage Ratings (60 Hz)*.

See 220.2(A) of the *NEC* for a list of *nominal voltages* used in computing branch-circuit and feeder loads.

**Voltage to Ground.** For grounded circuits, the voltage between the given conductor and that point or conductor of the circuit that is grounded; for ungrounded circuits, the greatest voltage between the given conductor and any other conductor of the circuit.

The *voltage to ground* of a 480/277-volt wye system would be 277 volts; of a 120/208-volt wye system, 120 volts; and of a 3-phase, 3-wire ungrounded 480-volt system, 480 volts.

For a 3-phase, 4-wire delta system with the center of one leg grounded, there are two voltages to ground. For example, on a 240-volt system, two legs each would have 120 volts to ground and the third, or "high" leg, would have 208 volts to ground. See 215.8, 230.56, and 408.3(E) of the *NEC* for special marking and arrangements on such circuit conductors.

**Watertight.** Constructed so that moisture will not enter the enclosure under specified test conditions.

Unless the enclosure is hermetically sealed and thus *watertight,* it is possible for moisture to enter the enclosure. See the commentary following the definition of *enclosure* and following Table 430.91 of the *NEC.*

**Weatherproof.** Constructed or protected so that exposure to the weather will not interfere with successful operation.

FPN: Rainproof, raintight, or watertight equipment can fulfill the requirements for weatherproof where varying weather conditions other than wetness, such as snow, ice, dust, or temperature extremes, are not a factor.

See the commentary following the definition of *enclosure.* Industry standards for enclosures are found in the commentary following 430.91 of the *NEC.*

**Working Near (live parts).** Any activity inside a Limited Approach Boundary.

The Limited Approach Boundary is the outermost boundary of an activity of a person *working near live parts* and cannot be crossed by an unqualified person unless escorted by a qualified person. Persons outside the Limited Approach Boundary might be within the Flash Protection Boundary, making them susceptible to a second-degree burn. Workers must consider all of the hazards and how they are related to be properly protected.

**Working On (live parts).** Coming in contact with live parts with the hands, feet, or other body parts, with tools, probes, or with test equipment, regardless of the personal protective equipment a person is wearing.

Any work that requires a person to cross the Prohibited Approach Boundary is considered to be working on a live part and is subject to all requirements associated with *working on live parts.* Note that measuring voltage requires that the Prohibited Approach Boundary be breached, suggesting that measuring voltage exposes a worker to an electrical hazard.

## II. Over 600 Volts, Nominal

Whereas the preceding definitions are intended to apply wherever the terms are used throughout this standard, the following definitions are applicable only to parts of this standard specifically covering installations and equipment operating at over 600 volts, nominal.

**Fuse.** An overcurrent protective device with a circuit-opening fusible part that is heated and severed by the passage of overcurrent through it.

FPN: A fuse comprises all the parts that form a unit capable of performing the prescribed functions. It may or may not be the complete device necessary to connect it into an electrical circuit.

**Switching Device.** A device designed to close, open, or both, one or more electric circuits.

*Circuit Breaker.* A switching device capable of making, carrying, and interrupting currents under normal circuit conditions, and also making, carrying for a specified time, and interrupting currents under specified abnormal circuit conditions, such as those of short circuit.

*Cutout.* An assembly of a fuse support with either a fuseholder, fuse carrier, or disconnecting blade. The fuseholder or fuse carrier may include a conducting element (fuse link), or may act as the disconnecting blade by the inclusion of a nonfusible member.

*Disconnecting (or Isolating) Switch (Disconnector, Isolator).* A mechanical switching device used for isolating a circuit or equipment from a source of power.

*Disconnecting Means.* A device, group of devices, or other means whereby the conductors of a circuit can be disconnected from their source of supply.

*Interrupter Switch.* A switch capable of making, carrying, and interrupting specified currents.

# ARTICLE 110
# General Requirements for Electrical Safety-Related Work Practices

The requirements defined within this article address how people interact with electrical equipment. The base expectation of Article 110 is that if an installation meets the requirements of the *National Electrical Code®* (NFPA 70, *NEC®*, or *Code*) and is installed according to the manufacturer's directions, that installation is safe when operating normally. The requirements of this article are intended to apply when something is wrong and the equipment is not operating normally. A second assumption is that the equipment is maintained adequately. The requirements in this article address only shock (electrocution) and the thermal aspects of an arc flash.

### 110.1 Scope

Chapter 1 covers electrical safety-related work practices and procedures for employees who work on or near exposed energized electrical conductors or circuit parts in workplaces that are included in the scope of this standard. Electric circuits and equipment not included in the scope of this standard might present a hazard to employees not qualified to work near such facilities. Requirements have been included in Chapter 1 to protect unqualified employees from such hazards.

Article 110 identifies work practices and procedures that can reduce or eliminate exposure of people to hazards associated with electrical energy. Although the scope statement limits application to exposed energized electrical conductors or circuit parts, exposure to circuit conductors or parts of an unknown state also is covered.

Electrical hazards are related to the physical behavior of electrical current and related energy, and they may be widely scattered across a site or facility. Electrical injuries are not related to employer or industrial segment. The scope statement indicates that this standard applies wherever an employee is exposed to an electrical hazard. The work practices contained in this article apply to all work tasks where exposure or potential exposure to exposed live parts exists. No industrial segment, including the construction industry, is excluded from the scope.

Persons who work with electrical conductors and circuit parts are especially exposed. However, all persons who use equipment connected to an electrical energy source are exposed.

NFPA 70E is all about work practices, whereas the *NEC* is primarily about facility installation. NFPA 70E provides for work practices that minimize exposure to electrical hazards associated with equipment or installations that *are* or *are not* covered by the *NEC*. The intent is for NFPA 70E to be all-inclusive where potential exposure to injury from an electrical energy source exists. The intent of NFPA 70E is to provide practices that are needed to protect both qualified and unqualified persons from exposure to hazards associated with electrical energy.

### 110.2 Purpose

These practices and procedures are intended to provide for employee safety relative to electrical hazards in the workplace.

Article 110 covers both work practices and procedures. Its sole purpose is to identify work practices that provide for employee safety from electrical hazards, and its requirements have historically proven to reduce exposure to hazards effectively. Requirements contained in Article 110 are the result of the merged collective experience of all Technical Committee members.

### 110.3 Responsibility

The safety-related work practices contained in Chapter 1 shall be implemented by employees. The employer shall provide the safety-related work practices and shall train the employee who shall then implement them.

Work practices and procedures must include input from, and be embraced by, both the workers (employees) and the line organization (employer). Employers are accountable for generating procedures, in accordance with 110.3. On the other hand, employees are accountable for executing work tasks as defined in the procedures published by the employer. The best working environment results when employers and employees work together in the area of safety. The requirements of this article are consistent with the regulations in OSHA 29 CFR 1910, Subpart S.

### 110.4 Multiemployer Relationship

**(A) Safe Work Practices.** On multiemployer worksites (in all industry sectors), more than one employer may be responsible for hazardous conditions that violate safe work practices.

The multiemployer relationship requirements of 110.4 are new to the 2004 edition of NFPA 70E. This section was added to define requirements for multiemployer worksites. The OSHA multiemployer citation policy can cite more than one employer for hazardous conditions that violate an OSHA regulation. The requirements in this section can help employers avoid such citations.

Employers could be categorized as on-site or outside employers. An "on-site" employer is one that might be landlord of a facility. Host (on-site) employers might have more than one division functioning at a particular facility location. For instance, a single location might have an exploration division and a downstream division at the same physical location. Although the exploration division might consider the downstream division to be "landlord," both divisions would be considered as the "on-site" employer from the perspective of consensus standards. An on-site employer can be a site owner or another employer acting as the site landlord.

An "outside" employer is one who is contractually obligated to the "on-site" employer. Service contractors, such as soda machine vendors, are outside employers, as are prime contractors and subcontractors. An outside employer can be a construction contractor, a maintenance contractor, an equipment vendor, or a service provider.

**(B) Outside Personnel (Contractors, etc.).** Whenever outside servicing personnel are to be engaged in activities covered by the scope and application of this standard, the on-site employer and the outside employer(s) shall inform each other of existing hazards, personal protective equipment/clothing requirements, safe work practice procedures, and emergency/evacuation procedures applicable to the work to be performed. This coordination shall include a meeting and documentation.

More than one employer commonly performs work on a single worksite during the same period of time. Generally, employers are responsible for managing exposure of their own employees to safety hazards. However, 110.4(B) requires that employers that recognize a hazard created by another employer must take steps to ensure that the employer that created the hazard is notified that a hazard exists. Employers generally are responsible for eliminating, guarding, or otherwise warning both employees and nonemployees that an electrical hazard exists. If the hazard presents an imminent threat, employers must protect employees from the threat either by correcting the hazard or by eliminating exposure to it. Generally, employers that create exposure to an electrical hazard are responsible for taking all steps necessary to ensure that no one is injured.

Other articles of this standard require an employer to generate and publish an electrical safety program and certain procedures. This article requires that all employers working in the same physical or electrical area exchange information about hazards associated with the work, communication methods, and all other processes that could impact the safety of employees of either employer.

A documented meeting must be held that enables the exchange of information between employees. The documentation should identify attendees and discussion agenda. Employers should maintain the record of the meeting(s) until the multiemployer relationship ceases to exist.

### 110.5 Organization

Chapter 1 of this standard is divided into three articles. Article 110 provides general requirements regarding the preparation for, and conduct of, work performed on or near electrical components regardless of whether such components are energized or not. Article 120 emphasizes working deenergized and describes the work practices used to deenergize electrical components to put them into an electrically safe work condition before attempting work on or near them. Article 130 provides requirements for working on or near electrical components that have not been placed into an electrically safe work condition.

This standard is organized to emphasize strategies sequentially, with the most important strategy considered first. Article 110 identifies requirements that are aligned with preparing to execute electrical work, including a requirement to establish an electrical safety program. Each requirement in this article should be addressed in an electrical safety program. For instance, the electrical safety program should establish electrical safety training requirements for both qualified and unqualified employees.

Article 120 identifies requirements associated with the primary NFPA 70E strategy to eliminate exposure to an electrical hazard by removing the source of energy and ensuring that the energy cannot reaccumulate, thus establishing an electrically safe work condition. The article continues by defining other mechanisms to mitigate or eliminate exposure to electrical safety hazards where an electrically safe work condition could not be established for valid reasons.

If the exposure to electrical hazards is not eliminated by implementing the requirements of Article 110 or Article 120, the requirements of Article 130 must be implemented to mitigate any exposure to the maximum extent possible.

### 110.6 Training Requirements

**(A) Safety Training.** The training requirements contained in this section shall apply to employees who face a risk of electrical hazard that is not reduced to a safe level by the electrical installation requirements of Chapter 4. Such employees shall be trained to understand the specific hazards associated with electrical energy. They shall be trained in safety-related work practices and procedural requirements as necessary to provide protection from the electrical hazards associated with their respective job or task assignments. Employees shall be trained to identify and understand the relationship between electrical hazards and possible injury.

A base expectation of Article 110 is that, when operating normally, an installation is safe if it meets both of the following conditions:

1. All the requirements in Chapter 4 of this standard (which come from the *NEC*)
2. All manufacturers' instructions

However, if the installation or equipment fails to meet these conditions, employees face an elevated risk of injury. For example, an equipment door or cover left with fasteners unlatched does not meet the requirements of Chapter 4. Equipment that is not adequately maintained does not meet these requirements and causes employees to be exposed to an electrical hazard.

Each employee who is or might be exposed to an elevated risk of injury by exposure to an electrical hazard must be trained to understand the specific hazards to which he or she might be exposed. To increase understanding, the training should include the following:

- What electrical hazards are present in the workplace
- How each electrical hazard affects body tissues
- How to determine the degree of each hazard
- How to avoid exposure to each hazard
- How to minimize risk by body position
- What personal protective equipment (PPE) is needed for the employee to execute his or her work assignment
- How to select and inspect PPE
- What employer-provided procedures, including specific work practices, must be implemented by the employee
- How increased duration of exposure to an electrical hazard results in a higher frequency of injuries
- How to perform a hazard/risk analysis

- How to determine Limited, Restricted, and Prohibited Approach Boundaries and recognize that these boundaries are related to protection from exposure to electrical shock and electrocution
- How to determine the Flash Protection Boundary and its relationship to the quantity of available energy

**(B) Type of Training.** The training required by this section shall be classroom or on-the-job type, or a combination of the two. The degree of training provided shall be determined by the risk to the employee.

Classroom training is effective for some objectives, and on-the-job training (OJT) is effective for others. In most instances, effective training makes use of both training processes. The qualification of the instructor is very important. Frequently, new employees learn by emulating a more experienced worker. If the experienced worker understands electrical safety requirements and practices them, the mentoring process is good. However, because a significant amount of new information is available, mentors might also need to be trained.

Some employees have minimal exposure to electrical hazards, especially with effective maintenance. For instance, arc flash training is not necessary for an office worker who is never exposed to an arc flash. Workers who are not exposed to medium and high voltages might not need to be trained to understand the characteristics of higher-voltage energy.

If any risk of injury from electrical energy exists, employees must be trained to recognize and deal with risk. Sometimes, risk of injury is of sufficient magnitude that the work task should not be executed.

**(C) Emergency Procedures.** Employees working on or near exposed energized electrical conductors or circuit parts shall be trained in methods of release of victims from contact with exposed energized conductors or circuit parts. Employees shall be regularly instructed in methods of first aid and emergency procedures, such as approved methods of resuscitation, if their duties warrant such training.

Employee training must ensure that each employee understand the steps necessary to release a victim who might be in contact with an energized conductor. The employee must be trained to know that the first action in responding to an electrical contact incident must be to remove the source of the electricity.

Employees must understand emergency first-aid procedures. Each worker must know the procedure necessary to reach emergency assistance. Employees who are frequently exposed to electrical hazards should be trained to provide cardiopulmonary resuscitation (CPR).

Employees who are qualified to work on or near exposed energized electrical conductors or circuit parts should be trained to perform CPR and emergency first aid. CPR and emergency first-aid training should be up-to-date. In some instances, unqualified employees might be expected to provide emergency first aid or CPR.

**(D) Employee Training.**

**(1) Qualified Person.** A qualified person shall be trained and knowledgeable of the construction and operation of equipment or a specific work method and be trained to recognize and avoid the electrical hazards that might be present with respect to that equipment or work method.

The definition of *qualified person* suggests that safety training is necessary, and 110.6(D)(1) further defines the meaning of *safety training.* The requirements expand and define training considered necessary for a person to be qualified to perform work on or near exposed, energized electrical conductors or circuit parts.

To avoid unexpected exposure to an electrical hazard, a person must be thoroughly familiar with and demonstrate familiarity with the details of how electrical equipment is constructed. For instance, the actions necessary to withdraw a motor control center (MCC) unit from an MCC structure might be dramatically different between equipment from two different manufacturers or from different models of the same manufacturer. A qualified person must be aware of the difference and be familiar with how to perform the task. A person also might be qualified only for a single work task.

Section 110.6(D)(1) requires that a qualified person must be knowledgeable and familiar with certain aspects of executing electrical work. Note that the terms *knowledgeable* and *familiar* are synonyms for each other. To establish that an employee is knowledgeable or familiar with details of executing electrical work, employee understanding must be established in some way. Employers should establish and maintain documents that provide evidence of the qualifications of each employee. The documentation also should define the limits of each employee's qualification.

A qualified person might understand the shock hazard and have only a passing knowledge of other electrical hazards. In that case, that person must not perform work tasks that expose him or her to the other hazards. For instance, an instrument technician who is exposed to a maximum circuit voltage of 120 volts and maximum circuit size of 30 amperes may not require arc flash training. However, if the same person performs work where an arc flash hazard exists, he or she must understand that exposure and be qualified for that task.

A qualified person must be able to recognize all electrical hazards within the realm of his or her work assignment. He or she must be trained to understand and implement the procedural requirements of the employer. He or she also must be able to select and use adequate protective equipment.

An unqualified person may work in an environment that is influenced by electrical hazards as long as he or she receives specific instruction regarding the existence, location, and degree of any electrical hazards. The unqualified person must be under the direct supervision of a qualified person for the duration of the exposure.

A qualified person must understand how to use a voltage detector to determine if electrical energy is present or not and understand how to use the specific voltage detector that will be used. He or she also must understand any limitations established by the manufacturer. If the manufacturer establishes a duty cycle, for instance, the qualified person must understand the meaning of *duty cycle* and how it applies to his or her work assignment.

(a) Such persons shall also be familiar with the proper use of the special precautionary techniques, personal protective equipment, including arc-flash, insulating and shielding materials, and insulated tools and test equipment. A person can be considered qualified with respect to certain equipment and methods but still be unqualified for others.

(b) An employee who is undergoing on-the-job training and who, in the course of such training, has demonstrated an ability to perform duties safely at his or her level of training and who is under the direct supervision of a qualified person shall be considered to be a qualified person for the performance of those duties.

(c) Such persons permitted to work within the Limited Approach Boundary of exposed live parts operating at 50 volts or more shall, at a minimum, be additionally trained in all of the following:

(1) The skills and techniques necessary to distinguish exposed energized parts from other parts of electrical equipment
(2) The skills and techniques necessary to determine the nominal voltage of exposed live parts
(3) The approach distances specified in Table 130.2(C) and the corresponding voltages to which the qualified person will be exposed
(4) The decision-making process necessary to determine the degree and extent of the hazard and the personal protective equipment and job planning necessary to perform the task safely

Section 110.6(D)(1)(c) establishes that only qualified persons are permitted to work within the Limited Approach Boundary. These qualified persons must possess the following:

- The skill and knowledge to determine whether an exposed electrical conductor is energized or not energized
- The ability and knowledge to determine the nominal voltage of an electrical circuit by reading drawings, signs, and labels
- A general knowledge of equipment construction
- A general knowledge that exposure to electrical hazards increases as he or she approaches an exposed live part

A qualified worker must be trained to do the following:

- Understand all electrical hazards and be capable of deciding whether his or her actions might result in a release of energy
- Determine whether PPE is necessary, what type of PPE is necessary, and how the PPE is rated
- Understand the protective characteristics of each PPE item
- Apply the decision-making process, which refers to a hazard/risk analysis
- Understand how to determine whether a hazard exists, if he or she could be exposed to the hazard when executing the work task, and the degree of the hazard, if one does exist
- Understand that all work on or near exposed live parts must be well planned
- Understand that each task in the work plan must be executed as planned
- Understand that if any step in the work plan cannot be executed according to the plan, the work must stop until it is replanned

**(2) Unqualified Persons.** Unqualified persons shall be trained in and be familiar with any of the electrical safety-related practices that might not be addressed specifically by Chapter 1 but are necessary for their safety.

In general, unqualified persons might be exposed to electrical hazards. If exposure to an electrical hazard exists or might exist, the unqualified person must be trained to understand how the exposure could occur and how to avoid injury that could result from the exposure. Examples of unqualified persons include office workers, janitors, equipment operators, apprentices, or workers from crafts other than electrical.

### 110.7 Electrical Safety Program

**(A) General.** The employer shall implement an overall electrical safety program that directs activity appropriate for the voltage, energy level, and circuit conditions.

FPN: Safety-related work practices are just one component of an overall electrical safety program.

Each employer must develop and implement an electrical safety program that directly addresses all employee exposure to each specific electrical hazard that exists on the work site. The electrical safety program implemented by "off-site" (contractor or vendor) employers must address each electrical hazard to which employees might be exposed. The electrical safety program (and related training) must be appropriate for the conditions that exist.

The electrical safety program must be written, published, and available to all employees. It may be part of a more extensive safety program, or it may be an independent program, but in either situation, it must cover all circuit conditions to which an employee might be exposed.

The program should contain all procedures that guide work activity for the employer.

**(B) Awareness and Self-Discipline.** The electrical safety program shall be designed to provide an awareness of the potential electrical hazards to employees who might from time to time work in an environment influenced by the presence of electrical energy. The program shall be developed to provide the required self-discipline for employees who occasionally must perform work on or near exposed energized electrical conductors and circuit parts. The program shall instill safety principles and controls.

For a worker to avoid exposure to an electrical hazard, he or she must be aware of all aspects of the work environment. The electrical safety program, including training, must emphasize employee self-discipline and require supervisors to discuss the safety aspects of work tasks. Discussing electrical incidents that illustrate various aspects of similar work tasks improves recognition that a similar incident or injury could occur and increases employee awareness.

**(C) Electrical Safety Program Principles.** The electrical safety program shall identify the principles upon which it is based.

FPN: For examples of typical electrical safety program principles, see Annex E.

**(D) Electrical Safety Program Controls.** An electrical safety program shall identify the controls by which it is measured and monitored.

FPN: For examples of typical electrical safety program controls, see Annex E.

**(E) Electrical Safety Program Procedures.** An electrical safety program shall identify the procedures for working on or near live parts operating at 50 volts or more or where an electrical hazard exists before work is started.

FPN: For an example of a typical electrical safety program procedure, see Annex E.

Procedures that provide guidance for employees to execute work tasks must be a significant part of the electrical safety program. All work tasks that involve exposure to an electrical hazard must be guided by a procedure.

The program might contain several procedures; however, it must contain a procedure that defines requirements and provides guidance for workers as they perform work on or near live parts.

See *The Electrical Safety Program Book* for procedures that might be necessary.

**(F) Hazard/Risk Evaluation Procedure.** An electrical safety program shall identify a hazard/risk evaluation procedure to be used before work is started on or near live parts operating at 50 volts or more or where an electrical hazard exists.

FPN: For an example of a hazard risk procedure, see Annex F.

One procedure in the published program must define the process that employees will use to assess both the hazard and the risk associated with each work task on or near live parts operating at 50 volts or greater. That hazard/risk evaluation procedure must identify a process that enables employees to determine whether a hazard exists. It must evaluate electrical shock and arc-flash. The process must provide sufficient information to enable both the employer and employee to make an informed decision on whether the risk of injury is acceptable or not acceptable.

Note: Written authorization to execute work on or near live parts (energized electrical work permit) is discussed in Article 130.1(A).

**(G) Job Briefing.**

**(1) General.** Before starting each job, the employee in charge shall conduct a job briefing with the employees involved. The briefing shall cover such subjects as hazards associated with the job, work procedures involved, special precautions, energy source controls, and personal protective equipment requirements.

A job briefing is a discussion of the work task. The briefing must be held prior to beginning each work task associated with work on or near live parts. The briefing must include a discussion of electrical hazards and how employees might be exposed to them. As a minimum, the discussion must include the following subjects:

- Electrical hazards associated with the work task
- Procedures that must be followed when executing the work task
- Any special precautions that are required by the working conditions
- Where and how to remove the source of energy
- Emergency response and emergency communications
- Required PPE
- Other work in the immediate physical area
- Other work associated with the same electrical circuits or equipment

**(2) Repetitive or Similar Tasks.** If the work or operations to be performed during the work day or shift are repetitive and similar, at least one job briefing shall be conducted before the start of the first job of the day or shift. Additional job briefings shall be held if significant changes that might affect the safety of employees occur during the course of the work.

If the work task is a repetitive task that is performed several times during the day, a single job briefing held before the worker performs the task for the first time is satisfactory. If significant changes that might affect the safety of employees occur during the day, however, a new job briefing is required.

**(3) Routine Work.** A brief discussion shall be satisfactory if the work involved is routine and if the employee, by virtue of training and experience, can reasonably be expected to recognize and avoid the hazards involved in the job. A more extensive discussion shall be conducted if either of the following apply:

(1) The work is complicated or particularly hazardous.
(2) The employee cannot be expected to recognize and avoid the hazards involved in the job.

FPN: For an example of a job briefing form and planning checklist, see Annex I.

The job briefing should be as extensive as necessary to ensure that workers have a complete understanding of their exposure to an electrical hazard. If the task is simple and routine (such as exchanging filters), only a brief discussion is necessary. If, however, the work task is complex or unfamiliar to the worker, a more complete job briefing is necessary. A worker from another employer or another area of the facility should not be expected to be familiar with a new work environment. Thus, the job briefing should be complete and ensure that the particulars of the environment are explained to the "new" worker.

The job briefing provides an opportunity for supervisors or lead persons to review plans that have been developed to cover the work.

### 110.8 Working On or Near Electrical Conductors or Circuit Parts

**(A) General.** Safety-related work practices shall be used to safeguard employees from injury while they are working on or near exposed electric conductors or circuit parts that are or can

**become energized. The specific safety-related work practice shall be consistent with the nature and extent of the associated electric hazards.**

Exposed electrical conductors or circuit parts that are energized at less than 50 volts do not normally present an electrical hazard. However, creating a short or open circuit in these conductors could result in a process upset and damage to the environment.

Employers are required to provide procedures that contain practices that will prevent injury to employees. Work that is performed on or near exposed live parts is particularly dangerous. Work on or near exposed conductors that are not in an electrically safe work condition expose workers to injury. Only qualified persons may perform work on or near these conductors.

Workers must select and use work practices that are consistent with the degree of the potential hazard. For instance, if the potential hazard includes possible exposure to an arc flash, the worker must select flame-resistant clothing and other PPE that is at least as protective as the potential hazard is dangerous. It is acceptable for the rating of the PPE to exceed the degree of exposure, but it is unacceptable for the rating of the PPE to be less than the degree of exposure.

**(1) Live Parts—Safe Work Condition.** **Live parts to which an employee might be exposed shall be put into an electrically safe work condition before an employee works on or near them, unless work on energized components can be justified according to 130.1.**

The primary protective strategy must be to establish an electrically safe work condition. After this strategy is executed, all electrical energy has been removed from all conductors and circuit parts to which the worker could be exposed. After the electrically safe work condition has been established, no PPE is necessary, and unqualified workers are permitted to execute the remainder of the work.

The only exception to this requirement is if deenergizing the circuit conductors or equipment cannot be justified as described in the following list. Justification for work on or near exposed live parts must be in writing. A qualified person may perform work on or near exposed live parts under the following conditions:

- Deenergizing the conductors or equipment could result in an increased hazard. For instance, a life support system might be dependent on the continuation of the electrical service.
- Deenergizing the conductors or equipment could require a complete shutdown of a continuous process. For instance, the design of the electrical circuit is such that a continuous processing facility must be completely shut down.

Experience shows a tendency for workers to err on the side of accepting exposure to electrical hazards, while managers and supervisors tend to be reluctant to accept increased exposure to hazards.

**(2) Live Parts—Unsafe Work Condition.** **Only qualified persons shall be permitted to work on electrical conductors or circuit parts that have not been put into an electrically safe work condition.**

An electrical hazard is considered to be present until an electrically safe work condition exists. Unqualified persons are not permitted to perform any task with potential exposure to an electrical hazard.

The acts of opening a disconnecting means, measuring for absence of voltage, visually verifying a physical break in the power conductors, and installing safety grounds all contain a risk of injury. These acts are necessary to create an electrically safe work condition, and, until they are completed, the worker should be wearing PPE based on the degree of potential hazard. PPE that is rated greater than 40 cal/cm$^2$ could be required until the electrically safe work condition exists. Workers who implement the requirements for an electrically safe work condition must be qualified for that work task.

**(B) Working On or Near Exposed Electrical Conductors or Circuit Parts that Are or Might Become Energized.** Prior to working on or near exposed electrical conductors and circuit parts operating at 50 volts or more, lockout/tagout devices shall be applied in accordance with 120.1, 120.2, and 120.3. If, for reasons indicated in 130.1, lockout/tagout devices cannot be applied, 130.2(A) through 130.2(D)(2) shall apply to the work.

Locks and tags both are required if they can be installed on circuits with an operating voltage of 50 volts or greater. The installation of locks and tags must be in accordance with a lockout/tagout program defined by the requirements in Article 120. Lockout is only one step in establishing an electrically safe work condition. When an electrically safe work condition cannot be established, all requirements associated with approach boundaries described in 130.2 must be observed.

**(1) Electrical Hazard Analysis.** If the live parts operating at 50 volts or more are not placed in an electrically safe work condition, other safety-related work practices shall be used to protect employees who might be exposed to the electrical hazards involved. Such work practices shall protect each employee from arc flash and from contact with live parts operating at 50 volts or more directly with any part of the body or indirectly through some other conductive object. Work practices that are used shall be suitable for the conditions under which the work is to be performed and for the voltage level of the live parts. Appropriate safety-related work practices shall be determined before any person approaches exposed live parts within the Limited Approach Boundary by using both shock hazard analysis and flash hazard analysis.

Qualified workers who are permitted to work on or near exposed energized conductors or circuit parts must select and use work practices that provide protection from shock, arc flash, and other electrical hazards. The work practices that are used must minimize any potential for injury. For instance, body position is one factor that a qualified person should recognize as an element of the analysis that could reduce exposure to electrical shock or arc flash.

The hazard/risk analysis must determine whether any conductor will remain energized for the duration of the work task. The analysis must determine the shock approach boundaries and the Flash Protection Boundary.

Both a shock analysis and an arc flash analysis are required before any person is permitted to approach the exposed live part. These analyses must answer the following questions:

- Does a shock hazard exist?
- Will the worker be exposed to the shock hazard at any point during the work task?
- What is the degree of the hazard?
- What protective equipment is necessary to minimize the exposure?
- Does an arc flash hazard exist?
- Will the worker be exposed to a thermal hazard at any point during the work task?
- What is the degree of the arc flash hazard?
- What protective equipment is necessary to minimize exposure to the thermal hazard?
- Does a co-occupancy hazard exist?
- What measures will be taken to minimize the impact of other work?
- Will other workers be exposed to an electrical hazard because of the work task?
- Will the worker be exposed to any other electrical hazard while executing the work task?
- What authorization is necessary to justify executing the work task while the exposed conductor(s) is (are) energized?

- What workers are required to be within an approach boundary?
- Are unqualified workers required?
- How will the voltage on the conductor (or nearby conductors) be determined?
- Is the risk of injury acceptable?

(a) Shock Hazard Analysis. A shock hazard analysis shall determine the voltage to which personnel will be exposed, boundary requirements, and the personal protective equipment necessary in order to minimize the possibility of electrical shock to personnel.

FPN: See 130.2 for the requirements of conducting a shock hazard analysis.

The shock hazard analysis is only one part of the hazard/risk analysis. The shock hazard analysis is intended to determine if a risk of electrocution or shock might exist when the work task is being executed. That risk increases as a worker approaches an exposed live part. Requirements that reduce the risk are associated with the Limited, Restricted, and Prohibited Approach Boundaries. The degree of the risk increases as the voltage increases. The shock hazard analysis must determine if work practices and protective equipment reduce the risk of electrocution to an acceptable level.

(b) Flash Hazard Analysis. A flash hazard analysis shall be done in order to protect personnel from the possibility of being injured by an arc flash. The analysis shall determine the Flash Protection Boundary and the personal protective equipment that people within the Flash Protection Boundary shall use.

FPN: See 130.3 for the requirements of conducting a flash hazard analysis.

Paragraph 110.8(B)(1)(b) requires an analysis of potential exposure to the thermal hazard associated with an arcing fault. The flash hazard analysis must determine whether the worker could be exposed to the extreme temperature generated by the electrical current during an arcing fault. The analysis has two purposes. First, it must determine the location of the arc flash boundary. Second, the analysis must determine the rating of flame-retardant clothing that must be worn by the worker.

Several methods are available to establish the degree of the potential arc flash hazard, but no specific method is required. No method to determine the intensity of a potential arc flash has been substantiated to the satisfaction of the Technical Committee, and no preferred calculating method exists. Methods of calculating arc flash intensity are evolving, and research is currently underway to help facilities determine the calculation method to select for their electrical safety program.

The product of the flash hazard analysis must provide sufficient information for a worker to select PPE that will protect him or her from injury as the result of the thermal effects of any potential arcing fault. If the flash hazard analysis suggests that the intensity of the arc flash could expose a worker to 40 calories per square centimeter (cal/cm$^2$), the work must not be performed unless an electrical safe work condition has been established. If the intensity is greater than 40 cal/cm$^2$, no protective equipment exists that can protect the worker from the intense pressure (arc blast) that also will be produced by the arcing fault.

**(2) Energized Electrical Work Permit.** If live parts are not placed in an electrically safe work condition (i.e., for the reasons of increased or additional hazards or infeasibility per 130.1), work to be performed shall be considered energized electrical work and shall be performed by written permit only.

Historically, electrical workers have accepted the increased risk associated with working on or near exposed live parts. Also historically, managers and supervisors have not recognized the increased risk of electrical injury. Experience suggests that if managers and supervisors are advised that a significant risk of injury exists, they are reluctant to accept that increased risk. If a manager or supervisor is requested to authorize work on or near exposed live parts by signing a permit, he or she will be more critical of the plan to execute such work.

The requirement for an energized electrical work permit is new to the 2004 edition of NFPA 70E. Use of the written permit helps to ensure that all parties have reviewed the work task and determined that an electrically safe work condition cannot be established [see OSHA 29 CFR 1910.333(a)(1)].

The intent of the written permit is to ensure that the decision to accept the increased risk of injury is shifted from the worker to a manager or supervisor, where the responsibility will reside at an appropriate level in the organization. The worker always has the option to refuse to perform any task that he or she believes has an unacceptable risk.

All work on or near exposed live parts must be authorized by a written document. The written authorization must be kept on file until the work is completed. High-level managers should review the number of energized work permits that are authorized over a specific period of time. For instance, if the number of permits is increasing, the high-level manager should question the process that accepts the increased exposure to electrical hazards.

FPN: See 130.1(A) for the requirements of an energized electrical work permit.

**(3) Unqualified Persons.** Unqualified persons shall not be permitted to enter spaces that are required under 400.16 to be accessible to qualified employees only, unless the electric conductors and equipment involved are in an electrically safe work condition.

Unqualified persons are not trained to recognize whether an electrical hazard exists, and they must not be closer to an exposed live part than the Limited Approach Boundary or the Flash Protection Boundary, whichever is greater. An unqualified person may be within the Limited Approach Boundary, provided he or she has been advised by a qualified person how to avoid contact with the live part and how to select and wear PPE that is necessary for protection from the effects of an arcing fault, shock, or electrocution.

**(4) Safety Interlocks.** Only qualified persons following the requirements for working inside the Restricted Approach Boundary as covered by 130.2(C) shall be permitted to defeat or bypass an electrical safety interlock over which the person has sole control, and then only temporarily while the qualified person is working on the equipment. The safety interlock system shall be returned to its operable condition when the work is completed.

Two types of safety interlocks exist. One is a mechanical interlock that prevents a door from being opened or a disconnect means from being closed while the equipment is energized. The second type is an electrical circuit contact that prevents a circuit component from being operated unless a specific circuit condition exists.

These interlocks are installed to prevent a specific hazardous condition. Only qualified workers have been trained to understand how defeating a safety interlock establishes an unsafe condition.

Any safety interlock that is defeated or bypassed creates an unsafe condition and must be attended for the duration of the unsafe condition.

### 110.9 Use of Equipment

**(A) Test Instruments and Equipment.**

**(1) Rating.** Test instruments, equipment, and their accessories shall be rated for circuits and equipment to which they will be connected.

Test instruments should be considered safety equipment. Although they have some characteristics of a tool, all test equipment and accompanying accessories should be purchased without artificial cost restraints. Test equipment must be selected on the basis of the intended use and expected voltage or current rating. Leads and probes are an integral part of the test equipment and must be rated at least as great as the instrument.

**(2) Design.** Test instruments, equipment, and their accessories shall be designed for the environment to which they will be exposed, and for the manner in which they will be used.

Instruments and similar equipment must be selected for the conditions of use. If they are to be used in an environment that could expose the instrument, leads, and probes to a lightning surge, they must be designed and rated for that situation.

**(3) Visual Inspection.** Test instruments and equipment and all associated test leads, cables, power cords, probes, and connectors shall be visually inspected for external defects and damage before the equipment is used on any shift. If there is a defect or evidence of damage that might expose an employee to injury, the defective or damaged item shall be removed from service, and no employee shall use it until repairs and tests necessary to render the equipment safe have been made.

All test instruments must be inspected for physical damage before each use. The inspection must include all leads, probes, and other attachments. If any damage, such as cracked cases, cut or pinched leads, or damaged probe tips, is observed, the instrument must be removed from service, repaired, and tested before it is used again.

**(B) Portable Electric Equipment.** This section applies to the use of cord-and-plug-connected equipment, including cord sets (extension cords).

**(1) Handling.** Portable equipment shall be handled in a manner that will not cause damage. Flexible electric cords connected to equipment shall not be used for raising or lowering the equipment. Flexible cords shall not be fastened with staples or hung in such a fashion as could damage the outer jacket or insulation.

**(2) Grounding-type Equipment.**

(a) A flexible cord used with grounding-type utilization equipment shall contain an equipment grounding conductor.

(b) Attachment plugs and receptacles shall not be connected or altered in a manner that would interrupt continuity of the equipment grounding conductor at the point where plugs are attached to receptacles. Additionally, these devices shall not be altered to allow the grounding pole of a plug to be inserted into slots intended for connection to the current-carrying conductors.

(c) Adapters that interrupt the continuity of the equipment grounding conductor shall not be used.

**(3) Visual Inspection of Portable Cord-and-Plug-Connected Equipment and Flexible Cord Sets.**

(a) Frequency of Inspection. Before use on any shift, portable cord-and-plug-connected equipment shall be visually inspected for external defects (such as loose parts, deformed and missing pins) and for evidence of possible internal damage (such as pinched or crushed outer jacket).

The grounding conductor is an integral part of the safety system that is built into the equipment by manufacturers for tools and devices that are not double insulated. The integrity of the grounding conductor is paramount to minimizing the chance that a worker will become a part of the electrical circuit providing power to the tool. Each person must visually inspect the tool before it is used. Any defect, such as a cord with a missing ground pin, is sufficient indication to cause the tool to be removed from use. If a defect is found in the visual inspection, the tool must be removed from service until it is repaired. See Exhibit 110.1 for a photo of a typical cord cap.

Grounding conductors are not required on tools that are rated by the manufacturer as double insulated. These tools are required to be inspected for indication of damage.

***EXHIBIT 110.1.*** *A cord cap. (Courtesy of Pass & Seymour/Legrand®)*

*Exception: Cord-and-plug-connected equipment and flexible cord sets (extension cords) that remain connected once they are put in place and are not exposed to damage shall not be required to be visually inspected until they are relocated.*

A cord-connected electrical device, such as a water cooler or computer terminal that is installed and not moved, is not required to be inspected each day. Although not required, less frequent inspection of the grounding conductor integrity is recommended.

(b) Defective Equipment. If there is a defect or evidence of damage that might expose an employee to injury, the defective or damaged item shall be removed from service, and no employee shall use it until repairs and tests necessary to render the equipment safe have been made.

The integrity of the grounding conductor of any portable or cord-connected tools and devices must be verified by tests prior to returning the tool or device to service.

The integrity of the insulating system for double-insulated tools must be verified by tests after repairs have been made.

(c) Proper Mating. When an attachment plug is to be connected to a receptacle, the relationship of the plug and receptacle contacts shall first be checked to ensure that they are of mating configurations.

(d) Conductive Work Locations. Portable electric equipment used in highly conductive work locations (such as those inundated with water or other conductive liquids) or in job locations where employees are likely to contact water or conductive liquids shall be approved for those locations. In job locations where employees are likely to contact or be drenched with water or conductive liquids, ground-fault circuit-interrupter protection for personnel shall also be used.

**(4) Connecting Attachment Plugs.**

(a) Employees' hands shall not be wet when plugging and unplugging flexible cords and cord-and-plug-connected equipment if energized equipment is involved.

When cord-and-plug-connected equipment is placed into service, the pin configuration of the plug must match the configuration of the receptacle (see Exhibit 110.2). Pins must not be damaged. Attachment devices that permit the installation of devices with pin configurations that do not match must ensure the integrity of all connecting pins of the cord cap.

Moisture can provide a conducting path from the hot conductor in a cord cap to the surface of the device. A person inserting a wet cord cap into an energized receptacle is exposed to a shock hazard. The person handling the wet cord cap must be wearing PPE that is rated at least as great as the circuit voltage.

If the cord-connected equipment is outdoors and subject to moisture ingress, the cord must be protected by a listed ground-fault circuit interrupter. Some work environments could contain another conductive compound. Any cord-connected equipment or device installed in these areas must be protected by a ground-fault circuit interrupter.

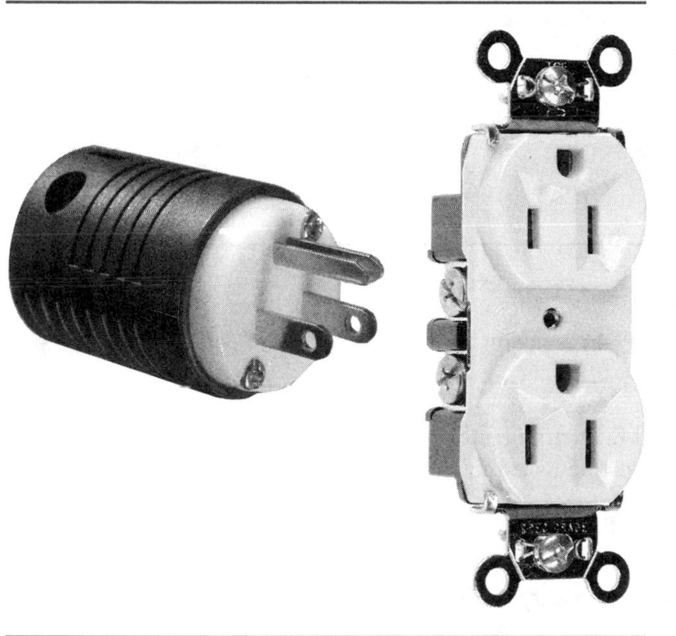

***EXHIBIT 110.2.*** *A three-pronged plug (left), which goes into a three-pronged receptacle (right). (Courtesy of Pass & Seymour/Legrand®)*

**(b) Energized plug and receptacle connections shall be handled only with insulating protective equipment if the condition of the connection could provide a conductive path to the employee's hand (if, for example, a cord connector is wet from being immersed in water).**

Any conductive path from a worker's hand to an energized conductor exposes that worker to shock or electrocution. A worker standing in water or other conductive material must wear adequate insulating gloves when removing the plug from an energized receptacle. If the environment contains highly conductive compounds, gloves with adequate voltage rating are required. If the receptacle or plug is wet from water or other conductive material, adequately rated gloves are required.

**(c) Locking-type connectors shall be secured after connection.**

Twistlock plugs and some other portable cord-connecting devices are designed to provide a connection that is secure from accidental withdrawal (see Exhibit 110.3). Devices that provide the additional security must be inserted so that the design intent of the device is complete. For instance, twistlock plugs must be turned to the "secure" position.

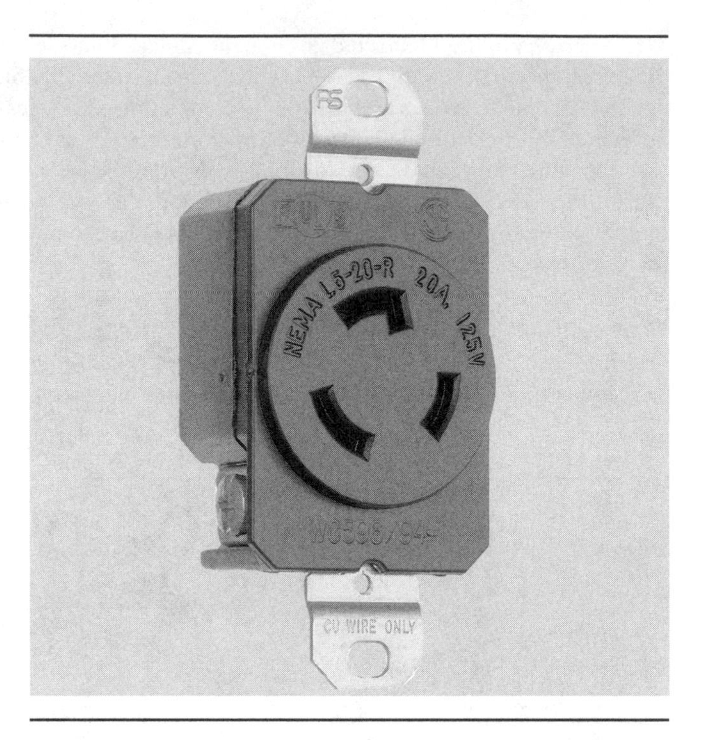


***EXHIBIT 110.3.*** *A Twistlock plug. (Courtesy of Pass & Seymour/Legrand®)*

**(C) GFCI Protection Devices.** GFCI protection devices shall be tested per manufacturer's instructions.

All manufacturers of listed ground-fault circuit interrupter (GFCI) devices provide installation and testing instructions. GFCIs must be tested in accordance with the instructions provided with each order.

**(D) Overcurrent Protection Modification.** Overcurrent protection of circuits and conductors shall not be modified, even on a temporary basis, beyond that permitted by 410.9(A) and 410.9(B).

Overcurrent devices are important components of a safety system to prevent conductors and devices from current flow that exceeds the conductor's ability to safely conduct current. Exceeding the conductor or equipment rating can result in overheating. One result of overheating can be to ignite nearby flammable material. Device and component overheating can also result in massive component failure.

Overcurrent devices are addressed in Article 410.9. Protection specified by that article must not be exceeded for any reason.

# ARTICLE 120
# Establishing an Electrically Safe Work Condition

The most effective way to prevent electrical injury is to completely remove the source of electrical energy and eliminate the possibility of its reappearance. To do that, workers must identify all possible sources of electricity and locate the disconnecting means of each source. Electricity sometimes can appear from the load side. The worker must identify all potential sources of energy, eliminate them, visually verify (if possible) that they are eliminated, lock them out, test to verify the absence of voltage, and ground the conductors or parts if necessary. An electrically safe work condition exists only after these actions have been completed.

### 120.1 Process of Achieving an Electrically Safe Work Condition

An electrically safe work condition shall be achieved when performed in accordance with the procedures of 120.2 and verified by the following process:

An electrically safe work condition does not exist until *all* of the six steps of 120.1(1) through 120.1(6) have been completed. Until then, workers might contact an exposed live part.

If an electrically safe work condition does exist, no electrical energy is in proximity of the work task(s). All danger of injury from an electrical hazard has been removed, and neither protective equipment nor special safety training is required. However, other hazards might remain.

(1) Determine all possible sources of electrical supply to the specific equipment. Check applicable up-to-date drawings, diagrams, and identification tags.

The first step in establishing an electrically safe work condition is to ensure that all sources or potential sources of electrical energy have been identified. A worker should rely on all possible sources of information to identify all sources of energy. In many cases, some electrical energy sources might be labeled well and operated frequently. Sometimes a sneak circuit exists that can be identified only by reviewing diagrammatic-type drawings. Diagrammatic-type drawings must be maintained in an up-to-date condition to provide accurate information.

Panel schedules that are complete and up-to-date can provide the same information that is gleaned from a diagram. For instance, lighting circuits with separate neutrals can be identified from an up-to-date panel schedule.

(2) After properly interrupting the load current, open the disconnecting device(s) for each source.

The rating of some disconnecting equipment is insufficient to interrupt the load current demanded by the circuit. These devices must not be operated unless the load has been removed by another action.

(3) Wherever possible, visually verify that all blades of the disconnecting devices are fully open or that drawout-type circuit breakers are withdrawn to the fully disconnected position.

Disconnecting means sometimes fail to open all phase conductors when the handle is operated. Ensuring that operating the handle of the device actually establishes a physical break in *all* conductors is critical. After operating the handle of the disconnecting device, the worker should open the door or cover and observe the physical opening in each blade of the disconnect switch. Opening a door or removing a cover could expose a worker to electrical hazards. Therefore, the worker must be protected from those hazards by PPE.

The physical opening is sometimes difficult or impossible to observe directly. In those instances, the worker should verify the physical opening by measuring voltage on the load side of the device after the handle has been operated. The voltage test must include measuring voltage to ground from each conductor and between each conductor and each of the other conductors (phase-to-ground and phase-to-phase). Taking these voltage measurements should be considered working on or near exposed live parts, regardless of the status of the operating handle.

(4) Apply lockout/tagout devices in accordance with a documented and established policy.

Employers must establish and implement a lockout/tagout procedure. Employees must implement all aspects of the published lockout/ragout procedure.

**(5) Use an adequately rated voltage detector to test each phase conductor or circuit part to verify they are deenergized. Test each phase conductor or circuit part both phase-to-phase and phase-to-ground. Before and after each test, determine that the voltage detector is operating satisfactorily.**

In each instance, the worker must determine that no voltage exists on each conductor to which he or she might be exposed. That determination must be made by measuring the voltage to ground and the voltage to all conductors from any potential power source with a voltage detector that is rated for the maximum voltage available from any potential source of energy. The voltage-detecting device must be functionally tested before taking the measurement and then again after the measurement is taken to ensure that the voltage detector is working satisfactorily.

A measurement must be made from each conductor to ground and between each conductor to each other conductor from a potential source of energy.

**(6) Where the possibility of induced voltages or stored electrical energy exists, ground the phase conductors or circuit parts before touching them. Where it could be reasonably anticipated that the conductors or circuit parts being deenergized could contact other exposed energized conductors or circuit parts, apply ground connecting devices rated for the available fault duty.**

Electrical conductors sometimes break and fall onto another conductor that is installed at a lower elevation. For instance, more than one outside overhead line might be installed on a single pole. Physical damage to a higher conductor could fall onto a lower installed conductor and result in reenergizing the lower conductor.

A set of safety grounds is necessary to protect workers from potentially hazardous voltage, such as in the following accidental instances:

- A long conductor installed in proximity to the deenergized conductor could induce a hazardous voltage onto the otherwise deenergized conductor.
- Another worker could inadvertently connect a conductor that is not locked out and add a source of energy to a circuit under repair.
- Other situations that could reintroduce voltage to a conductor under repair are identified by the hazard/risk analysis.

Safety grounds provide protection only if the rating is sufficiently great to conduct any potentially available energy. Inadequately rated safety grounds introduce a hazard that otherwise would not exist. The rating must be established by the manufacturer or another generally accepted rating process. See Exhibit 120.1 for an approved grounding connector.

### 120.2 Working On or Near Deenergized Electrical Conductors or Circuit Parts That Have Lockout/Tagout Devices Applied

**Each employer shall identify, document, and implement lockout/tagout procedures conforming to Article 120 to safeguard employees from exposure to electrical hazards while they are working on or near deenergized electrical conductors or circuit parts that are likely to result in injury from inadvertent or accidental contact or equipment failure. The lockout/tagout procedure shall be appropriate for the experience and training of the employees and conditions as they exist in the workplace.**

Section 120.2 establishes a requirement for each employer to implement a lockout/tagout procedure to ensure that each hazardous source of electricity is removed and controlled. The procedure must consider all possible sources of electrical energy and ensure that the electrical

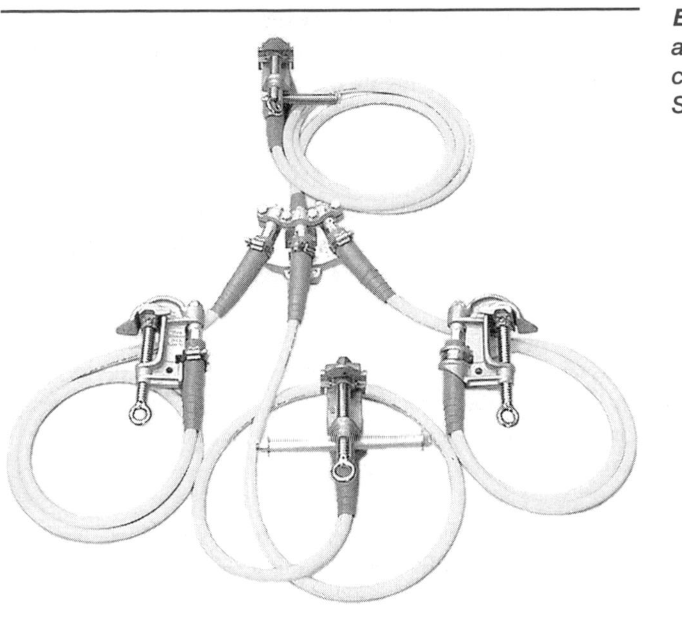

*EXHIBIT 120.1. An approved grounding conductor. (Courtesy of Salisbury)*

energy cannot reappear while the work task is being executed. The procedure should identify if and when employees may perform work on or near exposed live parts.

**(A) General.** All electrical circuit conductors and circuit parts shall be considered energized until the source(s) of energy is (are) removed, at which time they shall be considered deenergized. All electrical circuit conductors and circuit parts shall not be considered to be in an electrically safe condition until all sources of energy are removed, the disconnecting means is under lockout/tagout, the absence of voltage is verified by an approved voltage testing device, and, where exposure to energized facilities exists, are temporarily grounded. (*See 120.1 for the six-step procedure to establish an electrically safe work condition.*) Electrical conductors and circuit parts that have been disconnected, but not under lockout/tagout, tested, and grounded (where appropriate) shall not be considered to be in an electrically safe work condition, and safe work practices appropriate for the circuit voltage and energy level shall be used. Lockout/tagout requirements shall apply to fixed, permanently installed equipment, to temporarily installed equipment, and to portable equipment.

Lockout/tagout is only one step in the process of establishing an electrically safe working condition. Installing locks and tags does not ensure that electrical hazards have been removed. Workers must select and use work practices that are identical to working on or near exposed live parts until an electrically safe work condition has been established.

Each lockout/tagout procedure must apply to all exposed live parts, regardless of whether they are temporary, permanent, or portable. When implementing those procedural requirements, workers must select and use work practices (including PPE) that are appropriate for the voltage and energy level of the circuit as if it were known to be energized.

**(B) Principles of Lockout/Tagout Execution.**

**(1) Employee Involvement.** Each person who could be exposed directly or indirectly to a source of electrical energy shall be involved in the lockout/tagout process.

FPN: An example of direct exposure is the qualified electrician who works on the motor starter control, the power circuits, or the motor. An example of indirect exposure is the person who works on the coupling between the motor and compressor.

Each worker who could be exposed to an electrical hazard when executing the work task must be involved in the lockout/tagout process. Temporary workers and contract workers must also understand how the lockout/tagout procedure eliminates their exposure to electrical hazards and participate in the process.

**(2) Training.** All persons who could be exposed shall be trained to understand the established procedure to control the energy and their responsibility in executing the procedure. New (or reassigned) employees shall be trained (or retrained) to understand the lockout/tagout procedure as related to their new assignment.

Each person who is associated with executing the job must be trained to understand his or her role in the lockout/tagout process. Each person must understand and accept individual responsibility for the integrity of the lockout/tagout process. Employees that have been reassigned (either permanently or temporarily) must be retrained to understand their new role in the lockout/tagout procedure.

**(3) Plan.** A plan shall be developed on the basis of the existing electrical equipment and system and shall utilize up-to-date diagrammatic drawing representation(s).

A plan must be developed for each lockout/tagout. The planner might be the person in charge [see 120.2(C)(2)], the supervisor, or the worker. The plan must identify the location, both physically and electrically, that requires a lockout or tagout device and must be based on up-to-date diagrammatic information.

That information might be a single-line or three-line diagram. The drawing can be one that is hand marked to illustrate recent circuit modifications, provided the information accurately depicts the configuration of the circuit. The information on the drawing must match the information on the equipment labels. The plan should include identifying information that is found on both the drawing and the equipment label.

**(4) Control of Energy.** All sources of electrical energy shall be controlled in such a way as to minimize employee exposure to electrical hazards.

All disconnecting means must be opened (disconnected) and placed in a condition that minimizes the chance of being reclosed.

**(5) Identification.** The lockout/tagout device shall be unique and readily identifiable as a lockout/tagout device.

Lockout/tagout devices must be physically different from all other devices that are used on the site or facility. A worker must be able to recognize a lockout/tagout device by sight. No possibility of confusing lockout/tagout devices with locks or tags used for other purposes can exist. For instance, information tags and process control locks must not be confused with tags or locks used for lockout/tagout. See Exhibit 120.2 for a lockout station containing locks, tags, and devices.

**(6) Voltage.** Voltage shall be removed and absence of voltage verified.

All voltage sources must be removed, and the absence of voltage must be verified.

**(7) Coordination.** The established electrical lockout/tagout procedure shall be coordinated with all of the employer's procedures associated with lockout/tagout of other energy sources. The lockout/tagout procedure shall be audited for execution and completeness on an annual basis.

*EXHIBIT 120.2. A lockout station. (Courtesy of Salisbury)*

Some projects involve employers other than the facility owner, such as contractors. Because each employer is required to implement a lockout/tagout procedure, different requirements might exist. To ensure that the requirements of each procedure are observed, one or more procedures might need to have additional requirement(s).

A coordination meeting must be held, and each procedural requirement that is more or less restrictive than other employers' lockout/tagout procedure(s) must be reviewed and discussed. The result of the meeting should be that all lockout/tagout procedures in effect on the project are coordinated with one another.

Each employer must audit at least one work task where lockout/tagout has been applied. The audit must determine whether all requirements of the lockout/tagout procedure were observed. The audit must also determine whether the requirements contained in the published procedure are sufficient to ensure that the electrical energy is satisfactorily controlled.

**(C) Responsibility.**

**(1) Procedures.** The employer shall establish lockout/tagout procedures for the organization, provide training to employees, provide equipment necessary to execute the details of the procedure, audit execution of the procedures to ensure employee understanding/compliance, and audit the procedure for improvement opportunity and completeness.

Each employer must write and publish a lockout/tagout procedure that applies to all work involving exposure or potential exposure to an electrical hazard. If a tool or equipment is required to implement the details of the procedure, the employer must purchase and supply those tool(s) or equipment in sufficient quantity.

Each employer must train every employee to understand his or her role in implementing the details of the procedure. The training must ensure that each employee understands the enforcement aspects of respecting an installed lockout or tagout device. Each employer also is responsible for initiating audits, as necessary, to ensure that the lockout/tagout procedure is effective and that no changes are warranted.

**(2) Form of Control.** Three forms of hazardous electrical energy control shall be permitted: individual employee control, simple lockout/tagout, and complex lockout/tagout. *[See 120.2(D).]* For the individual employee control and the simple lockout/tagout, the qualified person shall be in charge. For the complex lockout/tagout, the person in charge shall have overall responsibility. (*See Annex* G *for a sample lockout/tagout procedure.*)

The basic idea behind 120.2(C)(2) for controlling electrical energy is to ensure that all possible sources of this energy are disconnected and cannot reappear unexpectedly. In some instances, workers might not understand the details of an installed electrical circuit, either the circuitry or the location of electrical equipment. Consequently, workers depend on the integrity of the information provided to them. Each employer is expected to assign to one person the responsibility of determining the integrity and completeness of each lockout/tagout. The person in charge need not be the same individual for all sites or for all lockouts or tagouts. Instead, that responsibility could change with each application of the procedure.

The lockout/tagout procedure should be as simple as possible to enable workers to understand the reasons for each requirement. For instance, if a worker is troubleshooting an individually mounted disconnect switch that is directly in from of him or her, the switch handle could not be operated while the person is standing directly in front of it. However, if the worker moves from the position directly in front of the switch, the handle could be operated without his or her knowledge. In this illustration, the worker would act as the person in charge.

Sometimes a single disconnect switch provides energy for a device such as a motor. A worker who intends to remove the motor should operate the disconnecting means for the motor. Because the potential hazard will not be continually in view, the worker should install a lockout device on the disconnect switch. Again, the worker would act as the person in charge.

In other instances, implementing the requirements of the lockout/tagout procedure can be complex. Adding any additional factor, such as more workers or more sources of energy, increases the complexity of the lockout/tagout. As that complexity increases, the need to understand the electrical circuit increases. A single person in charge can improve the chance of accurately locating and controlling all potential sources of electrical energy.

**(3) Audit Procedures.** An audit shall be conducted at least annually by a qualified person and shall cover at least one lockout/tagout in progress and the procedure details. The audit shall be designed to correct deficiencies in the procedure or in employee understanding.

Each employer must audit at least one work task where lockout/tagout has been applied. The audit must determine whether all requirements of the lockout/tagout procedure were observed and whether the requirements contained in the published procedure are sufficient to ensure that the electrical energy is satisfactorily controlled. The objective of the audit should be to identify weaknesses (or potential weaknesses) in the procedure, in employee training, or in enforcement of the requirements.

**(D) Hazardous Electrical Energy Control Procedures.**

**(1) Individual Qualified Employee Control Procedure.** The individual qualified employee control procedure shall be permitted when equipment with exposed conductors and circuit parts is deenergized for minor maintenance, servicing, adjusting, cleaning, inspection, operating conditions, and the like. The work shall be permitted to be performed without the placement of lockout/tagout devices on the disconnecting means, provided the disconnecting means is adjacent to the conductor, circuit parts, and equipment on which the work is performed, the disconnecting means is clearly visible to the individual qualified employee involved in the work, and the work does not extend beyond one shift.

When an employee is working on or within a disconnecting means with a single source of energy, such as replacing a motor fuse, the need to install a lockout/tagout device does not exist. The worker is physically positioned to ensure that he or she is in control of the operating mechanism of the switch, provided the worker does not leave the front of the switch. However, if the worker turns away from the switch or leaves the area, the lockout/tagout shifts from an individual qualified employee control procedure to a simple lockout/tagout, and a lockout/tagout device becomes necessary.

Some equipment, such as a disconnect switch, cannot be opened with a lock installed. If the work task is to measure voltage or change a fuse in this disconnecting means, a lockout device cannot be installed. The purpose of the individual qualified employee control procedure is to permit these tasks without a lockout/tagout device. *It is important to note that the individual qualified employee control procedure requires that the exposed conductor be continuously within sight and within arm's reach.*

***EXHIBIT 120.3.** Simple lockout devices. (Courtesy of Salisbury)*

**(2) Simple Lockout/Tagout Procedure.** All lockout/tagout procedures that are not under individual qualified employee control *[see 120.2(D)(1)]* or complex lockout/tagout *[see 120.2(D)(3)]* shall be considered to be simple lockout/tagout procedures. All lockout/tagout procedures that involve only a qualified person(s) deenergizing one set of conductors or circuit part source for the sole purpose of performing work on or near electrical equipment shall be considered to be a simple lockout/tagout. Simple lockout/tagout plans shall not be required to be written for each application. Each worker shall be responsible for his or her own lockout/tagout.

Some work tasks involve only a single source of electrical energy and a single disconnecting means. If the disconnecting means is capable of accepting a lockout/tagout device, the lockout/tagout can be considered to be a simple lockout/tagout procedure and no written lockout/tagout plan is necessary. A simple lockout/tagout must be a planned activity, although a written plan is not always needed (see Exhibit 120.3).

**(3) Complex Lockout/Tagout Procedure.**

(a) A complex lockout/tagout plan shall be permitted where one or more of the following exist:

(1) Multiple energy sources
(2) Multiple crews
(3) Multiple crafts
(4) Multiple locations
(5) Multiple employers
(6) Different disconnecting means
(7) Particular sequences
(8) A job or task that continues for more than one work period

If any of the conditions listed in 120.2(D)(3)(a) exist, the risk of exposing a worker to an electrical hazard increases. As the number of people involved in the lockout increases, the chance decreases that each person understands the location and purpose of any individual disconnecting means.

These conditions are known to increase the complexity and difficulty of a lockout/tagout. If one or more of these conditions exist, the lockout/tagout is defined as complex (see Exhibit 120.4).

(b) A person shall be in charge of a complex lockout/tagout procedure. Such person shall be a qualified individual who is specifically appointed with overall responsibility to ensure that all energy sources are under lockout/tagout and to account for all persons working on the job/task.

Each complex lockout/tagout must be under the direct control of a single person in charge. The person in charge must be assigned and assume the responsibility of ensuring that an electrically

***EXHIBIT 120.4.*** *Complex lockout. (Courtesy of Salisbury)*

safe work condition is established before any work task associated with the job is begun. The person in charge also must assume the responsibility of ensuring that all persons who are assigned to the job are accounted for before the electrically safe work condition is removed.

(c) The complex lockout/tagout procedure shall identify the person in charge. In this (these) instance(s), the person in charge shall be permitted to install locks/tags, or direct their installation, on behalf of other employees. The person in charge shall be held accountable for safe execution of the complex lockout/tagout. The complex lockout/tagout procedure shall address all the concerns of employees who might be exposed. All complex lockout/tagout procedures shall require a written plan of execution that identifies the person in charge. All complex lockout/tagout plans shall identify the method to account for all persons who might be exposed to electrical hazards in the course of the lockout/tagout.

Additional employees often are assigned to an area to provide sufficient manpower to complete all necessary work tasks while a shutdown is under way. Contract or other temporary employees might be unfamiliar with the location of electrical circuits and disconnecting means. The person in charge is responsible for ensuring that no temporary employee is unnecessarily exposed to an electrical hazard.

A written plan must identify each step required to install lockout and tagout devices, including the following:

- The disconnecting means
- Who will install lockout/tagout devices
- How the absence of voltage will be verified
- How employees will be accounted for before, during, and after the work is complete

The plan must be reviewed with or by all workers to establish clearly the authority of the person in charge.

**(4) Coordination.**

(a) The established electrical lockout/tagout procedure shall be coordinated with all other employer's procedures for control of exposure to electrical energy sources such that all employer's procedural requirements are adequately addressed on a site basis.

Some projects involve contractors and other employers other than the facility owner. Because each employer is required to implement a lockout/tagout procedure, different requirements might exist. To ensure that the requirements of each procedure are observed, one or more procedures might need to have additional requirement(s).

(b) The procedure for control of exposure to electrical hazards shall be coordinated with other procedures for control of other hazardous energy sources such that they are based on similar/identical concepts.

Other standards require an employer to implement a "control of hazard energy procedure" that covers all sources of energy *except* electrical energy. The electrical lockout/tagout procedure must be coordinated with any other control of hazardous energy procedure to ensure that the requirements of each procedure have a similar basis and similar requirements. For instance, the general control of hazardous energy procedure and the electrical lockout/tagout procedure must have similar or identical requirements for locks and tags.

The electrical lockout/tagout procedure could be integrated into an overall control of hazardous energy procedure; however, that procedure must address all the issues identified in this standard.

(c) The electrical lockout/tagout procedure shall always include voltage testing requirements where there might be direct exposure to electrical energy hazards.

The published lockout/tagout procedure must contain the requirements necessary to ensure that all employees know whether they are exposed to an exposed energized conductor. The procedure should identify acceptable voltage testing devices and contain a requirement to ensure that the voltage testing device is functioning properly, both before and after each use. Employee training must ensure that each qualified employee is familiar with the requirements for testing voltage.

(d) Electrical lockout/tagout devices shall be permitted to be similar to lockout/tagout devices for control of other hazardous energy sources, such as pneumatic, hydraulic, thermal, and mechanical, provided such devices are used only for control of hazardous energy and for no other purpose.

Devices used for control of hazardous energy must be easily recognizable. Lockout/tagout devices that are used to control exposure to electrical energy should have the same physical characteristics as devices used for control of other forms of energy so that workers are not confused. All lockout/tagout devices should be identical until personal information is added by a qualified person.

**(5) Training and Retraining.** Each employer shall provide training as required to ensure employees' understanding of the lockout/tagout procedure content and their duty in executing such procedures.

Employees who work at different job assignments must understand how their personal safety is impacted by the lockout/tagout. Each employee must accept helping their coworkers avoid injury as their duty. Workers that are temporarily reassigned to increase the number of workers available for a specific job or work task must be trained to understand their role in implementing and maintaining the lockout/tagout.

**(E) Equipment.**

**(1) Lock Application.** Energy isolation devices for machinery or equipment installed after January 2, 1990, shall be capable of accepting a lockout device.

All energy isolation devices that were installed in 1990 or later must be capable of being locked in the open position. Any disconnect switch or circuit breaker that is installed in the future also must be capable of being locked in the open position.

Any disconnecting means that cannot be locked in the disconnected (or open) position must not be used as an energy isolation device.

**(2) Lockout/Tagout Device.** Each employer shall supply, and employees shall use, lockout/tagout devices and equipment necessary to execute the requirements of 120.3(E). Locks and tags used for control of exposure to electrical energy hazards shall be unique, shall be readily identifiable as lockout/tagout devices, and shall be used for no other purpose.

Employers must provide all equipment necessary to implement the requirements of the lockout/tagout procedure. The lockout or tagout devices may be identical to lockout or tagout devices that are used for the control of other types of hazardous energy. The devices must be visually unique and easily recognizable. The lockout/tagout devices must not be used for any other purpose.

**(3) Lockout Device.**

(a) A lockout device shall include a lock (either keyed or combination).

(b) The lockout device shall include a method of identifying the individual who installed the lockout device.

(c) A lockout device shall be permitted to be only a lock, provided the lock is readily identifiable as a lockout device, in addition to a means of identifying the person who installed the lock.

(d) Lockout devices shall be attached to prevent operation of the disconnecting means without resorting to undue force or the use of tools.

(e) The tag used in conjunction with a lockout device shall contain a statement prohibiting unauthorized operation of the disconnecting means or unauthorized removal of the device.

(f) Lockout devices shall be suitable for the environment and for the duration of the lockout.

(g) Whether keyed or combination locks are used, the key or combination shall remain in the possession of the individual installing the lock or the person in charge, when provided by the established procedure.

A lockout device might include a tag, chain, cable tie, or other component; however, lockout devices are intended to mean "a lock." The lock may be used in conjunction with other components, but the basic lockout device is a lock. The lock can be operated either with a key or a combination. That key (or combination) must prevent unauthorized removal of the lock, and the lock's installer must be in control of the key (or combination).

The lockout device must include information that identifies the person who installed the lock and be installed in a manner that prevents operation of the energy isolation device. The lockout device also must contain information suggesting disciplinary action for unauthorized removal.

**(4) Tagout Device.**

(a) A tagout device shall include a tag together with an attachment means.

(b) The tagout device shall be readily identifiable as a tagout device and suitable for the environment and duration of the tagout.

(c) A tagout device attachment means shall be capable of withstanding at least 224.4 N (50 lb) of force exerted at a right angle to the disconnecting means surface. The tag attachment means shall be nonreusable, attachable by hand, self-locking, and nonreleasable, equal to an all-environmental tolerant nylon cable tie.

(d) Tags shall contain a statement prohibiting unauthorized operation of the disconnecting means or removal of the tag.

*Exception to (a), (b), and (c): A "hold card tagging tool" on an overhead conductor in conjunction with a hotline tool to install the tagout device safely on a disconnect that is isolated from the worker(s).*

A tagout device is intended to mean a tag and other equipment necessary to attach the complete assembly to an energy isolation device. The tagout device must be unique and easily recognizable as a tagout device. It must be attached to the energy isolation device with a component that ensures that the tag stays in place. The tagout device must contain information suggesting disciplinary action for unauthorized removal.

In an outside overhead line installation, a "hold card" installed by a utility in accordance with an established procedure is permissible.

**(5) Electrical Circuit Interlocks.** Up-to-date diagrammatic drawings shall be consulted to ensure that no electrical circuit interlock operation can result in reenergizing the circuit being worked on.

Section 120.2(E)(5) has two basic requirements. First, electrical circuit interlocks must be considered when developing the lockout/tagout plan. Second, diagrammatic drawings (single-line diagrams, schematic diagrams, or similar drawings) that are consulted must contain current information.

**(6) Control Devices.** Locks/tags shall be installed only on circuit disconnecting means. Control devices, such as pushbuttons or selector switches, shall not be used as the primary isolating device.

Energy isolation devices must directly open the conductors that supply energy to the work location. Control devices such as pushbuttons, selector switches, or other devices that do not directly disconnect the power conductors must not be used for lockout/tagout.

**(F) Procedures.** The employer shall maintain a copy of the procedures required by this section and shall make the procedures available to all employees.

The employer must implement a lockout/tagout procedure for the company or corporation. That procedure must contain the elements and requirements defined in Article 120.

**(1) Planning.** The procedure shall require planning, including 120.2(F)(1)(a) through 120.2(F)(2)(n).

The procedure must contain a requirement for a plan to exist whenever a lockout or tagout is implemented. That plan must consider the information in 120.2(F)(1)(a) through (F)(c).

(a) Locating Sources. Up-to-date single-line drawings shall be considered a primary reference source for such information. When up-to-date drawings are not available, the employer shall be responsible for ensuring that an equally effective means of locating sources of energy is employed.

Locating all possible sources of electrical energy accurately and completely is crucial. Section 120.2(F)(1)(a) is intended to apply to those times when up-to-date drawings are not available. For instance, small businesses might not be able to maintain drawings easily. Those employers must provide an alternative means of locating all possible sources of electrical energy accurately and completely. If changes to the electrical system are unlikely, the employer could prepare a document that lists all possible sources of energy for various system components.

(b) Exposed Persons. The plan shall identify persons who might be exposed to an electrical hazard during the execution of the job or task.

(c) Person In Charge. The plan shall identify the person in charge and his or her responsibility in the lockout/tagout.

The plan must identify the person in charge. The person in charge may be identified by name or by position.

(d) Individual Qualified Employee Control. Individual qualified employee control shall be in accordance with 120.2(D)(1).

The procedure must define the conditions under which individual qualified employee control may be used.

(e) Simple Lockout/Tagout. Simple lockout/tagout procedure shall be in accordance with 120.2(D)(2).

The procedure must define the conditions under which simple lockout/tagout may be used.

(f) Complex Lockout/Tagout. Complex lockout/tagout procedure shall be in accordance with 120.2(D)(3).

The procedure must define the conditions that establish a complex lockout/tagout for the company or corporation.

**(2) Elements of Control.** The procedure shall identify elements of control.

(a) Deenergizing Equipment (Shutdown). The procedure shall establish the person who performs the switching and where and how to deenergize the load.

The procedure must identify the person who will operate the switches. It also must define any special condition or requirement related to deenergizing the load.

(b) Stored Energy. The procedure shall include requirements for releasing stored electric or mechanical energy that might endanger personnel. All capacitors shall be discharged, and high capacitance elements shall also be short-circuited and grounded before the associated equipment is touched or worked on. Springs shall be released or physical restraint shall be applied when necessary to immobilize mechanical equipment and pneumatic and hydraulic pressure reservoirs. Other sources of stored energy shall be blocked or otherwise relieved.

Stored energy can exist in many different forms. Electrical energy can be stored in capacitors or other capacitive elements, such as a long cable. Stored energy, in all its forms, must be relieved or blocked. Electrical equipment frequently contains springs and sometimes pneumatic pressure components, which are sources of stored energy that must be relieved.

(c) Disconnecting Means. The procedure shall identify how to verify that the circuit is deenergized (open).

(d) Responsibility. The procedure shall identify the person who is responsible to verify that the lockout/tagout procedure is implemented and who is responsible to ensure that the task is completed prior to removing locks/tags. A mechanism to accomplish lockout/tagout for multiple (complex) jobs/tasks where required, including the person responsible for coordination, shall be included.

One person must be assigned the responsibility of making certain that the requirements of the lockout/tagout procedure are implemented. The procedure also should define the person responsible for verifying that the job or task is complete before the locks and tags are removed.

The procedure must define the process for executing a complex lockout/tagout. If more than one employer has work on the site, or if multiple procedures apply for any reason, the procedure must define the process for achieving coordination between or among the procedures.

(e) Verification. The procedure shall verify that equipment cannot be restarted. The equipment operating controls, such as pushbuttons, selector switches, and electrical interlocks, shall be operated or otherwise shall be verified that the equipment cannot be restarted.

The procedure must identify the process or processes for ensuring that all the correct disconnecting means have been locked in the "open" position.

(f) Testing. The procedure shall establish the following:

(1) What voltage detector will be used and who will use it to verify proper operation of the voltage detector before and after use
(2) A requirement to define the boundary of the work area
(3) A requirement to test before touching every exposed conductor or circuit part(s) within the defined boundary of the work area
(4) A requirement to retest for absence of voltage when circuit conditions change or when the job location has been left unattended
(5) Where there is no accessible exposed point to take voltage measurements, planning considerations shall include methods of verification.

Many injuries result from inadequate testing for absence of voltage. The procedure or plan must specifically identify the following:

- The testing device to be used
- Who will use the testing device
- The boundary of the safe zone established by the lockout/tagout

The procedure and/or plan also must define the following:

- A requirement for testing every conductor every time before a person touches them
- A requirement to ensure that the testing requirement applies each time the worker(s) leaves the work area for any reason and for any length of time
- The mechanism to be used to determine that the conductor is, in fact, deenergized if no points are exposed to test for voltage

(g) Grounding. Grounding requirements for the circuit shall be established, including whether the grounds shall be installed for the duration of the task or temporarily are established by the procedure. Grounding needs or requirements shall be permitted to be covered in other work rules and might not be part of the lockout/tagout procedure.

The plan must consider all possible ways that a voltage could reappear in the vicinity of the point of the work. The plan must define any requirement for any temporary grounds or safety grounds. If other work rules consider and address conductor grounding, the lockout/tagout procedure or plan is not required to address the same issue.

(h) Shift Change. A method shall be identified in the procedure to transfer responsibility for lockout/tagout to another person or person in charge when the job or task extends beyond one shift.

The procedure must define a formal method of transferring responsibility for the lockout/tagout from one person in charge to another if the work extends beyond one shift. The procedure also must define the method of communicating the transfer of responsibility to the employees.

(i) Coordination. The procedure shall establish how coordination is accomplished with other jobs or tasks in progress, including related jobs or tasks at remote locations, including the person responsible for coordination.

When more than one task or job is being executed in the same physical or electrical circuit area, the actions of one employee might have an impact on the actions of another employee. One person must be assigned responsibility for ensuring that work tasks by different trades or contractors are coordinated to minimize the possibility of one person's actions having a negative impact on another.

(j) Accountability for Personnel. A method shall be identified in the procedure to account for all persons who could be exposed to hazardous energy during the lockout/tagout.

The procedure must address a method to account for all workers who are involved in the work under lockout/tagout. Accounting for all workers at the conclusion of the work normally is assigned to the person in charge.

(k) Lockout/Tagout Application. The procedure shall clearly identify when and where lockout applies, in addition to when and where tagout applies, and shall address the following:

(1) Lockout shall be defined as installing a lockout device on all sources of hazardous energy such that operation of the disconnecting means is prohibited and forcible removal of the lock is required to operate the disconnect means.
(2) Tagout shall be defined as installing a tagout device on all sources of hazardous energy, such that operation of the disconnect means is prohibited. The tagout device shall be installed in the same position available for the lockout device.
(3) Where it is not possible to attach a lock to existing disconnecting means, the disconnecting means shall not be used as the only means to put the circuit in an electrically safe work condition.
(4) The use of tagout procedures without a lock shall be permitted only in cases where equipment design precludes the installation of a lock on a energy isolation device(s). When tagout is employed, at least one additional safety measure shall be employed. In such cases, the procedure shall clearly establish responsibilities and accountability for each person who might be exposed to electrical hazards.

Section 120.2(F)(2)(k) requires the procedure to define when lockout is acceptable and when tagout is acceptable. If tagout is permitted, the employer should be able to justify to a third party that a lockout device could not be installed. *Note that all disconnecting means installed since January 2, 1990, are required to accept a lockout device.*

If tagout is permitted, the procedure or plan must define clearly and unambiguously individual responsibility and accountability for each person potentially exposed to an electrical hazard.

(l) Removal of Lockout/Tagout Devices. The procedure shall identify the details for removing locks or tags when the installing individual is unavailable. When locks or tags are removed by other than the installer, the employer shall attempt to locate the person prior to removing the lock or tag. When the lock or tag is removed because the installer is unavailable, the installer shall be informed prior to returning to work.

The person who installed the lockout/tagout devices must remove them. However, if that person is not available, the lockout/tagout device may be removed by a member of the line organization, provided an attempt to locate the person fails and removing the lockout/tagout device will not expose a person to an electrical hazard. If a member of the line organization removes the lockout/tagout device, the person who installed the device must be informed that the lockout/tagout device has been removed before he or she returns to work.

The procedure must provide details of the process for emergency removal of a lockout/tagout device.

(m) Release for Return to Service. The procedure shall identify steps to be taken when the job or task requiring lockout/tagout is completed. Before electric circuits or equipment are reenergized, appropriate tests and visual inspections shall be conducted to verify that all tools, mechanical restraints and electrical jumpers, shorts, and grounds have been removed, so that the circuits and equipment are in a condition to be safely energized. Where appropriate, the employees responsible for operating the machines or process shall be notified when circuits and equipment are ready to be energized, and such employees shall provide assistance as necessary to safely energize the circuits and equipment. The procedure shall contain a statement requiring the area to be inspected to ensure that nonessential items have been removed. One such step shall ensure that all personnel are clear of exposure to dangerous conditions resulting from reenergizing the service and that blocked mechanical equipment or grounded equipment is cleared and prepared for return to service.

The procedure must define all steps necessary to ensure that the work task is complete and that no one will be exposed unexpectedly to an electrical hazard when the electrical service is restored.

(n) Temporary Release for Testing/Positioning. The procedure shall clearly identify the steps and qualified persons' responsibilities when the job or task requiring lockout/tagout is to be interrupted temporarily for testing or positioning of equipment; then the steps shall be identical to the steps for return to service. See 110.9 and 130.4 for requirements when using test instruments and equipment.

The procedure must define the steps for restoring the electrical service for testing or repositioning. Restoring the electrical service after the work task has begun is very hazardous and should be avoided.

### 120.3 Temporary Protective Grounding Equipment

**(A) Placement.** Temporary protective grounds shall be placed at such locations and arranged in such a manner as to prevent each employee from being exposed to hazardous differences in electrical potential.

When protective grounds (safety grounds) are necessary to avoid possibly reenergizing a conductor that would expose a worker to an unexpected electrical hazard, the grounds must be installed on the conductor at a point between the worker and the source of energy. If the unexpected source of electricity could be from both directions in the electrical circuit, safety grounds must be installed on both sides of the worker. The safety grounds should be installed in a manner that establishes a zone of equipotential where each employee is working.

**(B) Capacity.** Temporary protective grounds shall be capable of conducting the maximum fault current that could flow at the point of grounding for the time necessary to clear the fault.

A set of safety grounds that is exposed to a fault current is subjected to significant mechanical forces. The physical and electrical integrity of the conductors that comprise the ground set must be sufficient to withstand both the mechanical and electrical forces associated with the fault. The ground set must have a rating established by the manufacturer and must be applied within that rating. To establish the necessary rating, available fault current must be determined by a system analysis. The rating of the ground set must be at least as great as the available fault current.

Safety grounds can be fabricated in the field. However, employers who choose this option must be able to verify that the rating of the safety grounds exceeds the available fault current. This option is not recommended.

**(C) Equipment Approval.** Temporary protective grounding equipment shall meet the requirements of ASTM F 855, *Standard Specification for Temporary Protective Grounds to be Used on De-energized Electric Power Lines and Equipment,* 1997.

Safety grounds must be tested in accordance with ASTM F 855, *Standard Specification for Temporary Protective Grounds to Be Used on De-energized Electric Power Lines and Equipment.*

**(D) Impedance.** Temporary protective grounds shall have an impedance low enough to cause immediate operation of protective devices in case of accidental energizing of the electric conductors or circuit parts.

The objective of safety grounds is to provide a path to earth so that the overcurrent protective device can operate. The impedance of the grounding path must be low enough to permit a significant fault current to flow through the overcurrent device.

# ARTICLE 130
# Working On or Near Live Parts

Deciding to work on or near live parts should be a last resort in the workplace, after all other opportunities for establishing an electrically safe work condition have been exhausted. For an employer, accepting the risk of working on or near live parts has moral and legal implications. For a worker, accepting the risk of working on or near live parts significantly increases his or her chance for injury, electrocution, and equipment damage. This is an extremely serious decision that must be supported by a written work permit that is reviewed by several people [see 130.1(A)(1)].

No work must be performed on or near live parts unless the employee is trained and qualified to recognize and avoid contact with live parts. The employee must determine where a difference of potential greater than 49 volts exists between exposed parts within arm's reach of the work task. The work must be planned, and the plan must be shared with all employees who might be involved in or associated with any job task involving live work.

Employers are required to supply, and employees are required to wear, personal protective equipment (PPE) that is selected on the basis of the hazards associated with the overall job. The PPE to be used must be inspected before each use to ensure the integrity of the equipment and that it has been maintained in usable condition.

### 130.1 Justification for Work

Live parts to which an employee might be exposed shall be put into an electrically safe work condition before an employee works on or near them, unless the employer can demonstrate that deenergizing introduces additional or increased hazards or is infeasible due to equipment design or operational limitations. Energized parts that operate at less than 50 volts to ground shall not be not required to be deenergized if there will be no increased exposure to electrical burns or to explosion due to electric arcs.

The basic rule for justifying work on or near live parts is that exposed live parts energized at 50 volts or more must be placed in an electrically safe work condition before an employee works on them. Only two explanations for not creating an electrically safe work condition are satisfactory:

- If deenergizing the electrical circuit would result in an increased or additional hazard, the task may be performed with the circuit energized. An example of an additional hazard could be the loss of electrical power to life support equipment. An example of an increased hazard might be that loss of electrical power could result in an environmental spill.
- If deenergizing the electrical circuit is infeasible due to equipment design or operational limitations, the task may be performed with the circuit energized. An example of infeasible due to equipment design might be that removing the source of voltage for a single instrument circuit would require a complete shutdown of a continuous process.

However, the difference between *infeasible* and *inconvenient* is significant. In some instances, an employer treats the terms *infeasible* and *inconvenient* interchangeably. This paragraph is intended to ensure that *inconvenient* cannot serve to justify work on or near exposed live parts. If work is performed on or near an exposed energized electrical circuit, the employer must be able to document that the work task meets the criteria for one of the acceptable reasons for executing the work with the circuit energized.

The procedure should identify if and when a qualified person is permitted to work on or near exposed live parts.

FPN No. 1: Examples of increased or additional hazards include, but are not limited to, interruption of life support equipment, deactivation of emergency alarm systems, and shutdown of hazardous location ventilation equipment.

"Lack of illumination" has been removed from the list of examples of increased or additional hazards in FPN No. 1. This text was deleted to emphasize the fact that temporary lighting could be installed to enable an electrically safe work condition.

FPN No. 2: Examples of work that might be performed on or near exposed energized electrical conductors or circuit parts because of infeasibility due to equipment design or operational limitations include performing diagnostics and testing (e.g., start-up or troubleshooting) of electric circuits that can only be performed with the circuit energized and work on circuits that form an integral part of a continuous process that would otherwise need to be completely shut down in order to permit work on one circuit or piece of equipment.

FPN No. 3: For voltages of less than 50 volts, the decision to deenergize should include consideration of the capacity of the source and any overcurrent protection between the energy source and the worker.

Sources of electrical energy that are energized at less than 50 volts can be hazardous. For instance, control circuits that operate at less than 50 volts could impact process conditions and result in a release of another kind of energy. Even if the capacity of the energy source is limited, the integrity of the circuit could be critical.

**(A) Energized Electrical Work Permit.**

**(1) Where Required.** If live parts are not placed in an electrically safe work condition (i.e., for the reasons of increased or additional hazards or infeasibility per 130.1), work to be performed shall be considered energized electrical work and shall be performed by written permit only.

The requirement for an energized electrical work permit is new in the 2004 edition of NFPA 70E. Section 130.1(A)(1) correlates with OSHA requirements to deenergize equipment and circuits except when it is infeasible to do so.

Permits that cover routine work tasks to be performed by trained and qualified employees can be written to cover a long period of time. For instance, a worker might be trained and qualified to replace a fuse that involves an exposed live part. If the worker is trained to understand the electrical hazards associated with exchanging the fuse and is wearing any necessary PPE, a permit might be issued that covers, for instance, a three-month period.

If the work is unusual or other than routine from any perspective, the work task must be covered by an energized work permit.

Experience shows that electrical workers tend to accept risk of exposure, believing that they are capable of managing the risk. Experience also shows that managers are less likely to accept increased risk of injury. The energized work permit is intended to ensure that the increased risk (and increased possibility of injuries) associated with exposure to an exposed live part receives adequate consideration.

All work to be done on or near energized live parts must be authorized by a written permit. By signing the permit, the person authorizing the work is accepting some responsibility for the exposure. That individual therefore tends to be more creative in finding alternative ways to accomplish such work. Experience has shown that, when an energized work permit is required, work on or near exposed live parts is reduced more than 50 percent.

**(2) Elements of Work Permit.** The energized electrical work permit shall include, but not be limited to, the following items:

(1) A description of the circuit and equipment to be worked on and their location
(2) Justification for why the work must be performed in an energized condition (130.1)
(3) A description of the safe work practices to be employed [110.8(B)]
(4) Results of the shock hazard analysis [110.8(B)(1)(a)]
(5) Determination of shock protection boundaries [130.2(B) and Table 130.2(C)]
(6) Results of the flash hazard analysis (130.3)
(7) The Flash Protection Boundary [130.3(A)]
(8) The necessary personal protective equipment to safely perform the assigned task [130.3(B), 130.7(C)(9), and Table 130.7(C)(9)(a)]
(9) Means employed to restrict the access of unqualified persons from the work area [110.8(A)(2)]

(10) Evidence of completion of a job briefing, including a discussion of any job-specific hazards [110.7(G)]

(11) Energized work approval (authorizing or responsible management, safety officer, or owner, etc.) signature(s)

When the energized work permit is executed, the items listed in 130.1(A)(2) must be considered, and the permit should provide evidence of their consideration. Note that these items evidence the fact that consideration of electrical hazards is of primary importance.

**(3) Exemptions to Work Permit.** Work performed on or near live parts by qualified persons related to tasks such as testing, troubleshooting, voltage measuring, etc., shall be permitted to be performed without an energized electrical work permit, provided appropriate safe work practices and personal protective equipment in accordance with Chapter 1 are provided and used.

FPN: For an example of an acceptable energized electrical work permit, see Annex J.

Workers who are authorized to perform these work tasks must be trained and qualified to execute those task(s). They must understand the elevated exposure and risk of injury. Before executing these tasks, these workers must perform a hazard analysis, wear any PPE that is required by the analysis, and implement the safe work practices identified in Chapter 1.

### 130.2 Approach Boundaries to Live Parts

**(A) Shock Hazard Analysis.** A shock hazard analysis shall determine the voltage to which personnel will be exposed, boundary requirements, and the personal protective equipment necessary in order to minimize the possibility of electric shock to personnel.

More than one electrical hazard exists to employees working on or near live parts. Section 130.2 is intended to minimize exposure to shock and electrocution. Shock approach boundaries are identified as Limited, Restricted, and Prohibited. Crossing one of these approach boundaries increases the chance that a worker might contact an exposed live part. A person must be qualified before he or she can cross the Limited Approach Boundary. To cross the Restricted Approach Boundary, in addition to being a qualified person, the person must also wear PPE for protection from shock.

The Limited Approach Boundary is intended to restrict the approach of unqualified persons. The Restricted Approach Boundary is intended to restrict the approach of qualified persons (see Exhibit 130.1).

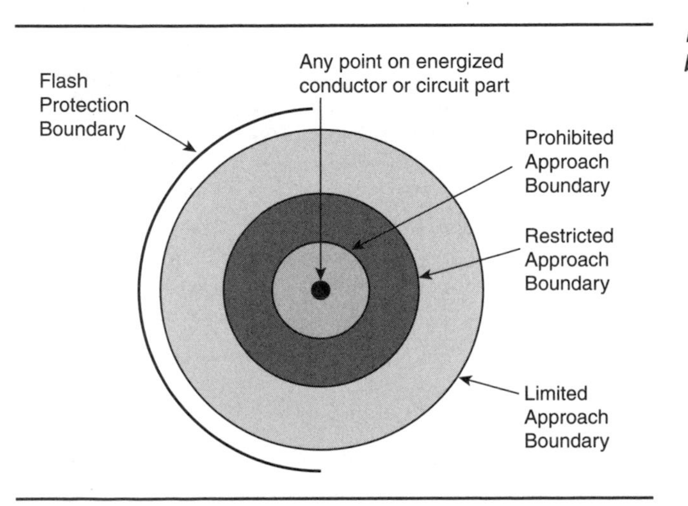


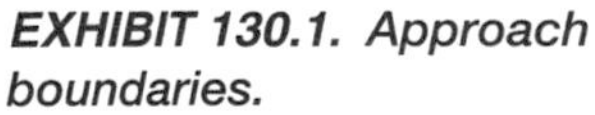
***EXHIBIT 130.1.* Approach boundaries.**

**(B) Shock Protection Boundaries.** The shock protection boundaries identified as Limited, Restricted, and Prohibited Approach Boundaries are applicable to the situation in which approaching personnel are exposed to live parts. See Table 130.2(C) for the distances associated with various system voltages.

FPN: In certain instances, the Flash Protection Boundary might be a greater distance from the exposed live parts than the Limited Approach Boundary.

**(C) Approach to Exposed Live Parts Operating at 50 Volts or More.** No qualified person shall approach or take any conductive object closer to exposed live parts operating at 50 volts or more than the Restricted Approach Boundary set forth in Table 130.2(C), unless any of the following apply:

(1) The qualified person is insulated or guarded from the live parts operating at 50 volts or more (insulating gloves or insulating gloves and sleeves are considered insulation only with regard to the energized parts upon which work is being performed), and no uninsulated part of the qualified person's body crosses the Prohibited Approach Boundary set forth in Table 130.2(C).

(2) The live part operating at 50 volts or more is insulated from the qualified person and from any other conductive object at a different potential.

(3) The qualified person is insulated from any other conductive object as during live-line bare-hand work.

***TABLE 130.2(C)*** *Approach Boundaries to Live Parts for Shock Protection. (All dimensions are distance from live part to employee.)*

| (1) | (2) | (3) | (4) | (5) |
|---|---|---|---|---|
| | **Limited Approach Boundary**[1] | | **Restricted Approach Boundary[1]; Includes Inadvertent Movement Adder** | |
| **Nominal System Voltage Range, Phase to Phase** | **Exposed Movable Conductor** | **Exposed Fixed Circuit Part** | | **Prohibited Approach Boundary**[1] |
| Less than 50 | Not specified | Not specified | Not specified | Not specified |
| 50 to 300 | 3.05 m (10 ft 0 in.) | 1.07 m (3 ft 6 in.) | Avoid contact | Avoid contact |
| 301 to 750 | 3.05 m (10 ft 0 in.) | 1.07 m (3 ft 6 in.) | 304.8 mm (1 ft 0 in.) | 25.4 mm (0 ft 1 in.) |
| 751 to 15 kV | 3.05 m (10 ft 0 in.) | 1.53 m (5 ft 0 in.) | 660.4 mm (2 ft 2 in.) | 177.8 mm (0 ft 7 in.) |
| 15.1 kV to 36 kV | 3.05 m (10 ft 0 in.) | 1.83 m (6 ft 0 in.) | 787.4 mm (2 ft 7 in.) | 254 mm (0 ft 10 in.) |
| 36.1 kV to 46 kV | 3.05 m (10 ft 0 in. ) | 2.44 m (8 ft 0 in.) | 838.2 mm (2 ft 9 in.) | 431.8 mm (1 ft 5 in.) |
| 46.1 kV to 72.5 kV | 3.05 m (10 ft 0 in.) | 2.44 m (8 ft 0 in.) | 965.2 mm (3 ft 2 in.) | 635 mm (2 ft 1 in.) |
| 72.6 kV to 121 kV | 3.25 m (10 ft 8 in.) | 2.44 m (8 ft 0 in.) | 991 mm (3 ft 3 in.) | 812.8 mm (2 ft 8 in.) |
| 138 kV to 145 kV | 3.36 m (11 ft 0 in.) | 3.05 m (10 ft 0 in.) | 1.093 m (3 ft 7 in.) | 939.8 mm (3 ft 1 in.) |
| 161 kV to 169 kV | 3.56 m (11 ft 8 in.) | 3.56 m (11 ft 8 in.) | 1.22 m (4 ft 0 in.) | 1.07 m (3 ft 6 in.) |
| 230 kV to 242 kV | 3.97 m (13 ft 0 in.) | 3.97 m (13 ft 0 in.) | 1.6 m (5 ft 3 in.) | 1.45 m (4 ft 9 in.) |
| 345 kV to 362 kV | 4.68 m (15 ft 4 in.) | 4.68 m (15 ft 4 in.) | 2.59 m (8 ft 6 in.) | 2.44 m (8 ft 0 in.) |
| 500 kV to 550 kV | 5.8 m (19 ft 0 in.) | 5.8 m (19 ft 0 in.) | 3.43 m (11 ft 3 in.) | 3.28 m (10 ft 9 in.) |
| 765 kV to 800 kV | 7.24 m (23 ft 9 in.) | 7.24 m (23 ft 9 in.) | 4.55 m (14 ft 11 in.) | 4.4 m (14 ft 5 in.) |

Note: For Flash Protection Boundary, see 130.3(A).

[1]See definition in Article 100 and text in 130.2(D)(2) and Annex C for elaboration.

Section 130.2(C) defines requirements that must be observed before a qualified person can cross the Restricted Approach Boundary. This section recognizes that a tool or other object is considered to be an extension of the person's body. If a worker is holding an object in his or her hand, the requirements of this section apply to both the worker and the object. The effect of these requirements is that the worker (or extended body part) is prevented from being exposed to any difference of potential greater than 50 volts.

The purpose of columns 2 and 3 of Table 130.2(C) is to recognize that the possibility of contacting an electrical conductor that can move is greater than the possibility of contacting an electrical conductor that is fixed in place with no possibility of moving. This risk is associated with the relative position of the conductor and the worker. If that distance can vary because the conductor can move (such as a bare overhead conductor or a conductor installed on racks in a manhole), or if the distance can vary because the platform (articulating arm) on which the employee is standing can move, then column 2 applies.

The Restricted Approach Boundary accounts for some inadvertent movement by the worker. In other words, a worker might move his or her hand unintentionally. The inadvertent movement adder accounts for this unintended movement.

**(D) Approach by Unqualified Persons.** Unqualified persons shall not be permitted to enter spaces that are required under 400.16(A) to be accessible to qualified employees only, unless the electric conductors and equipment involved are in an electrically safe work condition.

The spaces mentioned in 130.2(D) contain live parts that are exposed if the door or cover is opened.

**(1) Working At or Close to the Limited Approach Boundary.** Where one or more unqualified persons are working at or close to the Limited Approach Boundary, the designated person in charge of the work space where the electrical hazard exists shall cooperate with the designated person in charge of the unqualified person(s) to ensure that all work can be done safely. This shall include advising the unqualified person(s) of the electrical hazard and warning him or her to stay outside of the Limited Approach Boundary.

Workers who are not associated with electrical work in progress could be exposed to an electrical hazard. For instance, a painter could be working in the same room with an exposed live part. The electrical supervisor and the painter's supervisor must establish a communication path to ensure that the painter is advised of the location of the exposed live part and how to avoid exposure to any associated electrical hazard(s).

**(2) Entering the Limited Approach Boundary.** Where there is a need for an unqualified person(s) to cross the Limited Approach Boundary, a qualified person shall advise him or her of the possible hazards and continuously escort the unqualified person(s) while inside the Limited Approach Boundary. Under no circumstance shall the escorted unqualified person(s) be permitted to cross the Restricted Approach Boundary.

Unqualified persons must not cross the Limited Approach Boundary. Should it become necessary for an unqualified person to cross that boundary, a qualified person must escort him or her and advise the unqualified person about how to avoid exposure to an associated electrical hazard.

### 130.3 Flash Hazard Analysis

A flash hazard analysis shall be done in order to protect personnel from the possibility of being injured by an arc flash. The analysis shall determine the Flash Protection Boundary and the personal protective equipment that people within the Flash Protection Boundary shall use.

The purpose of a flash hazard analysis is to determine if a thermal hazard exists and to select protective equipment necessary to mitigate exposure to the hazard.

An arc flash analysis is a review of an electrical circuit to determine its capacity to deliver sufficient energy to cause a thermal burn from an arcing fault. The primary purpose of the analysis is to determine the distance from the potential arcing fault point that will expose a person to a second-degree burn (Flash Protection Boundary). Any body part that is closer to the potential arcing fault must be protected from the thermal effects of that fault. The analysis must determine that point. The Flash Protection Boundary can be determined by individual calculations at a specific point in a circuit or with the use of labels (or another communication method) that indicate the Flash Protection Boundary.

The Flash Protection Boundary is intended to trigger the need for PPE that can protect the worker from potential thermal injury. After the Flash Protection Boundary is determined, the worker must be able to select PPE that will minimize the possibility of a second-degree burn.

Determining the degree of possible exposure is complex and involves the interaction of as many as 14 variables. An arcing fault is a transformation of electrical energy to other forms of energy, such as heat and pressure. Heat energy is the hazard that is addressed in the arc flash analysis. The release of thermal energy in an arcing fault is dependent on many variables. Although the variables are known, how they interact is currently unknown.

Methods to calculate the degree of possible thermal exposure are currently evolving. Several have been developed and are currently available, and all produce a result that is stated in terms of calories per square centimeter. Flame-resistant (FR) protective clothing and equipment are also rated in calories per square centimeter. FR clothing that has a rating equal to or higher than the potential arc intensity must be worn.

**(A) Flash Protection Boundary.** For systems that are 600 volts or less, the Flash Protection Boundary shall be 4.0 ft, based on the product of clearing times of 6 cycles (0.1 second) and the available bolted fault current of 50 kA or any combination not exceeding 300 kA cycles (5000 ampere seconds). For clearing times and bolted fault currents other than 300 kA cycles, or under engineering supervision, the Flash Protection Boundary shall alternatively be permitted to be calculated in accordance with the following general formula:

$$D_c = [2.65 \times MVA_{bf} \times t]^{1/2}$$

or

$$D_c = [53 \times MVA \times t]^{1/2}$$

where:

$D_c$ = distance in feet from an arc source for a second-degree burn

$MVA_{bf}$ = bolted fault capacity available at point involved (in mega volt-amps)

$MVA$ = capacity rating of transformer (mega volt-amps). For transformers with $MVA$ ratings below 0.75 MVA, multiply the transformer $MVA$ rating by 1.25

$t$ = time of arc exposure (in seconds)

At voltage levels above 600 volts, the Flash Protection Boundary is the distance at which the incident energy equals 5 J/cm$^2$(1.2 cal/cm$^2$). For situations where fault-clearing time is 0.1 second (or faster), the Flash Protection Boundary is the distance at which the incident energy level equals 6.24 J/cm$^2$(1.5 cal/cm$^2$).

Section 130.3(A) is intended to provide guidance in how to determine the Flash Protection Boundary. Determining the thermal intensity of an arcing fault is difficult and complex. Many small businesses have an electrical system that does not exceed 600 volts. Small businesses are less likely to have access to engineering supervision needed to execute complex calculations, and their electrical installations usually have limited capacity and relatively simple circuitry. A default Flash Protection Boundary of 4 ft has been established, provided the system capacity does not exceed 300 kA cycles (5000 ampere seconds). The available bolted-fault current must not exceed 50 kA, with duration of 6 cycles. (The product of 50 kA and 6 cycles is 300 kA cycles.) If the product of the available bolted-fault current (in thousands of amperes) and the clearing time (in cycles) of the overcurrent device is greater than 300, the default Flash Protection Boundary must not be used.

The equation provided in this section was published in an IEEE paper authored by R. H. Lee in 1982. The equation produces a distance that will expose a surface to 1.2 calories per square centimeter.

A thermal burn occurs because the temperature of cells in skin tissue is raised to a level that damages their structure. Some time is required for the transfer of heat from the source (arcing fault) into the surface of the skin. If the duration of the arcing fault can be limited to less than 6 cycles, the injury limit can be increased to 1.5 calories per square centimeter. The information needed to perform this calculation normally is available from a facility coordination study.

After the Flash Protection Boundary is determined, the worker must be provided with information that enables him or her to select FR clothing or equipment. Table 130.7(C)(9)(a) provides a list of work tasks intended to suggest protective equipment based on standard configurations and conditions. The employer could choose to use this table to determine the amount of protection that is necessary.

Arcing faults rarely occur in open air. In the majority of instances, arcing faults occur within an enclosure. An arcing fault is most likely to occur when movement is occurring within the enclosure (for instance, movement occurs when a switch contact is closed or opened, when a contactor operates, when a door or cover is removed or installed). Many arcing faults are associated with movement initiated by a person, such as when a person reaches inside a cubicle with a conductive object in his or her hand.

This section also recognizes that other methods can be used to determine the Flash Protection Boundary and the degree of protection, and it permits their use under engineering supervision. Annex D exhibits some other methods of determining the Flash Protection Boundary and the rating of protective equipment.

**(B) Protective Clothing and Personal Protective Equipment for Application with a Flash Hazard Analysis.** Where it has been determined that work will be performed within the Flash Protection Boundary by 130.3(A), the flash hazard analysis shall determine, and the employer shall document, the incident energy exposure of the worker (in calories per square centimeter). The incident energy exposure level shall be based on the working distance of the employee's face and chest areas from a prospective arc source for the specific task to be performed. Flame-resistant (FR) clothing and personal protective equipment (PPE) shall be used by the employee based on the incident energy exposure associated with the specific task. Recognizing that incident energy increases as the distance from the arc flash decreases, additional PPE shall be used for any parts of the body that are closer than the distance at which the incident energy was determined As an alternative, the PPE requirements of 130.7(C)(9) shall be permitted to be used in lieu of the detailed flash hazard analysis approach described in 130.3(A).

FPN: For information on estimating the incident energy, see Annex D.

In addition to determining the Flash Protection Boundary, the flash hazard analysis per 130.3(B) must provide information necessary for a worker to select flame-resistant (FR) clothing that can mitigate exposure to a potential arc flash.

Protective clothing constructed and tested in accordance with the test procedures developed by the ASTM F 18 technical committee is rated in calories per square centimeter. In choosing protective clothing, purchasers must compare the degree of arc flash hazard with the ratings of clothing construction from different manufacturers. Therefore, the degree of the potential arc flash hazard must be determined in calories per square centimeter or converted to that measure. If the clothing rating is equal to or greater than the degree of arc flash hazard, the clothing can protect the worker from a second-degree burn in most exposures. Note that although most second-degree burns will be mitigated, protection is not absolute. This section suggests that body parts closer to the potential arc source should have additional protection.

The term *incident energy* is intended to be the amount of *thermal* energy that could be received (incident upon) by either clothing or exposed skin. FR clothing must be rated at least as great as the incident energy potentially available in the arcing fault.

Because understanding of the thermal hazard is evolving, an employer might choose to define protective clothing by several different methods. PPE requirements also could be defined by calculating the available incident energy by one of several different methods. Appropriate PPE could be determined on the basis of the tables included in NFPA 70E, 2004 edition. The employer procedure that defines the flash hazard analysis must define the process selected by the employer.

The ASTM F 18 committee has established a standard dimension of 18 in. between the potential arc source and the worker's clothing. The 18-in. dimension is based loosely on the worker's arm being extended approximately 12 in. and the potential arc being approximately 6 in. from the front of the equipment. The distance between an arcing fault and a worker is a very important consideration. The intensity of the incident energy decreases exponentially as that distance increases. If a worker could increase the distance of his or her body from the exposed live part, he or she certainly should do so.

Parts of a worker's body that are closer to the potential arc (such as the hands) should have increased protection. No method has been identified to suggest the amount of thermal protection needed, but a worker's hands probably are exposed to much higher temperatures than other body parts.

### 130.4 Test Instruments and Equipment Use

Only qualified persons shall perform testing work on or near live parts operating at 50 volts or more.

Workers must be trained to understand that they are exposed to shock and electrocution when performing work tasks involving testing. Each qualified person must be trained to understand how to use the specific meter (see Exhibit 130.2) and to understand and interpret its indication(s). (Of course, the meter must be in good working condition, appropriate for the task, and tested before use.)

All employees who are qualified persons shall be trained to test for the absence of voltage. Each qualified person must be able to operate each meter that he or she might be expected to use and to interpret any possible meter indication. No voltage test device should be available for use until each qualified person has been trained to use it.

### 130.5 Work On or Near Uninsulated Overhead Lines

**(A) Uninsulated and Energized.** Where work is performed in locations containing uninsulated energized overhead lines that are not guarded or isolated, precautions shall be taken to prevent employees from contacting such lines directly with any unguarded parts of their body or indirectly through conductive materials, tools, or equipment. Where the work to be performed is such that contact with uninsulated energized overhead lines is possible, the lines shall be deenergized and visibly grounded at the point of work, or suitably guarded.

Electrical conductors that are not fully insulated for the circuit voltage are considered to have the same potential for shock and electrocution as conductors that are completely bare. Some overhead conductors have a covering as protection from environmental degradation. Exhibit 130.3 illustrates the danger of operating equipment in close proximity of overhead power lines and the need to treat these lines as energized, unless they have been deenergized and visibly grounded.

No work task should be performed unless the worker is protected from unintentional contact with any overhead lines. Workers carrying long objects must exercise caution and avoid crossing the Limited Approach Boundary. When long objects are moved, a worker should be assigned to each end of the object to maintain control of each end. Any object that is not fully insulated for the circuit voltage is considered to be conductive. To be insulated for the voltage,

*EXHIBIT 130.2. A meter typical of those used to test for voltage. (Courtesy of Fluke)*

an object must have a rating established by the manufacturer and rated according to a standard testing method.

Unqualified workers must not approach an overhead line that is not in an electrically safe work condition. Qualified workers must observe and comply with the approach boundaries identified in Table 130.2(C), Approach Boundaries to Live Parts for Shock Protection.

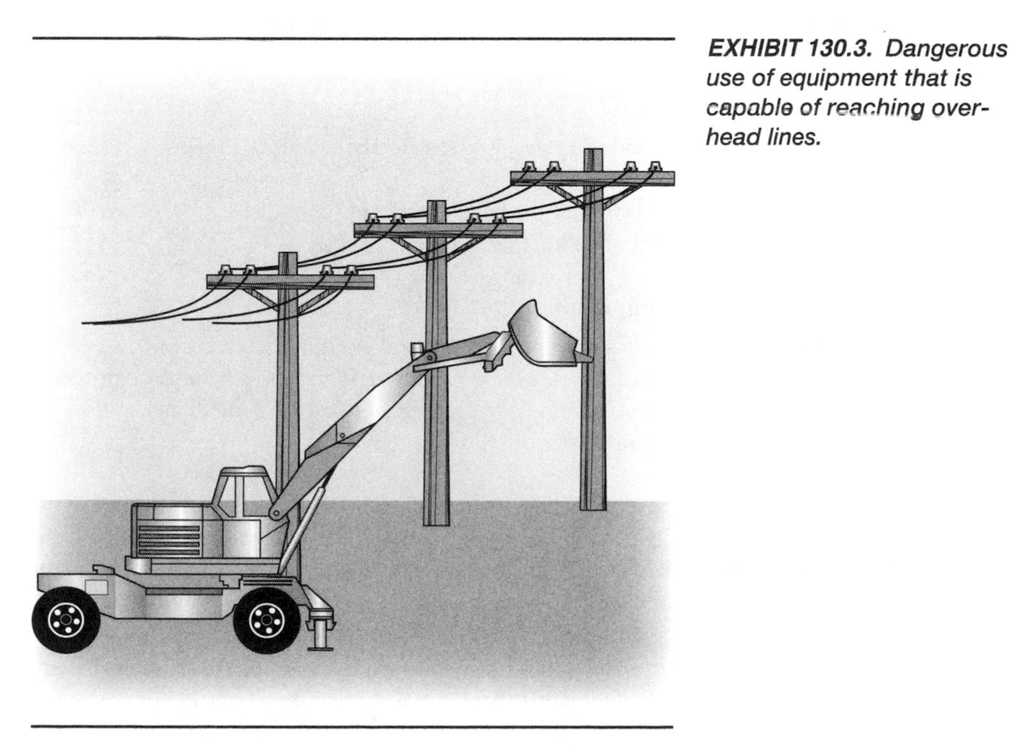

*EXHIBIT 130.3. Dangerous use of equipment that is capable of reaching overhead lines.*

**(B) Deenergizing or Guarding.** If the lines are to be deenergized, arrangements shall be made with the person or organization that operates or controls the lines to deenergize them and visibly ground them at the point of work. If arrangements are made to use protective measures, such as guarding, isolating, or insulation, these precautions shall prevent each employee from contacting such lines directly with any part of his or her body or indirectly through conductive materials, tools, or equipment.

In many cases, responsibility for operation and maintenance of outside overhead lines is assigned to a person or crew. Responsibility for operation and maintenance of transmission and distribution lines might be assigned to a utility or other similar group. The person responsible for operation and maintenance of the affected conductors must be consulted and directly involved in deenergizing and grounding the overhead conductors.

Suitable guards could be installed to prevent accidental contact with the overhead lines. However, the guards must be of sufficient strength to control the approach or any possible movement of the person or object, to eliminate the chance of unintentional contact. In most instances, where an object is being handled in proximity to an uninsulated and ungrounded overhead line, line hose *is not satisfactory* to prevent unintentional contact.

**(C) Employer and Employee Responsibility.** The employer and employee shall be responsible for ensuring that guards or protective measures are satisfactory for the conditions. Employees shall comply with established work methods and the use of protective equipment.

Both employers and employees are responsible for ensuring that any installed guards are adequate for the conditions. That shared responsibility involves procedures and/or work methods provided by the employer and implemented by the employee. The employer and employee must work together to ensure that effective procedures exist, are stringently applied, and are reviewed frequently.

**(D) Approach Distances for Unqualified Persons.** When employees without electrical training are working on the ground or in an elevated position near overhead lines, the location shall be such that the employee and the longest conductive object the employee might contact cannot come closer to any unguarded, energized overhead power line than the Limited Approach Boundary. If the voltage on the line exceeds 50 kV, the distance shall be 3.04 m (10 ft) plus 100 mm (4 in.) for every 10 kV over 50 kV.

FPN: Objects that are not insulated for the voltage involved should be considered to be conductive.

Section 130.5(D) underscores the importance of the Limited Approach Boundary. Unqualified workers must observe the specified dimension [see Table 310.2(C)].

**(E) Vehicular and Mechanical Equipment.**

**(1) Elevated Equipment.** Where any vehicle or mechanical equipment structure will be elevated near energized overhead lines, they shall be operated so that the Limited Approach Boundary distance of Table 130.2(C), Column 2, is maintained. However, under any of the following conditions, the clearances shall be permitted to be reduced:

(1) If the vehicle is in transit with its structure lowered, the Limited Approach Boundary to overhead lines in Table 130.2(C), Column 2, shall be permitted to be reduced by 1.83 m (6 ft). If insulated barriers, rated for the voltages involved, are installed and they are not part of an attachment to the vehicle, the clearance shall be permitted to be reduced to the design working dimensions of the insulating barrier.
(2) If the equipment is an aerial lift insulated for the voltage involved, and if the work is performed by a qualified person, the clearance (between the uninsulated portion of the aerial

lift and the power line) shall be permitted to be reduced to the Restricted Approach Boundary given in Table 130.2(C), Column 4.

A significant number of electrocutions occur when a vehicle contacts an energized overhead conductor. Section 130.5(E)(1) limits the approach of vehicular and other mechanical equipment to outside overhead lines that are not in an electrically safe work condition. This section also establishes conditions under which the approach dimensions for this type of equipment may be reduced. Only two valid reasons to reduce the approach limit are permissible: (1) If the equipment is insulated for at least the circuit voltage, the approach distance may be reduced to the Restricted Approach Boundary; (2) if the equipment is in motion, the approach boundary may be reduced.

The requirements of this section apply to any type of mechanical equipment. For instance, these requirements apply to operating a man lift or any other articulating equipment.

**(2) Equipment Contact.** Employees standing on the ground shall not contact the vehicle or mechanical equipment or any of its attachments, unless either of the following conditions apply:

(1) The employee is using protective equipment rated for the voltage.
(2) The equipment is located so that no uninsulated part of its structure (that portion of the structure that provides a conductive path to employees on the ground) can come closer to the line than permitted in 130.5(E)(1).

Workers who are outside the equipment that is in contact with an energized conductor have greater exposure to electrocution than workers who are inside the equipment cab. Workers must not enter the physical area surrounding equipment operating in proximity to overhead conductors. Such workers are exposed to touch potential that is effectively equal to the operating circuit voltage if they are in contact with the equipment when it touches an energized conductor. Workers are exposed to electrocution by step potential if they are standing on the ground near the equipment when it makes contact with an overhead line.

A barricade should be erected around the physical area to surround equipment that could contact an overhead line. Signs should be installed to warn people to stay out of the area. The barricaded area should be no smaller than the Limited Approach Boundary.

**(3) Equipment Grounding.** If any vehicle or mechanical equipment capable of having parts of its structure elevated near energized overhead lines is intentionally grounded, employees working on the ground near the point of grounding shall not stand at the grounding location whenever there is a possibility of overhead line contact. Additional precautions, such as the use of barricades or insulation, shall be taken to protect employees from hazardous ground potentials (step and touch potential), which can develop within a few feet or more outward from the grounded point.

In some instances, equipment operating in proximity to an overhead line can have a temporary grounding conductor installed to ensure that the overcurrent device would operate if needed. Such a grounding conductor would expand the touch- and step-potential hazard to include the ground rod (or other ground point) in addition to the physical area surrounding the conductor and the ground rod.

Barricade and warning signs should be erected that consider the grounding conductor and ground rod when the Limited Approach Boundary is established.

### 130.6 Other Precautions for Personnel Activities

#### (A) Alertness.

**(1) When Hazardous.** Employees shall be instructed to be alert at all times when they are working near live parts operating at 50 volts or more and in work situations where unexpected electrical hazards might exist.

Although employees must be instructed to be alert, the electrical safety program should also emphasize alertness. The supervisor should evaluate the condition of each employee when the job line-up instruction is given.

**(2) When Impaired.** Employees shall not knowingly be permitted to work in areas containing live parts operating at 50 volts or more or other electrical hazards while their alertness is recognizably impaired due to illness, fatigue, or other reasons.

When the job line-up instruction is given, in addition to alertness, the supervisor must evaluate whether any employee is impaired for any reason. In addition to illness and fatigue, impairment might be the result of substance abuse or legal drugs. An employee's ability to act or react could be affected by conditions surrounding his or her personal life. The supervisor should look for signs that the employee's ability to think or act might be impaired for any reason.

**(B) Blind Reaching.** Employees shall be instructed not to reach blindly into areas that might contain exposed live parts where an electrical hazard exists.

"Reaching blindly" is reaching to contact any point that is not directly visible. Using a mirror as an aid to see behind a device or object is still considered to be reaching blindly. If an exposed live part exists, employees must not reach into the area unless they have direct visual observation. If the hazard/risk analysis indicates that an electrical hazard exists, reaching blindly must be avoided.

**(C) Illumination.**

**(1) General.** Employees shall not enter spaces containing live parts unless illumination is provided that enables the employees to perform the work safely.

**(2) Obstructed View of Work Area.** Where lack of illumination or an obstruction precludes observation of the work to be performed, employees shall not perform any task near live parts operating at 50 volts or more or where an electrical hazard exists.

Section 130.6(C)(2) suggests that if a worker cannot see the work area and discern all aspects of the physical equipment, the work must not be performed if the hazard/risk analysis indicates existence of an electrical hazard.

**(D) Conductive Articles Being Worn.** Conductive articles of jewelry and clothing (such as watchbands, bracelets, rings, key chains, necklaces, metalized aprons, cloth with conductive thread, metal headgear, or metal frame glasses) shall not be worn where they present an electrical contact hazard with exposed live parts.

Workers who are or might be exposed to an exposed live part must not wear jewelry. Note that eyewear with metal frames, body piercing jewelry, and conductive clothing all are prohibited.

**(E) Conductive Materials, Tools, and Equipment Being Handled.**

**(1) General.** Conductive materials, tools, and equipment that are in contact with any part of an employee's body shall be handled in a manner that prevents accidental contact with live parts. Such materials and equipment include, but are not limited to, long conductive objects, such as ducts, pipes and tubes, conductive hose and rope, metal-lined rules and scales, steel tapes, pulling lines, metal scaffold parts, structural members, bull floats, and chains.

A conductive object being held by an employee extends the reach of the employee. In the context of 130.6(E)(1), the word *conductive* means any object that does not have an assigned rating. Manufacturers (and sometimes testing laboratories) test materials and equipment to determine the effective insulation level and resultant voltage rating. All objects that do not have a voltage rating must be considered to be conductive.

Long objects are difficult to control and should be handled by two employees, one on each end of the object. This work practice enables the object to be moved without crossing the Limited Approach Boundary.

**(2) Approach to Live Parts.** Means shall be employed to ensure that conductive materials approach exposed live parts no closer than that permitted by Table 130.2(C).

Unless an object has been assigned a rating that describes its insulating qualities, the object should be considered to be conductive. Conductive objects must be handled or secured in a way that ensures the object does not breach the shock approach boundary identified in Table 130.2(C).

**(F) Confined or Enclosed Work Spaces.** When an employee works in a confined or enclosed space (such as a manhole or vault) that contains exposed live parts operating at 50 volts or more or an electrical hazard exists, the employer shall provide, and the employee shall use, protective shields, protective barriers, or insulating materials as necessary to avoid inadvertent contact with these parts. Doors, hinged panels, and the like shall be secured to prevent their swinging into an employee and causing the employee to contact exposed live parts operating at 50 volts or more or where an electrical hazard exists.

Manholes, hand holes, vaults, and large sections of equipment could enable an employee to enter an area that has exposed conductors that might be energized. Employees should not enter those areas.

If it is determined that a task cannot be delayed, a hazard/risk analysis must be performed and documented. If the analysis determines that the risks could be reduced to an acceptable level by installing barriers, shields, or other isolating devices, the task may be performed, provided all hazards identified in the hazard/risk analysis are mitigated. The work task must be covered by an energized work permit.

Doors, hinged panels, and similar covers must be held open by a secure means to avoid the possibility that the door or cover could swing and surprise the worker while he or she is exposed to a shock hazard.

In addition to the electrical hazards, confined spaces might contain other hazards. Toxic gases or low oxygen content within the confined space could expose a worker to a life-threatening situation. OSHA has specific requirements for work performed in confined spaces that are not related to an electrical hazard.

**(G) Housekeeping Duties.** Where live parts present an electrical contact hazard, employees shall not perform housekeeping duties inside the Limited Approach Boundary where there is a possibility of contact, unless adequate safeguards (such as insulating equipment or barriers) are provided to prevent contact. Electrically conductive cleaning materials (including conductive solids such as steel wool, metalized cloth, and silicone carbide, as well as conductive liquid solutions) shall not be used inside the Limited Approach Boundary unless procedures to prevent electrical contact are followed.

Section 130.6(G) permits only qualified workers to perform housekeeping duties within the Limited Approach Boundary provided by Table 130.2(C). No conductive cleaning materials may be present within the Limited Approach Boundary unless they are being used by a qualified person. The conductive material must be removed from within the Limited Approach Boundary immediately after use. If all live parts are covered in such a way as to prevent unintentional contact, the work task may be executed.

**(H) Occasional Use of Flammable Materials.** Where flammable materials are present only occasionally, electric equipment capable of igniting them may not be used, unless measures are taken to prevent hazardous conditions from developing. Such materials include, but are not limited to, flammable gases, vapors, or liquids; combustible dust; and ignitible fibers or flyings.

FPN: Electrical installation requirements for locations where flammable materials are present on a regular basis are contained in Article 440.

Some cleaning materials are flammable. An area could become a Class I hazardous area on a temporary basis due to the possibility of accumulating vapors. Flammable materials must not be used near energized electrical equipment unless the area is classified or the possibility of generating an explosive environment is avoided.

**(I) Anticipating Failure.** When there is evidence that electric equipment could fail and injure employees, the electric equipment shall be deenergized unless the employer can demonstrate that deenergizing introduces additional or increased hazards or is infeasible because of equipment design or operational limitation. Until the equipment is deenergized or repaired, employees shall be protected from hazards associated with the impending failure of the equipment.

Electrical equipment frequently offers indications that failure is impending, and workers should be capable of recognizing these indications. If an enclosure feels hot to the touch, for instance, the chance of equipment failure is increased. If unusual noises or sounds are heard, the chance of equipment failure is increased. Unusual smells also are an indication of impending failure.

If these or other indications of impending failure exist, the equipment should be deenergized from a remote location. A circuit breaker or switch that supplies voltage to the suspect equipment should be opened.

Disconnecting means located in equipment that has an indication of impending failure should not be operated unless the worker is protected from the effects of equipment failure.

**(J) Routine Opening and Closing of Circuits.** Load-rated switches, circuit breakers, or other devices specifically designed as disconnecting means shall be used for the opening, reversing, or closing of circuits under load conditions. Cable connectors not of the load-break type, fuses, terminal lugs, and cable splice connections shall not be permitted to be used for such purposes, except in an emergency.

Only equipment that has been rated to serve as load-break equipment may be used for routine control of electrical equipment or circuits. The manufacturer establishes the load-break rating after testing to ensure that the equipment can safely interrupt the full-load current expected in the circuit. Unless the rating equals or exceeds the full-load current of the circuit, the disconnecting device must not be operated unless the circuit has been deenergized.

With the exception of pole-mounted fuse-cutout devices, fuses are not designed to be removed from an energized circuit. Cable connectors are not designed to be opened under load. Removing a fuse or opening a cable connector that is conducting current can initiate an arc that could escalate into an arcing fault. These and similar devices must not be disconnected or removed unless the load current has been interrupted by another means.

Disconnect switches that are not load rated can initiate an arcing fault if operated under load. These devices may be operated under load only if an emergency condition exists.

**(K) Reclosing Circuits After Protective Device Operation.** After a circuit is deenergized by a circuit protective device, the circuit shall not be manually reenergized until it has been determined that the equipment and circuit can be safely energized. The repetitive manual reclosing of circuit breakers or reenergizing circuits through replaced fuses shall be prohibited. When it is determined from the design of the circuit and the overcurrent devices involved that the automatic operation of a device was caused by an overload rather than a fault condition, examination of the circuit or connected equipment shall not be required before the circuit is reenergized.

Overcurrent protective devices operate if the rated current flow is exceeded. If an overcurrent device (fuse, circuit breaker, relay, overload, and similar devices) operates, a reasonable assumption is that the rated current has been exceeded. The rated current possibly was exceeded because of an expected condition, such as a motor overload. A motor could experience an overloaded condition due to a temporary mechanical condition in the driven equipment.

An electrical fault condition also could be the reason an overcurrent device operates. Repeatedly reapplying electrical energy to an electrical fault can result in a massive failure and introduce unexpected hazardous conditions. However, massive failure is unlikely if the energy is reapplied only once.

A circuit breaker or overload relay that trips may be reset only once. A fuse that blows also may be replaced only once. If the device trips a second time, a qualified person must determine the reason for the overloaded condition before the circuit is reenergized.

### 130.7 Personal and Other Protective Equipment

**(A) General.** Employees working in areas where electrical hazards are present shall be provided with, and shall use, protective equipment that is designed and constructed for the specific part of the body to be protected and for the work to be performed.

PPE generally is designed to afford protection from a single hazard and for a specific body part. For instance, voltage-rated gloves are designed for protection from electrocution. A flame-resistant shirt is designed to protect the upper torso from a thermal hazard. Although a specific item of PPE might offer protection from more than one hazard, PPE is not designed for that purpose.

When selecting PPE, the worker must identify the degree of each hazard. Next, the worker must determine which part of his or her body will be within the boundary associated with the hazard. If a worker's face is within the Flash Protection Boundary, for instance, a thermally rated face shield is satisfactory. If the worker's entire head is within the Flash Protection Boundary, protection for the back of the worker's head is required in addition to protection for the worker's face.

**(B) Care of Equipment.** Protective equipment shall be maintained in a safe, reliable condition. The protective equipment shall be visually inspected before each use.

FPN: Specific requirements for periodic testing of electrical protective equipment are given in 130.7(C)(8) and 130.7(F).

To ensure that PPE is effective when needed, it must be inspected, maintained, and stored at defined intervals. The electrical safety program should define intervals and inspection methods. If inspection intervals and test methodology are defined in a national consensus standard, the electrical safety program inspection methods and intervals must be at least as frequent as those stated in the consensus standard. Tables 130.7(C)(8) and 130.7(F) list national consensus standards that cover electrical PPE.

Flame-resistant clothing must be cleaned and maintained as defined by the clothing manufacturer. Such clothing can lose its flame-resistant characteristics when cleaned. Manufacturer's instructions must be implemented with regard to restoring the flame-resistant characteristics.

**(C) Personal Protective Equipment.**

**(1) General.** When an employee is working within the Flash Protection Boundary he/she shall wear protective clothing and other personal protective equipment in accordance with 130.3.

Article 130 establishes a general requirement for an employee to be protected from the thermal effects of an exposure to an arcing fault. Section 130.7(C)(1) requires the protection from thermal exposure to be based on a flash hazard analysis or selected from Table 130.7(C)(9).

**(2) Movement and Visibility.** When flame-resistant (FR) clothing is worn to protect an employee, it shall cover all ignitible clothing and shall allow for movement and visibility.

When FR clothing is required by a flash hazard analysis, FR clothing must cover all clothing that is easily ignitible, such as untreated cotton. Some clothing, such as polyester, melts before igniting and also must be completely covered by FR clothing.

FR protection must not physically restrain worker movement or unnecessarily restrict the worker's ability to see.

**(3) Head, Face, Neck, and Chin Protection.** Employees shall wear nonconductive head protection wherever there is a danger of head injury from electric shock or burns due to contact with live parts or from flying objects resulting from electrical explosion. Employees shall wear nonconductive protective equipment for the face, neck, and chin whenever there is a danger of injury from exposure to electric arcs or flashes or from flying objects resulting from electrical explosion.

FPN: See 130.7(C)(13)(b) for arc flash protective requirements.

Section 130.7(C)(3) requires protection from shock, arc flash, and flying parts and pieces that might be expelled during an arcing fault. The PPE must protect all parts of the head when the entire head is within the arc flash boundary. The protective equipment must be nonconductive and offer some protection from the blast effects associated with the arcing fault. If a worker's head is completely within the Flash Protection Boundary, all parts of the head must be covered by FR clothing.

**(4) Eye Protection.** Employees shall wear protective equipment for the eyes whenever there is danger of injury from electric arcs, flashes, or from flying objects resulting from electrical explosion.

Electromagnetic energy is radiated from an arc. The pressure wave associated with the rapid heating expels parts of molten material from the arc. The worker must wear PPE to protect his or her eyes from damage. Eyeglasses (known in consensus standards as spectacles) that meet the requirements of ASTM Z 87.1, *Occupational and Educational Personal Eye and Face Protection Devices,* offer protection from impact and also filter a significant portion of the damaging ultraviolet energy.

Eyeglasses, commonly referred to as coverall glasses or goggles, must not be worn where the potential for exposure to an arcing fault exists. This type of eye protection normally is provided for protection from chemical splashes and can melt or burn easily.

**(5) Body Protection.** Employees shall wear FR clothing wherever there is possible exposure to an electric arc flash above the threshold incident-energy level for a second-degree burn, 5 J/cm$^2$ (1.2 cal/cm$^2$).

*Exception: For incident-energy exposures 8.36 J/cm$^2$ (2 cal/cm$^2$) and below, employees may wear non-melting clothing described in Hazard/Risk Category 0 in Table 130.7(C)(11).*

Employees who are within the Flash Protection Boundary must wear FR clothing for protection from the thermal effects of an arcing fault. The requirement is that all ignitible clothing must be protected by at least one layer of clothing that has an established incident-energy rating. Clothing could be shirt and pants, coveralls, or any other assembly that provides protection for ignitible clothing or exposed skin.

Some FR clothing is rated for use with incident energy exposures up to 100 calories per centimeter squared. The 70E Technical Committee determined that although it might be possible to protect a person from such extreme thermal exposure, the clothing would be unlikely to protect the worker from the effects of the accompanying pressure wave. If the arc flash analysis indicates an exposure of more than 40 calories per centimeter squared, the task must not be performed until an electrically safe work condition exists. FR clothing with a very high incident energy rating might be needed to perform the steps necessary to establish an electrically safe work condition. However, this is the only task that should be accomplished with the equipment energized.

FPN: Such clothing can be provided as shirt and trousers, or as coveralls, or as a combination of jacket and trousers, or, for increased protection, as coveralls with jacket and trousers. Various weight fabrics are available. Generally, the higher degree of protection is provided by heavier weight fabrics and/or by layering combinations of one or more layers of FR clothing. In some cases one or more layers of FR clothing are worn over flammable, non-melting clothing. Non-melting, flammable clothing, used alone, can provide protection at low incident energy levels of 8.36 J/cm$^2$ (2.0 cal/cm$^2$) and below.

Several manufacturers make FR clothing using different materials. Generally, as the weight of the material increases, the degree of protection also increases. If FR clothing is worn in layers, some air is trapped between the layers, providing extra thermal insulation. Layering FR clothing increases the amount of protection afforded by the overall material.

FR clothing may be worn over flammable or non-melting materials, provided that it would keep the flammable clothing below its ignition point.

Ordinary clothing is categorized by the weight of the material per unit of area. Although FR clothing also can be categorized by the same method, it is categorized in calories per square centimeter. If FR clothing rated only by weight is on hand, the manufacturer should be consulted to determine the protective characteristics of the material.

**(6) Hand and Arm Protection.** Employees shall wear rubber insulating gloves where there is danger of hand and arm injury from electric shock due to contact with live parts. Hand and arm protection shall be worn where there is possible exposure to arc flash burn. The apparel described in 130.7(C)(13)(c) shall be required for protection of hands from burns. Arm protection shall be accomplished by apparel described in 130.7(C)(5).

If an electrically safe work condition has not been established, the qualified worker must wear protection for his or her hands and arms. The protection must eliminate the possibility that the worker might be electrocuted. The same clothing worn for body protection must provide flash protection for the worker's arms. Clothing selected and worn to protect the upper torso from thermal exposure must have long sleeves, and the sleeves must not be shortened (not rolled up).

**(7) Foot and Leg Protection.** Where insulated footwear is used as protection against step and touch potential, dielectric overshoes shall be required. Insulated soles shall not be used as primary electrical protection.

If the hazard/risk analysis indicates that the worker's feet and legs would be exposed to an arc flash, FR clothing worn to protect the lower torso must protect the worker's legs from exposure. The worker's feet also must be covered by heavy-duty leather work shoes.

The integrity of the insulating quality of shoes with insulated soles cannot be established easily after the worker has been wearing them in a work environment. Therefore, shoes with insulated soles must not serve as the primary protection from touch and step potential. If such protection is warranted, the worker must wear dielectric overshoes (boots).

**(8) Standards for Personal Protective Equipment.** Personal protective equipment shall conform to the standards given in Table 130.7(C)(8).

FPN: Non-FR or flammable fabrics are not covered by a standard in Table 130.7(C)(8). See 130.7(C)(14)(a), 130.7(C)(14)(b), and 130.7(C)(15).

***TABLE 130.7(C)(8)*** *Standards on Protective Equipment*

| Subject | Number and Title |
|---|---|
| Head protection | ANSI Z89.1, *Requirements for Protective Headwear for Industrial Workers,* 1997 |
| Eye and face protection | ANSI Z87.1, *Practice for Occupational and Educational Eye and Face Protection,* 1998 |
| Gloves | ASTM D 120-02, *Standard Specification for Rubber Insulating Gloves,* 2002 |
| Sleeves | ASTM D 1051-02, *Standard Specification for Rubber Insulating Sleeves,* 2002 |
| Gloves and sleeves | ASTM F 496-02, *Standard Specification for In-Service Care of Insulating Gloves and Sleeves,* 2002 |
| Leather protectors | ASTM F 696-02, *Standard Specification for Leather Protectors for Rubber Insulating Gloves and Mittens,* 2002 |
| Footwear | ASTM F 1117-98, *Standard Specification for Dielectric Overshoe Footwear,* 1998 |
| | ANSI Z41, *Standard for Personnel Protection, Protective Footwear,* 1999 |
| Visual inspection | ASTM F 1236-01, *Standard Guide for Visual Inspection of Electrical Protective Rubber Products,* 2001 |
| Apparel | ASTM F 1506-02a, *Standard Performance Specification for Textile Material for Wearing Apparel for Use by Electrical Workers Exposed to Momentary Electric Arc and Related Thermal Hazards,* 2002a |
| Raingear | ASTM F 1891-02a, *Standard Specification for Arc and Flame Resistant Rainwear,* 2002a |
| Face protective products | ASTM F 2178-02, *Standard Test Method for Determining the Arc Rating of Face Protective Products,* 2002 |

The national consensus standards listed in Table 130.7(C)(8) describe tests necessary for equipment to be considered acceptable. Equipment that is not ordinarily considered PPE is not covered by the table.

Two protective equipment standards were added to Table 130.7(C)(8). ASTM F 1891, *Standard Specification for Arc and Flame Resistant Rainwear,* was added on the basis of advances in the manufacturing technology of FR clothing. This standard complements other standards (listed in the table) to provide comprehensive protection for workers. The melting or flammability characteristics of non-FR rainwear could exacerbate an injury to a worker. ASTM F 2178, *Standard Test Method for Determining the Arc Rating of Face Protective Products,* was also added to provide selectivity for a worker for face protection when the exposure is Category 2 or lower.

**(9) Selection of Personal Protective Equipment.**

(a) When Required for Various Tasks. When selected in lieu of the flash hazard analysis of 130.3(A), Table 130.7(C)(9)(a) shall be used to determine the hazard/risk category for a task. The assumed short-circuit current capacities and fault clearing times for various tasks are listed in the text and notes to Table 130.7(C)(9)(a). For tasks not listed, or for power systems with greater than the assumed short-circuit current capacity or with longer than the assumed fault clearing times, a flash hazard analysis shall be required in accordance with 130.3.

Table 130.7(C)(9)(a) lists common work tasks and may be used to determine a hazard/risk category. A hazard/risk, including an arc flash analysis, has been performed for each common task listed in the table. The hazard/risk analysis (arc flash analysis) is based on parameters that commonly are found in industrial workplaces and are identified as notes at the bottom of the table. An arc flash analysis must be performed for fault-clearing times and short-circuit capacities that exceed the information contained in the notes.

The work tasks and protective equipment identified in Table 130.7(C)(9)(a) were identified by a task group, and the protective equipment selected was based on the collective experience of the task group. The protective equipment is not necessarily based on calculations, but it is considered to be reasonable, based on the consensus judgment of the full 70E Technical Committee.

The protective equipment identified in Table 130.7(C)(9)(a) considers both electrical circuit parameters and the physical attributes of the equipment and work task. Work tasks that are not listed in the table must be subjected to a hazard/risk analysis that considers both shock and arc flash hazards.

FPN No. 1: Both larger and smaller available short-circuit currents could result in higher available arc-flash energies. If the available short-circuit current increases without a decrease in the opening time of the overcurrent protective device, the arc-flash energy will increase. If the available short-circuit current decreases, resulting in a longer opening time for the overcurrent protective device, arc-flash energies could also increase.

Dual-function overcurrent devices, such as current-limiting fuses or circuit breakers, do not limit the current when the fault current is less than the trigger point of the current-limiting element of the device. Until the fault current reaches the lower end of the current-limiting characteristic, very high incident energy exposures can exist. Fault current normally is dependent on the impedance of the ground-fault current return path. The integrity of the ground-fault current return path is crucial to ensuring that any fault current will be sufficient to reach the current-limiting range of the device.

FPN No. 2: Energized parts that operate at less than 50 volts are not required to be deenergized to satisfy an "electrically safe work condition." Consideration should be given to the capacity of the source, any overcurrent protection between the energy source and the worker, and whether the work task related to the source operating at less than 50 volts increases exposure to electrical burns or to explosion from an electric arc.

The generally accepted lowest voltage that will result in a significant shock is 50 volts. That voltage level is selected by several standards as the lower limit of the danger level. Employers and employees should be aware that electrical burns also could be received at lower voltages.

Notes 1 through 6 to Table 130.7(C)(9)(a) define the available short-circuit current and fault-clearing time that was assumed to determine the hazard/risk category for the table. The hazard/risk category for work tasks that have higher short-circuit current and/or higher fault-clearing times must be determined by a different process. Note that the clearing time is well within the current-limiting range of current-limiting devices. If the fault current is below the current-limiting range of the overcurrent device, a different process should be used to determine incident energy.

*TABLE 130.7(C)(9)(A) Hazard/Risk Category Classifications*

| Task (Assumes Equipment Is Energized, and Work Is Done Within the Flash Protection Boundary) | Hazard/ Risk Category | V-rated Gloves | V-rated Tools |
|---|---|---|---|
| **Panelboards Rated 240 V and Below—Notes 1 and 3** | | | |
| Circuit breaker (CB) or fused switch operation with covers on | 0 | N | N |
| CB or fused switch operation with covers off | 0 | N | N |
| Work on energized parts, including voltage testing | 1 | Y | Y |
| Remove/install CBs or fused switches | 1 | Y | Y |
| Removal of bolted covers (to expose bare, energized parts) | 1 | N | N |
| Opening hinged covers (to expose bare, energized parts) | 0 | N | N |
| **Panelboards or Switchboards Rated >240 V and up to 600 V (with molded case or insulated case circuit breakers)—Notes 1 and 3** | | | |
| CB or fused switch operation with covers on | 0 | N | N |
| CB or fused switch operation with covers off | 1 | N | N |
| Work on energized parts, including voltage testing | 2* | Y | Y |
| **600 V Class Motor Control Centers (MCCs)—Notes 2 (except as indicated) and 3** | | | |
| CB or fused switch or starter operation with enclosure doors closed | 0 | N | N |
| Reading a panel meter while operating a meter switch | 0 | N | N |
| CB or fused switch or starter operation with enclosure doors open | 1 | N | N |
| Work on energized parts, including voltage testing | 2* | Y | Y |
| Work on control circuits with energized parts 120 V or below, exposed | 0 | Y | Y |
| Work on control circuits with energized parts >120 V, exposed | 2* | Y | Y |
| Insertion or removal of individual starter "buckets" from MCC—Note 4 | 3 | Y | N |
| Application of safety grounds, after voltage test | 2* | Y | N |
| Removal of bolted covers (to expose bare, energized parts) | 2* | N | N |
| Opening hinged covers (to expose bare, energized parts) | 1 | N | N |
| **600 V Class Switchgear (with power circuit breakers or fused switches) — Notes 5 and 6** | | | |
| CB or fused switch operation with enclosure doors closed | 0 | N | N |
| Reading a panel meter while operating a meter switch | 0 | N | N |
| CB or fused switch operation with enclosure doors open | 1 | N | N |
| Work on energized parts, including voltage testing | 2* | Y | Y |
| Work on control circuits with energized parts 120 V or below, exposed | 0 | Y | Y |
| Work on control circuits with energized parts >120 V, exposed | 2* | Y | Y |
| Insertion or removal (racking ) of CBs from cubicles, doors open | 3 | N | N |
| Insertion or removal (racking) of CBs from cubicles, doors closed | 2 | N | N |
| Application of safety grounds, after voltage test | 2* | Y | N |
| Removal of bolted covers (to expose bare, energized parts) | 3 | N | N |
| Opening hinged covers (to expose bare, energized parts) | 2 | N | N |
| **Other 600 V Class (277 V through 600 V, nominal) Equipment—Note 3** | | | |
| Lighting or small power transformers (600 V, maximum) | — | — | — |
| Removal of bolted covers (to expose bare, energized parts) | 2* | N | N |
| Opening hinged covers (to expose bare, energized parts) | 1 | N | N |
| Work on energized parts, including voltage testing | 2* | Y | Y |
| Application of safety grounds, after voltage test | 2* | Y | N |
| Revenue meters (kW-hour, at primary voltage and current) | — | — | — |
| Insertion or removal | 2* | Y | N |
| Cable trough or tray cover removal or installation | 1 | N | N |
| Miscellaneous equipment cover removal or installation | 1 | N | N |
| Work on energized parts, including voltage testing | 2* | Y | Y |
| Application of safety grounds, after voltage test | 2* | Y | N |

| Task (Assumes Equipment Is Energized, and Work Is Done Within the Flash Protection Boundary) | Hazard/ Risk Category | V-rated Gloves | V-rated Tools |
|---|---|---|---|
| **NEMA E2 (fused contactor) Motor Starters, 2.3 kV Through 7.2 kV** | | | |
| Contactor operation with enclosure doors closed | 0 | N | N |
| Reading a panel meter while operating a meter switch | 0 | N | N |
| Contactor operation with enclosure doors open | 2* | N | N |
| Work on energized parts, including voltage testing | 3 | Y | Y |
| Work on control circuits with energized parts 120 V or below, exposed | 0 | Y | Y |
| Work on control circuits with energized parts >120 V, exposed | 3 | Y | Y |
| Insertion or removal (racking ) of starters from cubicles, doors open | 3 | N | N |
| Insertion or removal (racking) of starters from cubicles, doors closed | 2 | N | N |
| Application of safety grounds, after voltage test | 3 | Y | N |
| Removal of bolted covers (to expose bare, energized parts) | 4 | N | N |
| Opening hinged covers (to expose bare, energized parts) | 3 | N | N |
| **Metal Clad Switchgear, 1 kV and Above** | | | |
| CB or fused switch operation with enclosure doors closed | 2 | N | N |
| Reading a panel meter while operating a meter switch | 0 | N | N |
| CB or fused switch operation with enclosure doors open | 4 | N | N |
| Work on energized parts, including voltage testing | 4 | Y | Y |
| Work on control circuits with energized parts 120 V or below, exposed | 2 | Y | Y |
| Work on control circuits with energized parts >120 V, exposed | 4 | Y | Y |
| Insertion or removal (racking ) of CBs from cubicles, doors open | 4 | N | N |
| Insertion or removal (racking) of CBs from cubicles, doors closed | 2 | N | N |
| Application of safety grounds, after voltage test | 4 | Y | N |
| Removal of bolted covers (to expose bare, energized parts) | 4 | N | N |
| Opening hinged covers (to expose bare, energized parts) | 3 | N | N |
| Opening voltage transformer or control power transformer compartments | 4 | N | N |
| **Other Equipment 1 kV and Above** | | | |
| Metal clad load interrupter switches, fused or unfused | — | — | — |
| Switch operation, doors closed | 2 | N | N |
| Work on energized parts, including voltage testing | 4 | Y | Y |
| Removal of bolted covers (to expose bare, energized parts) | 4 | N | N |
| Opening hinged covers (to expose bare, energized parts) | 3 | N | N |
| Outdoor disconnect switch operation (hookstick operated) | 3 | Y | Y |
| Outdoor disconnect switch operation (gang-operated, from grade) | 2 | N | N |
| Insulated cable examination, in manhole or other confined space | 4 | Y | N |
| Insulated cable examination, in open area | 2 | Y | N |

Note: *V-rated Gloves* are gloves rated and tested for the maximum line-to-line voltage upon which work will be done.

*V-rated Tools* are tools rated and tested for the maximum line-to-line voltage upon which work will be done.

2* means that a double-layer switching hood and hearing protection are required for this task in addition to the other Hazard/Risk Category 2 requirements of Table 130.7(C)(10).

Y = yes (required)

N = no (not required)

Notes:

1. 25 kA short circuit current available, 0.03 second (2 cycle) fault clearing time.
2. 65 kA short circuit current available, 0.03 second (2 cycle) fault clearing time.
3. For < 10 kA short circuit current available, the hazard/risk category required may be reduced by one number.
4. 65 kA short circuit current available, 0.33 second (20 cycle) fault clearing time.
5. 65 kA short circuit current available, up to 1.0 second (60 cycle) fault clearing time.
6. For < 25 kA short circuit current available, the hazard/risk category required may be reduced by one number.

Standards Council decision SC-04-4-9/Log 799 issued TIA 04-1 (NFPA 70E) revises Notes 1, 2, 4, and 5 to read as follows:

1. Maximum of 25 kA short-circuit current available, 0.03 second (2 cycle) fault clearing time.
2. Maximum of 65 kA short-circuit current available, 0.03 second (2 cycle) fault clearing time.
3. Maximum of 42 kA short-circuit current available, 0.33 second (20 cycle) fault clearing time.
4. Maximum of 35 kA short-circuit current available, 0.5 second (30 cycle) fault clearing time.

A Tentative Interim Amendment is tentative because it has not been processed through the entire standards-making procedures. It is interim because it is effective only between editions of the standards. A TIA automatically becomes a proposal of the proponent of the next edition of the standard; as such, it then is subject to all of the procedures of the standards-making process. See full text of TIA 04-1 on p. 98.

**(10) Protective Clothing and Personal Protective Equipment Matrix.** Once the Hazard/Risk Category has been identified, Table 130.7(C)(10) shall be used to determine the required personal protective equipment (PPE) for the task. Table 130.7(C)(10) lists the requirements for protective clothing and other protective equipment based on Hazard/Risk Category numbers 0 through 4. This clothing and equipment shall be used when working on or near energized equipment within the Flash Protection Boundary.

The protective clothing matrix is intended to provide helpful information. FR clothing is available in many different constructions. Using FR clothing in layers is one way to achieve a higher level of protection. Table 130.7(C)(10) suggests acceptable combinations of clothing items to achieve a desired hazard/risk category. Other combinations might also be possible. Section 130.7 requires that any body part within the Flash Protection Boundary be protected from the thermal effects of an arcing fault. Defining a hazard/risk category is a viable way to translate incident energy exposure to selection of FR clothing.

Table 130.7(C)(10) may be used to determine FR clothing requirements in conjunction with any method of determining thermal exposure.

FPN No. 1: See Annex H for a suggested simplified approach to ensure adequate PPE for electrical workers within facilities with large and diverse electrical systems.

To simplify administration of FR clothing exposure, an employer might implement a procedure that has two or three different assemblies of arc flash protection. A facility could define a general requirement for workers to wear a minimum level of protective clothing. The procedure could define a secondary level of protection that is easily recognized by employees, and still a third level of protection that is required in special situations. The keys to this type of requirement are that the protection is adequate for the greatest exposure for each level of protective clothing and that each worker can recognize when it is necessary to wear a higher level of protection.

FPN No. 2: The PPE requirements of this section are intended to protect a person from arc-flash and shock hazards. While some situations could result in burns to the skin, even with the protection described in Table 130.7(C)(10), burn injury should be reduced and survivable. Due to the explosive effect of some arc events, physical trauma injuries could occur. The PPE requirements of this section do not provide protection against physical trauma other than exposure to the thermal effects of an arc flash.

The nature of thermal energy enables the wearing of FR clothing to reduce the chance of thermal injury. However, because the nature of an arcing fault is unpredictable, determining the degree of each hazard associated with an arcing fault is difficult, and the effectiveness of the

*TABLE 130.7(C)(10)* *Protective Clothing and Personal Protective Equipment (PPE) Matrix*

| **Protective Clothing and Equipment** | **Protective Systems for Hazard/Risk Category** | | | | | |
|---|---|---|---|---|---|---|
| **Hazard/Risk Category Number** | **–1 (Note 3)** | **0** | **1** | **2** | **3** | **4** |
| **Non-melting (according to ASTM F 1506-00) or Untreated Natural Fiber** | | | | | | |
| a. T-shirt (short-sleeve) | X | | | X | X | X |
| b. Shirt (long-sleeve) | | X | | | | |
| c. Pants (long) | X | X | X (Note 4) | X (Note 6) | X | X |
| **FR Clothing (Note 1)** | | | | | | |
| a. Long-sleeve shirt | | | X | X | X (Note 9) | X |
| b. Pants | | | X (Note 4) | X (Note 6) | X (Note 9) | X |
| c. Coverall | | | (Note 5) | (Note 7) | X (Note 9) | (Note 5) |
| d. Jacket, parka, or rainwear | | | AN | AN | AN | AN |
| **FR Protective Equipment** | | | | | | |
| a. Flash suit jacket (multilayer) | | | | | | X |
| b. Flash suit pants (multilayer) | | | | | | X |
| c. Head protection | | | | | | |
| 1. Hard hat | | | X | X | X | X |
| 2. FR hard hat liner | | | | | AR | AR |
| d. Eye protection | | — | — | — | — | — |
| 1. Safety glasses | X | X | X | AL | AL | AL |
| 2. Safety goggles | | | | AL | AL | AL |
| e. Face and head area protection | | — | — | — | — | — |
| 1. Arc-rated face shield, or flash suit hood | | | | X (Note 8) | | |
| 2. Flash suit hood | | | | | X | X |
| 3. Hearing protection (ear canal inserts) | | | | X (Note 8) | X | X |
| f. Hand protection | | | — | — | — | — |
| Leather gloves (Note 2) | | | AN | X | X | X |
| g. Foot protection | | | | | | |
| Leather work shoes | | | AN | X | X | X |

AN = As needed AR = As required

AL = Select one in group X = Minimum required

Notes:

1. See Table 130.7(C)(11). Arc rating for a garment is expressed in cal/cm$^2$.

2. If voltage-rated gloves are required, the leather protectors worn external to the rubber gloves satisfy this requirement.

3. Hazard/Risk Category Number "-1" is only defined if determined by Notes 3 or 6 of Table 130.7(C)(9)(a).

4. Regular weight (minimum 12 oz/yd$^2$ fabric weight), untreated, denim cotton blue jeans are acceptable in lieu of FR pants. The FR pants used for Hazard/Risk Category 1 shall have a minimum arc rating of 4.

5. Alternate is to use FR coveralls (minimum arc rating of 4) instead of FR shirt and FR pants.

6. If the FR pants have a minimum arc rating of 8, long pants of non-melting or untreated natural fiber are not required beneath the FR pants.

7. Alternate is to use FR coveralls (minimum arc rating of 4) over non-melting or untreated natural fiber pants and T-shirt.

8. A faceshield with a minimum arc rating of 8, with wrap-around guarding to protect not only the face, but also the forehead, ears, and neck (or, alternatively, a flash suit hood), is required.

9. Alternate is to use two sets of FR coveralls (the inner with a minimum arc rating of 4 and outer coverall with a minimum arc rating of 5) over non-melting or untreated natural fiber clothing, instead of FR coveralls over FR shirt and FR pants over non-melting or untreated natural fiber clothing.

determination might not be complete. The effects of arc blast also are not predictable. Equipment for protection from the pressure wave or other conditions associated with an arcing fault is not available. Any work performed on or near exposed live parts exposes a worker to an elevated risk of injury.

**(11) Protective Clothing Characteristics.** Table 130.7(C)(11) lists examples of protective clothing systems and typical characteristics including the degree of protection for various clothing. The protective clothing selected for the corresponding hazard/risk category number shall have an arc rating of at least the value listed in the last column of Table 130.7(C)(11).

FPN: The arc rating for a particular clothing system can be obtained from the FR clothing manufacturer.

Table 130.7(C)(11) is intended to provide general information that could assist a worker understand the process for selecting clothing based on a hazard/risk category designation. This table describes the protective nature of clothing that meets a specific hazard/risk category. Although the table is intended for use in conjunction with Table 130.7(C)(9)(a), it can be used with other methods of determining the necessary protective clothing.

When FR clothing meeting the requirements of ASTM F 1959, *Standard Test Method for Determining the Arc Rating of Face Protective Products,* 2002, is exposed to an electrical arc to establish its rating, the rating is the failure mode in which the material chars and breaks open. Charring causes the fibers of the material to close the mesh in the material weave. The charred fabric could break open and expose the material (or skin) underneath. The arc rating of FR clothing is either the arc-thermal performance value (ATPV—ASTM F 1959) or the energy level ($E_{BT}$—ASTM F 1959) that caused breakopen to occur when the fabric was tested.

***TABLE 130.7(C)(11)*** *Protective Clothing Characteristics*

| | Typical Protective Clothing Systems | |
|---|---|---|
| **Hazard/Risk Category** | **Clothing Description (Typical number of clothing layers is given in parentheses)** | **Required Minimum Arc Rating of PPE [$J/cm^2(cal/cm^2)$]** |
| 0 | Non-melting, flammable materials (i.e., untreated cotton, wool, rayon, or silk, or blends of these materials) with a fabric weight at least 4.5 $oz/yd^2$ (1) | N/A |
| 1 | FR shirt and FR pants or FR coverall (1) | 16.74 (4) |
| 2 | Cotton underwear—conventional short sleeve and brief/shorts, plus FR shirt and FR pants (1 or 2) | 33.47 (8) |
| 3 | Cotton underwear plus FR shirt and FR pants plus FR coverall, or cotton underwear plus two FR coveralls (2 or 3) | 104.6 (25) |
| 4 | Cotton underwear plus FR shirt and FR pants plus multilayer flash suit (3 or more) | 167.36 (40) |

Note: Arc rating is defined in Article 100 and can be either ATPV or $E_{BT}$. ATPV is defined in ASTM F 1959-99 as the incident energy on a fabric or material that results in sufficient heat transfer through the fabric or material to cause the onset of a second-degree burn based on the Stoll curve. EBT is defined in ASTM F 1959-99 as the average of the five highest incident energy exposure values below the Stoll curve where the specimens do not exhibit breakopen. $E_{BT}$ is reported when ATPV cannot be measured due to FR fabric breakopen.

**(12) Factors in Selection of Protective Clothing.** Clothing and equipment that provide worker protection from shock and arc flash hazards shall be utilized. Clothing and equipment required for the degree of exposure shall be permitted to be worn alone or integrated with flammable, nonmelting apparel. If FR clothing is required, it shall cover associated parts of the body as well as all flammable apparel while allowing movement and visibility. All personal protective equipment shall be maintained in a sanitary and functionally effective condition. Personal protective equipment items will normally be used in conjunction with one another as a system to provide the appropriate level of protection.

FPN: Protective clothing includes shirts, pants, coveralls, jackets, and parkas worn routinely by workers who, under normal working conditions, are exposed to momentary electric arc and related thermal hazards. Flame-resistant rainwear worn in inclement weather is included in this category of clothing.

Materials that have an established arc flash rating are permitted to be used either alone or in combination with materials without an established rating. Materials from one manufacturer also are permitted to be used with materials from a different manufacturer, provided both sets of clothing have an established rating.

Clothing must be maintained in a clean and sanitary condition. It must be stored in a way that has no deleterious effect on the protective characteristics of the clothing.

(a) Layering. Nonmelting, flammable fiber garments shall be permitted to be used as underlayers in conjunction with FR garments in a layered system for added protection. If nonmelting, flammable fiber garments are used as underlayers, the system arc rating shall be sufficient to prevent breakopen of the innermost FR layer at the expected arc exposure incident energy level to prevent ignition of flammable underlayers.

FPN: A typical layering system might include cotton underwear, a cotton shirt and trouser, and a FR coverall. Specific tasks might call for additional FR layers to achieve the required protection level.

Layering increases the overall protective characteristics of FR clothing. If ignitible fabrics are used as underlayers, the system arc flash rating must not permit the outer layer to break open and directly expose the ignitible material to the arcing fault. The outer layers must limit the temperature rise of the ignitible underlayers to no more than 1.2 calories per square centimeter.

(b) Outer Layers. Garments worn as outer layers over FR clothing, such as jackets or rainwear, shall also be made from FR material.

If the hazard/risk analysis and arc flash analysis require a worker to use FR clothing, and the FR clothing is used in conjunction with ignitible or meltable clothing, the outer layer must be FR fabric.

Note: FR rainwear also is available for purchase.

(c) Underlayers. Meltable fibers such as acetate, nylon, polyester, polypropylene, and spandex shall not be permitted in fabric underlayers (underwear) next to the skin.

*Exception: An incidental amount of elastic used on nonmelting fabric underwear or socks shall be permitted.*

Clothing made from meltable fabrics must not be worn next to the skin. However, an incidental amount of elastic used in these fabrics is acceptable.

FPN No. 1: FR garments (e.g., shirts, trousers, and coveralls) worn as underlayers that neither ignite nor melt and drip in the course of an exposure to electric arc and related thermal hazards generally provide a higher system arc rating than nonmelting, flammable fiber underlayers.

FPN No. 2: FR underwear or undergarments used as underlayers generally provide a higher system arc rating than nonmelting, flammable fiber underwear or undergarments used as underlayers.

(d) Coverage. Clothing shall cover potentially exposed areas as completely as possible. Shirt sleeves shall be fastened at the wrists, and shirts and jackets shall be closed at the neck.

When required by the hazard/risk analysis, FR clothing must completely cover all body areas within the arc flash boundary. Shirt sleeves must be fastened at the wrists, and the top button of shirts and/or jackets must be fastened to minimize the chance that heated air could reach below the FR clothing.

(e) Fit. Tight-fitting clothing shall be avoided. Loose-fitting clothing provides additional thermal insulation because of air spaces. FR apparel shall fit properly such that it does not interfere with the work task.

When the surface of FR clothing is heated, heat is conducted through the material. If the FR clothing is touching skin, the heat energy that is conducted through the FR clothing could result in a burn. To minimize this chance, FR clothing must be loose fitting to provide additional thermal insulation. However, FR clothing must not be so loose as to interfere with the worker's movements.

(f) Interference. The garment selected shall result in the least interference with the task but still provide the necessary protection. The work method, location, and task could influence the protective equipment selected.

The plan for the work task must define the protective garments that will be worn by the worker. As the plan is developed, the location and position of the worker must be considered to provide the best chance that the PPE will not interfere with the worker's movements as he or she executes the work task.

**(13) Arc Flash Protective Equipment.**

(a) Flash Suits. Flash suit design shall permit easy and rapid removal by the wearer. The entire flash suit, including the hood's face shield, shall have an arc rating that is suitable for the arc flash exposure. When exterior air is supplied into the hood, the air hoses and pump housing shall be either covered by FR materials or constructed of nonmelting and nonflammable materials.

Should an emergency condition develop, a worker might need to remove the PPE rapidly, and the design of the PPE must enable the worker to do so. In addition, the rating of the flash suit must at least be equal to the potential incident energy exposure (see Exhibit 130.4).

When a worker wears a hood with a viewing window, the concentration of oxygen can decrease if the hood fits tightly. Air may be supplied to the worker inside the hood. However, the air hoses and pump (if necessary) must be protected from ignition by a covering of FR materials. Any air hoses should be on the opposite side of the worker from the potential thermal exposure.

(b) Face Protection. Face shields shall have an arc rating suitable for the arc flash exposure. Face shields without an arc rating shall not be used. Eye protection (safety glasses or goggles) shall always be worn under face shields or hoods.

***EXHIBIT 130.4.*** *A flash suit. (Courtesy of Salisbury)*

Face shields and viewing windows must have an arc rating similar to an arc rating for FR clothing. Face shields and viewing windows also must meet the impact requirements of ANSI Z87.1, *Occupational and Educational Personal Eye and Face Protection Devices* (see Exhibit 130.5).

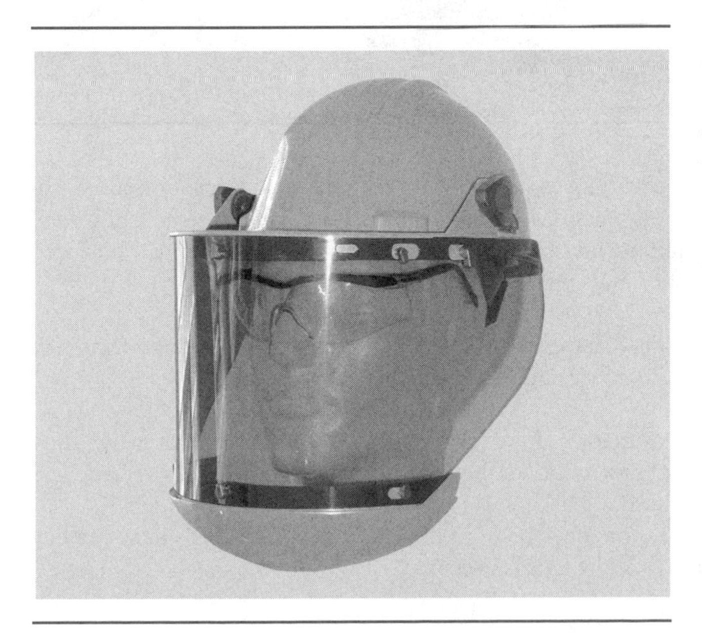

***EXHIBIT 130.5*** *An arc-rated face shield with a rating of 8 cal/cm$^2$ applicable for Category 2 tasks. This shield also has a protective coating to filter the UV characteristics of an arc flash. (Courtesy of Salisbury)*

Safety glasses or goggles (called *spectacles* in ANSI Z87.1) must be worn under the face shield or viewing window. Goggles, as considered by ANSI Z87.1, should not be worn, because they are permitted to have an ignitible and meltable component.

FPN: Face shields made with energy-absorbing formulations that can provide higher levels of protection from the radiant energy of an arc flash are available, but these shields are tinted and can reduce visual acuity. Additional illumination of the task area might be necessary when these types of arc protective face shields are used.

(c) Hand Protection. Leather or FR gloves shall be worn where required for arc flash protection. Where insulating rubber gloves are used for shock protection, leather protectors shall be worn over the rubber gloves.

FPN: Insulating rubber gloves and gloves made from layers of flame-resistant material provide hand protection against the arc flash hazard. Heavy-duty leather (e.g., greater than 12 oz/yd$^2$) gloves provide protection suitable up to Hazard/Risk Category 2. The leather protectors worn over insulating rubber gloves provide additional arc flash protection for the hands. During high arc flash exposures leather can shrink and cause a decrease in protection.

The hands probably are the most exposed part of a worker's body. Arc flash analyses generally are based on exposure at a distance of 18 in. or 24 in. Because the worker's hands are much closer, the thermal exposure is much greater to the hands. Additional thermal protection is warranted; however, no method exists for determining the degree of exposure for a worker's hands.

Gloves made from FR materials are available. The worker must wear voltage-rated gloves (see Exhibit 130.6) with heavy-duty leather over-protectors. The combination of rubber gloves and leather over-protectors provides significant protection.

***EXHIBIT 130.6.*** *Voltage-rated gloves. (Courtesy of Salisbury)*

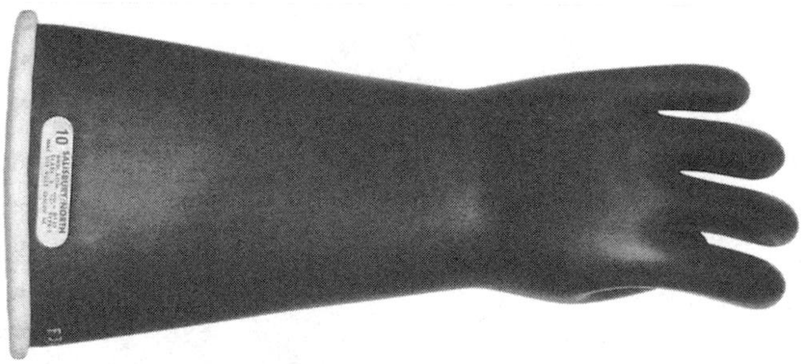

Experience has shown that leather gloves stitched with cotton thread offer significant protection if the overcurrent protection functions as intended by the *National Electrical Code*® (NFPA 70, *NEC*®, or *Code*) requirements. The cotton stitching might disintegrate; however, the leather offers protection for the hands underneath.

(d) Foot Protection. Heavy-duty leather work shoes provide some arc flash protection to the feet and shall be used in all tasks in Hazard/Risk Category 2 and higher.

Shoes with an arc rating are not available. However, experience has shown that heavy-duty leather work shoes offer significant protection for the feet. Normally, the worker's feet are less exposed than his or her hands or head.

Only heavy-duty leather work shoes must be worn if the hazard/risk analysis indicates that arc flash protection is necessary while the task is executed.

**(14) Clothing Material Characteristics.** FR clothing shall meet the requirements described in 130.7(C)(14)(a) through 130.7(C)(15).

FPN: FR materials, such as flame-retardant treated cotton, meta-aramid, para-aramid, and poly-benzimidazole (PBI) fibers, provide thermal protection. These materials can ignite but will not continue to burn after the ignition source is removed. FR fabrics can reduce burn injuries during an arc flash exposure by providing a thermal barrier between the arc flash and the wearer. In aramid and PBI blends, para-aramid adds strength to a fabric to prevent the fabric from breaking open due to the blast shock wave and high thermal energy of the arc.

(a) Melting. Clothing made from flammable synthetic materials that melt at temperatures below 315°C (600°F), such as acetate, nylon, polyester, polypropylene, and spandex, either alone or in blends, shall not be used.

FPN: These materials melt as a result of arc flash exposure conditions, form intimate contact with the skin, and aggravate the burn injury.

*Exception: Fiber blends that contain materials that melt, such as acetate, nylon, polyester, polypropylene, and spandex, shall be permitted if such blends in fabrics meet the requirements of ASTM F 1506, Standard Performance Specification for Textile Material for Wearing Apparel for Use by Electrical Workers Exposed to Momentary Electric Arc and Related Thermal Hazards, and if such blends in fabrics do not exhibit evidence of a melting and sticking hazard during arc testing according to ASTM F 1959 [see also 130.7(C)(15)].*

Section 130.7(C)(14) describes requirements that must be met by FR clothing. FR clothing that has an arc rating based on testing defined by ASTM F 1506, *Standard Performance Specification for Flame Resistant Textile Materials for Wearing Apparel for Use by Electrical Workers Exposed to Momentary Electric Arc and Related Thermal Hazards,* and ASTM F 1959, *Standard Test Method for Determining the Arc Thermal Performance Value of Materials for Clothing,* adheres to these requirements.

(b) Flammability. Clothing made from nonmelting flammable natural materials, such as cotton, wool, rayon, or silk, shall be permitted for Hazard/Risk Categories 0 and –1 considered acceptable if it is determined by flash hazard analysis that the exposure level is 8.36 J/cm$^2$ (2.0 cal/cm$^2$) or less, and that the fabric will not ignite and continue to burn under the arc exposure hazard conditions to which it will be exposed (using data from tests done in accordance with ASTM F 1958.) See also 130.7(C)(12)(a) for layering requirements.

FPN No. 1: Non-FR cotton, polyester-cotton blends, nylon, nylon-cotton blends, silk, rayon, and wool fabrics are flammable. These fabrics could ignite and continue to burn on the body, resulting in serious burn injuries.

FPN No. 2: Rayon is a cellulose-based (wood pulp) synthetic fiber that is a flammable but non-melting material.

Clothing that does not melt may be used for low-level arc flash exposures [see Table 130.7(10)].

**(15) Clothing Not Permitted.** Clothing made from materials that do not meet the requirements of 130.7(C)(14)(a) regarding melting, or made from materials that do not meet the flammability requirements of 130.7(C)(14)(b), shall not be permitted to be worn.

FPN: Some flame-resistant fabrics, such as non-FR modacrylic and nondurable flame-retardant treatments of cotton, are not recommended for industrial electrical or utility applications.

*Exception: Non-melting, flammable (non-FR) materials shall be permitted to be used as underlayers to FR clothing, as described in 130.7(C)(14)(a) and also shall be permitted to be used for Hazard/Risk Category 0 and 1 as described in Table 130.7(C)(10).*

Clothing that is likely to increase the extent of an arc flash injury must be avoided. Meltable fabrics or fabrics that continue to burn after the flame source has been removed must not be worn. Protective clothing should have an arc rating [see Table 130.7(10)].

**(16) Care and Maintenance of FR Clothing and FR Flash Suits.**

(a) Inspection. FR apparel shall be inspected before each use. Work clothing or flash suits that are contaminated, or damaged to the extent their protective qualities are impaired, shall not be used. Protective items that become contaminated with grease, oil, or flammable liquids or combustible materials shall not be used.

(b) Manufacturer's Instructions. The garment manufacturer's instructions for care and maintenance of FR apparel shall be followed.

The qualified worker must inspect his or her FR clothing before wearing it. The qualified worker should be trained to understand that if any flammable substance is on the surface of the FR clothing, the rating of the FR clothing is voided. The clothing must be free from tears or rips.

Maintenance instructions provided by the manufacturer must be followed to ensure that the protective characteristics of the clothing are maintained.

**(D) Other Protective Equipment.**

**(1) Insulated Tools and Equipment.** Employees shall use insulated tools and/or handling equipment when working inside the Limited Approach Boundary of exposed live parts where tools or handling equipment might make accidental contact. Insulated tools shall be protected from damage to the insulating material.

FPN: See 130.2(B) for working on exposed live parts.

When working inside the Limited Approach Boundary, workers must select and use work practices, including insulated tools that provide maximum protection from a release of energy. If contact with the exposed live part is likely, the worker must use insulated tools only. The term *insulated* means that the tool manufacturer has assigned a voltage rating to the insulating material. If the task requires the worker to cross or work in the vicinity of the Prohibited Approach Boundary, insulated tools are required.

Workers must be able to inspect the insulated tools for potential damage. Workers also must be able to determine whether the voltage rating remains intact.

(a) Requirements for Insulated Tools. The following requirements shall apply to insulated tools:

(1) Insulated tools shall be rated for the voltages on which they are used.
(2) Insulated tools shall be designed and constructed for the environment to which they are exposed and the manner in which they are used.

(b) Fuse or Fuse Holding Equipment. Fuse or fuse holder handling equipment, insulated for the circuit voltage, shall be used to remove or install a fuse if the fuse terminals are energized.

If the fuse terminals are energized, fuses should not be installed or removed in an industrial setting. When fuses are removed in an emergency, the fuse-handling equipment must be rated for the voltage.

Fuses that are mounted on poles and removed and installed by using a "hot stick" may be exchanged routinely with the fuse terminals energized.

(c) Ropes and Handlines. Ropes and handlines used near exposed live parts operating at 50 volts or more, or used where an electrical hazard exists, shall be nonconductive.

(d) Fiberglass-Reinforced Plastic Rods. Fiberglass-reinforced plastic rod and tube used for live line tools shall meet the requirements of ASTM F 711, *Standard Specification for Fiberglass-Reinforced Plastic (FRP) Rod and Tube Used; in Live Line Tools,* 1989 (R 1997).

(e) Portable Ladders. Portable ladders shall have nonconductive side rails if they are used where the employee or ladder could contact exposed live parts operating at 50 volts or more or where an electrical hazard exists. Nonconductive ladders shall meet the requirements of ANSI standards for ladders listed in Table 130.7(F).

(f) Protective Shields. Protective shields, protective barriers, or insulating materials shall be used to protect each employee from shock, burns, or other electrically related injuries while that employee is working near live parts that might be accidentally contacted or where dangerous electric heating or arcing might occur. When normally enclosed live parts are exposed for maintenance or repair, they shall be guarded to protect unqualified persons from contact with the live parts.

(g) Rubber Insulating Equipment. Rubber insulating equipment used for protection from accidental contact with live parts shall meet the requirements of the ASTM standards listed in Table 130.7(F).

(h) Voltage Rated Plastic Guard Equipment. Plastic guard equipment for protection of employees from accidental contact with live parts, or for protection of employees or energized equipment or material from contact with ground, shall meet the requirements of the ASTM standards listed in Table 130.7(F).

(i) Physical or Mechanical Barriers. Physical or mechanical (field fabricated) barriers shall be installed no closer than the restricted approach distance given in Table 130.2(C). While the barrier is being installed, the restrictive approach distance specified in Table 130.2(C) shall be maintained, or the live parts shall be placed in an electrically safe work condition.

**(E) Alerting Techniques.**

**(1) Safety Signs and Tags.** Safety signs, safety symbols, or accident prevention tags shall be used where necessary to warn employees about electrical hazards that might endanger them. Such signs and tags shall meet the requirements of ANSI Standard Z535 given in Table 130.7(F).

Signs that are required to warn people that an electrical hazard exists must be clear and legible. Legibility must be maintained and the signs kept clean. Signs that are no longer required must be removed.

**(2) Barricades.** Barricades shall be used in conjunction with safety signs where it is necessary to prevent or limit employee access to work areas containing live parts. Conductive barricades shall not be used where it might cause an electrical hazard. Barricades shall be placed no closer than the Limited Approach Boundary given in Table 130.2(C).

When workers not involved in the work task might be able to approach an exposed live part, the area defined by either the Limited Approach Boundary or the Flash Protection Boundary must be defined by a barricade to prevent inadvertent approach. The area may be protected by an attendant in lieu of installing a barricade.

**(3) Attendants.** If signs and barricades do not provide sufficient warning and protection from electrical hazards, an attendant shall be stationed to warn and protect employees. The primary duty and responsibility of an attendant providing manual signaling and alerting shall be to keep unqualified employees outside a work area where the unqualified employee might be exposed to electrical hazards. An attendant shall remain in the area as long as there is a potential for employees to be exposed to the electrical hazards.

An attendant may be assigned to warn people who approach the area where a live part is exposed. The attendant must remain alert to the approach by any person not directly involved in the work task. The attendant must remain on duty and must not leave the area while a live part is exposed.

**(F) Standards for Other Protective Equipment.** Other protective equipment required in 130.7(D) shall conform to the standards given in Table 130.7(F).

***TABLE 130.7(F)*** *Standards on Other Protective Equipment*

| Subject | Number and Title |
|---|---|
| Ladders | ANSI A14.1, *Safety Requirements for Portable Wood Ladders,* 1994 |
| | ANSI A14.3, *Safety Requirements for Fixed Ladders,* 2002 |
| | ANSI A14.4, *Safety Requirements for Job-Made Ladders,* 1992 |
| | ANSI A14.5, *Safety Requirement for Portable Reinforced Plastic Ladders,* 2000 |
| Safety signs and tags | ANSI Z535, *Series of Standards for Safety Signs and Tags,* 1998 |
| Blankets | ASTM D 1048, *Standard Specification for Rubber Insulating Blankets,* 1999 |
| Covers | ASTM D 1049, *Standard Specification for Rubber Covers,* 1998 |
| Line hoses | ASTM D 1050, *Standard Specification for Rubber Insulating Line Hoses,* 1990 |
| Line hoses and covers | ASTM F 478, *Standard Specification for In-Service Care of Insulating Line Hose and Covers,* 1999 |
| Blankets | ASTM F 479, *Standard Specification for In-Service Care of Insulating Blankets,* 1995 |
| Fiberglass tools/ ladders | ASTM F 711, *Standard Specification for Fiberglass-Reinforced Plastic (FRP) Rod and Tube Used; in Line Tools,* 1989 (R 1997) |
| Plastic guards | ASTM F 712, *Standard Test Methods for Electrically Insulating Plastic Guard Equipment for Protection of Workers,* 1995 |
| Temporary grounding | ASTM F 855, *Standard Specification for Temporary Protective Grounds to Be Used on De-energized Electric Power Lines and Equipment,* 1997 |
| Insulated hand tools | ASTM F 1505, *Standard Specification for Insulated and Insulating Hand Tools,* 2001 |

Table 130.7(F) identifies national consensus standards that define requirements for specific equipment. All equipment listed in the table impacts safe work practices that a qualified worker should implement.

## REFERENCES CITED IN COMMENTARY

ANSI Z87.1, *Occupational and Educational Personal Eye and Face Protection Devices,* American National Standards Institute, Inc., New York, NY, 2003.

ASTM F 855, *Standard Specification for Temporary Protective Grounds to be Used on Deenergized Electric Power Lines and Equipment,* American Society for Testing and Materials, West Conshohocken, PA, 1997.

ASTM F 1506, *Standard Performance Specification for Flame Resistant Textile Materials for Wearing Apparel for Use by Electrical Workers Exposed to Momentary Electric Arc and Related Thermal Hazards,* American Society for Testing and Materials, West Conshohocken, PA, 2001.

ASTM F 1891, *Standard Specification for Arc and Flame Resistant Rainwear,* American Society for Testing and Materials, West Conshohocken, PA, 2002.

ASTM F 1959, *Standard Test Method for Determining the Arc Thermal Performance Value of Materials for Clothing,* American Society for Testing and Materials, West Conshohocken, PA, 1999.

ASTM F 2178, *Standard Test Method for Determining the Arc Rating of Face Protective Products,* American Society for Testing and Materials, West Conshohocken, PA, 2002.

*Authoritative Dictionary of IEEE Standards Terms.* 7th edition, IEEE, Piscataway, NJ, 2000.

*General Information Directory* ("White Book"), Underwriters Laboratories Inc., Northbrook, IL, 2003.

Lee, Ralph, Life Fellow IEEE, "The Other Electrical Hazard: Electrical Arc Blast Burns," 1982.

*Manual of Style for NFPA Technical Committee Documents,* 2003 edition, National Fire Protection Association, Quincy, MA [available online at www.nfpa.org].

Mastrullo, K. G., Jones, R. A., and Jones, J. G., *The Electrical Safety Program Book,* National Fire Protection Association, Quincy, MA, 2004.

*National Electrical Code (NEC®) Style Manual,* 2003 edition, National Fire Protection Association, Quincy, MA [available online at www.nfpa.org].

NEMA WD 6, *Wiring, Devices—Dimensional Requirement,* National Electrical Manufacturers Association, Rosslyn, VA, 1997.

NFPA 70, *National Electrical Code®*, 2002 edition, National Fire Protection Association, Quincy, MA.

NFPA 2112, *Standard on Flame-Resistant Garments for Protection of Industrial Personnel Against Flash Fire,* 2001 edition, National Fire Protection Association, Quincy, MA.

OSHA 29 CFR 1910, Subpart S, U.S. Government Printing Office, Washington, DC.

*Standard Dictionary for Electrical and Electronic Engineers,* Institute of Electrical and Electronic Engineers, New York, NY, 2000.

UL 83, *Thermoplastic-Insulated Wires and Cables,* Underwriters Laboratories Inc., Northbrook, IL, 2003.

## Tentative Interim Amendment
**NFPA 70E**
**Standard for Electrical Safety Requirements for Employee Workplaces**
**2004 Edition**

Reference: 130.7 (C) (9) (a)
TIA 04-1 (NFPA 70E)
*(SC-04-4-9/Log 779)*

Pursuant to Section 5 of the NFPA Regulations Governing Committee Projects, the National Fire Protection Association has issued the following Tentative Interim Amendment to NFPA 70E, *Standard for Electrical Safety Requirements for Employee Workplaces,* 2004 edition. The TIA was processed by the Electrical Safety Requirements for Employee Workplaces Committee, and was issued by the Standards Council on April 15, 2004, with an effective date of May 5, 2004.

A Tentative Interim Amendment is tentative because it has not been processed through the entire standards-making procedures. It is interim because it is effective only between editions of the standard. A TIA automatically becomes a proposal of the proponent for the next edition of the standard; as such, it then is subject to all of the procedures of the standards-making process.

*1. Revise existing Notes 1, 2, 4, and 5 for Part II, Table 130.7 (C) (9) (a), as follows:*

**1.** *Maximum of* 25 kA short circuit current available, 0.03 second (2 cycle) fault clearing time.

**2.** *Maximum of* 65 kA short circuit current available, 0.03 second (2 cycle) fault clearing time

**4.** *Maximum of* 42 kA short circuit current available, 0.33 second (20 cycle) fault clearing time.

**5.** *Maximum of* 35 kA short circuit current available, up to 0.5 second (30 cycle) fault clearing time.

*2. Add references to Notes 4 and 5 at the following locations (highlighted) within the table:*

| Task (Assumes Equipment Is Energized, and Work Is Done Within the Flash Protection Boundary) | Hazard/Risk Category | V-rated Gloves | V-rated Tools |
|---|---|---|---|
| **Panelboards rated 240 V and below—Notes 1 and 3** | — | — | — |
| Circuit breaker (CB) or fused switch operation with covers on | 0 | N | N |
| CB or fused switch operation with covers off | 0 | N | N |
| Work on energized parts, including voltage testing | 1 | Y | Y |
| Remove/install CBs or fused switches | 1 | Y | Y |
| Removal of bolted covers (to expose bare, energized parts) | 1 | N | N |
| Opening hinged covers (to expose bare, energized parts) | 0 | N | N |

| | | | |
|---|---|---|---|
| **Panelboards or Switchboards rated >240 V and up to 600 V (with molded case or insulated case circuit breakers)—Notes 1 and 3** | — | — | — |
| CB or fused switch operation with covers on | 0 | N | N |
| CB or fused switch operation with covers off | 1 | N | N |
| Work on energized parts, including voltage testing | 2* | Y | Y |
| **600 V Class Motor Control Centers (MCCs)—Notes 2 (except as indicated) and 3** | — | — | — |
| CB or fused switch or starter operation with enclosure doors closed | 0 | N | N |
| Reading a panel meter while operating a meter switch | 0 | N | N |
| CB or fused switch or starter operation with enclosure doors open | 1 | N | N |
| Work on energized parts, including voltage testing | 2* | Y | Y |
| Work on control circuits with energized parts 120 V or below exposed | 0 | Y | Y |
| Work on control circuits with energized parts >120 V, exposed | 2* | Y | Y |
| Insertion or removal of individual starter "buckets" from MCC—Note 4 | 3 | Y | N |
| Application of safety grounds, after voltage test | 2* | Y | N |
| Removal of bolted covers (to expose bare, energized parts)—Note 4 | 2* | N | N |
| Opening hinged covers (to expose bare, energized parts) | 1 | N | N |
| **600 V Class Switchgear (with power circuit breakers or fused switches)—Notes 5 and 6** | — | — | — |
| CB or fused switch operation with enclosure doors closed | 0 | N | N |
| Reading a panel meter while operating a meter switch | 0 | N | N |
| CB or fused switch operation with enclosure doors open | 1 | N | N |
| Work on energized parts, including voltage testing | 2* | Y | Y |
| Work on control circuits with energized parts 120 V or below, exposed | 0 | Y | Y |
| Work on control circuits with energized parts >120 V, exposed | 2* | Y | Y |
| Insertion or removal (racking) of CBs from cubicles, doors open | 3 | N | N |
| Insertion or removal (racking) of CBs from cubicles, doors closed | 2 | N | N |
| Application of safety grounds, after voltage test | 2* | Y | N |
| Removal of bolted covers (to expose bare, energized parts) | 3 | N | N |
| Opening hinged covers (to expose bare, energized parts) | 2 | N | N |
| **Other 600 V Class (277 V through 600 V, Nominal) Equipment—Notes *2 (except as indicated) and* 3** | — | — | — |
| Lighting or small power transformers (600 V, maximum) | — | — | — |

| Task (Assumes Equipment Is Energized, and Work Is Done Within the Flash Protection Boundary) | Hazard/Risk Category | V-rated Gloves | V-rated Tools |
|---|---|---|---|
| Removal of bolted covers (to expose bare, energized parts) | 2* | N | N |
| Opening hinged covers (to expose bare, energized parts) | 1 | N | N |
| Work on energized parts, including voltage testing | 2* | Y | Y |
| Application of safety grounds, after voltage test | 2* | Y | N |
| Revenue meters (kW-hour, at primary voltage and current) | — | — | — |
| Insertion or removal | 2* | Y | N |
| Cable trough or tray cover removal or installation | 1 | N | N |
| Miscellaneous equipment cover removal or installation | 1 | N | N |
| Work on energized parts, including voltage testing | 2* | Y | Y |
| Application of safety grounds, after voltage test | 2* | Y | N |

*(Notes 1, 2, 4, and 5 are not applicable to the remainder of the table, so the rest of the table is not shown.)*

Copyright © 2004 All Rights Reserved
NATIONAL FIRE PROTECTION ASSOCIATION

CHAPTER 2

# Safety-Related Maintenance Requirements

An electrical work environment consists of three interrelated components: installation, safe work practices, and maintenance. Whether the work environment is safe depends on the quality of implementation of these contributing elements. For instance, safe work practices, defined in Chapter 1, are effective when the installation is *Code* compliant and the equipment is appropriately maintained. Any deficiency in the installation or maintenance adversely impacts safe work practices.

Chapter 2 addresses maintenance of the electrical equipment, which provides for the reliability and predictability necessary for the accurate operation of electrical equipment. Historically, electrical equipment has been very reliable and, in many instances, that reliability is taken for granted. NFPA 70E relies on the fact that if equipment is installed according to NFPA 70 (the *National Electrical Code*®, *NEC*®, or *Code*) and the manufacturer's directions, it is considered to be safe for operation. However, a minor deviation in the operating range of devices can significantly change the hazard exposure to a worker.

## ARTICLE 200
## Introduction

A comprehensive electrical equipment maintenance program provides two benefits to a company: increased reliability of the electrical systems, which avoids electrical outages and malfunctions, and decreased exposure of workers to electrical hazards. Equipment that operates accurately and within its design parameters requires fewer interactions. Historically, one-third of all electrical incidents are caused by a combination of unsafe equipment and unsafe conditions. This information suggests that adequately maintained and operated equipment significantly affects worker safety.

### 200.1 Scope

Chapter 2 addresses the following requirements:

(1) Chapter 2 covers practical safety-related maintenance requirements for electrical equipment and installations in workplaces as included in 90.1. These requirements identify only that maintenance directly associated with employee safety.

(2) Chapter 2 does not prescribe specific maintenance methods or testing procedures. It is left to the employer to choose from the various maintenance methods available to satisfy the requirements of Chapter 2.

The National Electrical Testing Association (NETA) maintains a set of testing procedures that are produced via a consensus process. These procedures can be used for acceptance testing and periodic maintenance testing to ensure system reliability. Acceptance testing verifies that the equipment functions as intended by the design specifications and provides a benchmark for comparing future test results. The comparison can identify any change in reliability for the equipment being tested. Maintenance tests enable a company to identify potential failures before they occur. A shutdown, then, can be scheduled, and repairs can be made with minimum exposure to workers that eliminates equipment damage.

(3) For the purpose of Chapter 2, maintenance shall be defined as preserving or restoring the condition of electrical equipment and installations, or parts of either, for the safety of employees who work on, near, or with such equipment. Repair or replacement of individual portions or parts of equipment shall be permitted without requiring modification or replacement of other portions or parts that are in a safe condition.

FPN: Refer to NFPA 70B, *Recommended Practice for Electrical Equipment Maintenance,* for specific maintenance methods and tests.

All electrical equipment has a predictable life cycle. Knowing the usable life of equipment is crucial if a facility is to predict the reliability and safe operation of its equipment. Repair parts for old equipment might be unavailable when a defect is discovered. Recognizing these two factors allows for planning, budgeting, and replacement of equipment in an orderly and safe manner.

Maintenance often is the most neglected component of a strategy to provide a safe work environment. NFPA 70B, *Recommended Practice for Electrical Equipment Maintenance,* is a widely used consensus-based document that provides prescribed maintenance methods and intervals to maximize the reliability of electrical equipment and systems. This narrative-based document describes electrical maintenance subjects and issues surrounding maintenance of electrical equipment. NFPA 70B provides workers with solutions, techniques, and testing intervals for adequate maintenance of electrical equipment.

# ARTICLE 205
# General Maintenance Requirements

Maintenance tasks, although routine, are key to equipment reliability. Rigid scheduling and documentation of maintenance tasks can help to ensure equipment safety as well as reliability.

## 205.1 Qualified Persons

Employees who perform maintenance on electrical equipment and installations shall be qualified persons as required in Chapter 1 and shall be trained in, and familiar with, the specific maintenance procedures and tests required.

A worker qualified for electrical equipment maintenance work must be familiar with the operation, maintenance, and history of the equipment as well as the safety training defined in Chapter 1. Familiarity with equipment and maintenance tools directly affects a worker's ability to recognize and avoid electrical hazards. A worker who is trained for a task and who does not perform that task for more than one year should no longer be considered to be qualified for the task. Employees should be provided with supplemental training or instruction to ensure that they continue to be qualified for the task.

## 205.2 Single Line Diagram

A single line diagram, where provided, for the electrical system shall be maintained.

Single line (S/L) diagrams are the best source of information for workers to locate the electrical hazards that might be encountered in their daily work routines. Maintaining these drawings in an up-to-date condition provides the following valuable information:

- All of the sources of power to a specific piece of equipment
- The interrupting capacity of devices at each point in the system
- All possible paths of potential backfeed

An up-to-date single line diagram enables an electrically safe work condition to be implemented. It provides the correct rating for overcurrent devices and enables calculation of available fault current.

Overcurrent protective devices that are not rated appropriately for the available fault current could malfunction or explode in a short-circuit situation. Shrapnel from this type of incident creates another hazard for the worker.

All qualified workers must have the ability to read and understand single line diagrams of systems they work on.

### 205.3 Spaces About Electrical Equipment

All working space and clearances required in Chapter 4 shall be maintained.

Adequate working space allows workers to perform their tasks in an unencumbered manner. The ability to use tools and equipment within adequate clearances provides the necessary safety factor to prevent an inadvertent movement, which could result in an electrical incident and injury. Working space must be kept clear of temporary obstructions, such as stored equipment.

### 205.4 Grounding and Bonding

Equipment, raceway, cable tray, and enclosure bonding and grounding shall be maintained to ensure electrical continuity.

Grounding and bonding provide the electrical continuity to enable any fault current to return to the energy source. Adequate grounding and bonding of electrical equipment provides personnel and equipment safety benefits by enabling the overcurrent device to operate. In a short-circuit or overload condition, the overcurrent device relies on a clear and effective grounding path to operate within its designed range. Without effective grounding and bonding, the tripping time might be extended, thus increasing the amount of incident energy to which a worker might be exposed.

### 205.5 Guarding of Live Parts

Enclosures shall be maintained to guard against accidental contact with live parts and other electrical hazards.

Electrical equipment is approved for the purpose on the basis of product standards. The base premise for worker safety is that no live parts are exposed. In some installations, live parts might be guarded by a fence or similar barrier. If the gate is kept locked, access is restricted only to authorized and qualified personnel who have the key to the lock.

Unused openings in enclosures must be covered in a manner that prevents inadvertent contact with live parts. Covering these openings prevents or minimizes the chance of introducing contamination, including conductive dust, that could result in an arc flash incident.

### 205.6 Safety Equipment

Locks, interlocks, and other safety equipment shall be maintained in proper working condition to accomplish the control purpose.

Locks and interlocks assist in providing safety for workers and equipment. Locks and interlocks ensure that only authorized persons have access to areas that contain exposed live parts. Maintaining these locks and interlocks in good working condition helps to minimize exposure to electrical hazards.

Some equipment arrangements use an interlocking system of keys to control the flow of electrical power through the system and to control the sequence of switch operations. The system's keying mechanism must work smoothly and without incident to accomplish this sequence. This keyed system is typically used to redirect power upon interruption from an incoming feeder or to provide system maintenance while minimizing the amount of energized equipment. Duplicate keys for these interlocked key systems must be destroyed or rigidly controlled.

Instructions should be readily available to authorized workers so they can execute the required steps for operating the equipment. A worker who does not perform this procedure on a regular basis might be unable to remember the procedure steps to transfer the power.

### 205.7 Clear Spaces

Access to working space and escape passages shall be kept clear and unobstructed.

Good housekeeping is always essential to a safe work environment. Maintaining adequate access to equipment is essential if a worker is to operate equipment in a timely manner. Material storage that blocks access or prevents safe work practices must be avoided at all times.

### 205.8 Identification of Components

Identification of components, where required, and safety-related instructions (operating or maintenance), if posted, shall be securely attached and maintained in legible condition.

Up-to-date operating and maintenance instructions are vital to ensure worker safety. Affected workers must be aware that instructions exist and where they are located, and the instructions must be readily accessible.

### 205.9 Warning Signs

Warning signs, where required, shall be visible, securely attached, and maintained in legible condition.

Warning signs are required to inform both qualified and unqualified workers of potential hazards that they might encounter. Section 110.16 of the 2002 *NEC* requires a warning label for equipment that has a potential arc flash hazard. The marking must be located so as to be clearly visible to qualified persons before examination, adjustment, servicing, or maintenance of the equipment.

### 205.10 Identification of Circuits

Circuit or voltage identification shall be securely affixed and maintained in updated and legible condition.

The *NEC* and OSHA require that circuits be identified and the identification securely affixed to the equipment. Mislabeled equipment sets a trap for workers who assume that they have deenergized the appropriate circuit feeding the equipment. However, circuit identification does not change the requirement to verify the absence of voltage when establishing an electrically safe work condition.

### 205.11 Single and Multiple Conductors and Cables

Electrical cables and single and multiple conductors shall be maintained free of damage, shorts, and ground that would present a hazard to employees.

Single- and multiconductor cables have a protective shield to protect conductors from physical damage under certain conditions. Age of equipment, environment, and work practices all are factors that could affect the integrity of that protection. Conductor insulation and outer coverings can become dry and brittle over time, and movement around these cables could create a shock and flash hazard. For example, if a company modifies an electrical installation by additional cables to a cable tray, workers crawling along the cable tray to install the new cables or to move existing cable might damage insulation on existing cables and create a shock or flash hazard.

### 205.12 Flexible Cords and Cables

Flexible cords and cables shall be maintained to avoid strain and damage.

Incorrect terminations of flexible cords and cables at an enclosure or device is a common problem. Tension placed on the cable easily can damage the outer covering, allowing conductors to be exposed and subjecting the worker to a hazard.

A damaged or missing ground prong is the most common problem with extension cords. The grounding path provided by the ground prong is necessary to prevent electrical shock or electrocution of the worker. Before each use, extension cords should be inspected to ensure that the ground prong has not been damaged.

**(1) Damaged Cords and Cables.** Cords and cables shall not have worn, frayed, or damaged areas that present an electrical hazard to employees.

**(2) Strain Relief.** Strain relief of cords and cables shall be maintained to prevent pull from being transmitted directly to joints or terminals.

# ARTICLE 210
# Substations, Switchgear Assemblies, Switchboards, Panelboards, Motor Control Centers, and Disconnect Switches

Inspection and appropriate maintenance of the components addressed by Article 210 are imperative, since undetected deterioration can cause severe electrical safety hazards to workers.

### 210.1 Enclosures

Enclosures shall be kept free of material that would create a hazard.

### 210.2 Area Enclosures

Fences, physical protection, enclosures, or other protective means, where required to guard against unauthorized access or accidental contact with exposed live parts, shall be maintained.

Fences, their gates, and other enclosures should be inspected to ensure that their integrity has not deteriorated. Fences and enclosures must continue to guard against entry of unauthorized personnel or animals. The gates or doors, especially where equipped with panic hardware, should be checked for security and appropriate operation. Keys to the locking devices for these areas must be carefully controlled to prevent unauthorized personnel from entering areas and, conversely, to enable timely access in the case of an emergency. Repairing any defects or damage to these area enclosures must be completed promptly; otherwise, the ability of the equipment to prevent entry by personnel might be compromised.

### 210.3 Conductors

Current-carrying conductors (buses, switches, disconnects, joints, and terminations) and bracing shall be maintained to:

(1) Conduct rated current without overheating

Discoloration of copper conductors or terminals is evidence of overheating. One cost-effective method of investigating overheating problems is to perform an infrared scan of the equipment. This scan can be performed while the system is operating. The task of performing an infrared scan must be considered a hazardous task, however. Personal protective equipment, as determined by a hazard analysis, must be selected and worn while the scan is being executed. If evidence of overheating is found, the equipment should be deenergized and the problem investigated and repaired in accordance with manufacturers' specifications.

(2) Withstand available fault current

Short circuits or fault currents represent a significant amount of destructive energy that could be released into electrical systems under abnormal conditions. During normal system operation, electrical energy is controlled and does useful work. However, under fault conditions, short-circuit current can cause serious damage to electrical systems and equipment and create the potential for serious injury to personnel.

### 210.4 Insulation Integrity

Insulation integrity shall be maintained to support the voltage impressed.

### 210.5 Protective Devices

Protective devices shall be maintained to adequately withstand or interrupt available fault current.

The protective devices are designed to operate within a prescribed range to disconnect the power to equipment or sections of equipment in a timely manner. These devices are intended to minimize the damage to equipment and injury to personnel. When a protective device is incorrectly sized or maintained, any arcing fault could result in injury and equipment damage.

## ARTICLE 215
## Premises Wiring

Article 215 addresses the maintenance needs of all wiring present in a facility.

### 215.1 Covers for Wiring System Components

Covers for wiring system components shall be in place with all associated hardware, and there shall be no unprotected openings.

Section 110.12(A) of the 2002 *NEC* requires that all unused openings be closed to afford protection substantially equivalent to the wall of the equipment. These requirements require closing or covering openings to protect workers from contact with live parts with a body part or conductive object and also by preventing the introduction of contamination or conductive dust in equipment that could result in an explosion.

### 215.2 Open Wiring Protection

Open wiring protection, such as location or barriers, shall be maintained to prevent accidental contact.

Articles 225, 230, and 398 of the 2002 *NEC* provide spacing and clearance installation requirements for open wiring.

### 215.3 Raceways and Cable Trays

Raceways and cable trays shall be maintained to provide physical protection and support for conductors.

## ARTICLE 220 Controller Equipment

Article 220 addresses maintenance of control equipment and defines which types of equipment are considered to be controllers.

### 220.1 Scope

This article shall apply to controllers, including electrical equipment that governs the starting, stopping, direction of motion, acceleration, speed, and protection of rotating equipment and other power utilization apparatus in the workplace.

A controller can be a remote-controlled magnetic contactor, switch, circuit breaker, or device that normally is used to start and stop motors and other apparatuses. In the case of motors, the controller must be capable of interrupting the locked-rotor current of the motor. Stop-and-start stations and similar control circuit components that do not open the power conductors to the motor are not considered to be controllers.

### 220.2 Protection and Control Circuitry

Protection and control circuitry used to guard against accidental contact with live parts and to prevent other electrical or mechanical hazards shall be maintained.

## ARTICLE 225 Fuses and Circuit Breakers

Precise operation of fuses and circuit breakers in an electrical system is imperative. Therefore, their adequate maintenance is essential to maintaining a safe work environment in a facility.

### 225.1 Fuses

Fuses shall be maintained free of breaks or cracks in fuse cases, ferrules, and insulators. Fuse clips shall be maintained to provide adequate contact with fuses.

Fuse terminals and fuse clips should be examined for discoloration caused by heat from poor contact or corrosion. Early detection of overheating is possible through the use of infrared examination.

Fuses should have an interrupting rating equal to or greater than the maximum fault current available at their point of application. The interrupting rating of fuses, ranging from 10,000 amperes to 300,000 amperes, should be clearly visible on the fuse label.

Many different types of fuses are used in power distribution systems and utilization equipment. Fuses differ by performance, characteristics, and physical size. Their ratings must be verified to ensure that the circuit design is maintained for the life of the equipment. When replacing fuses, a worker should never alter the fuseholder or force it to accept fuses that do not readily fit. Stocking an adequate supply of spare fuses with appropriate ratings minimizes replacement problems.

### 225.2 Molded-Case Circuit Breakers

Molded-case circuit breakers shall be maintained free of cracks in cases and cracked or broken operating handles.

Maintenance of molded-case circuit breakers generally can be divided into two categories: mechanical and electrical. Mechanical maintenance consists of inspection for good housekeeping, maintenance of appropriate mechanical mounting and electrical connections, and manual operation of the circuit breakers.

Molded-case circuit breakers should be kept free of external contamination so that internal heat can be dissipated normally. A clean circuit breaker enclosure reduces potential arcing conditions between live conductors and between live conductors and ground. The structural strength of the case is important in withstanding the stresses imposed during fault-current interruptions. Therefore, an inspection should be made for cracks in the case, and replacements made if necessary.

Excessive heat in a circuit breaker can cause a malfunction in the form of nuisance tripping and possibly an eventual failure. Loose connections are the most common cause of excessive heat. Periodic maintenance checks should involve checking for loose connections or evidence of overheating. All connections should be maintained in accordance with manufacturers' recommendations.

Molded-case circuit breakers can be in service for extended periods and yet never be called on to perform their overload- or short-circuit-tripping functions. Manual operation of the circuit breaker helps keep the contacts clean, but it does not exercise the tripping mechanism. Although manual operations exercise the breaker mechanisms, none of the mechanical linkages in the tripping mechanisms are moved with this exercise. Even though manual operation does not completely check molded-case circuit breakers, it must be completed as the best-case testing scenario available.

### 225.3 Circuit Breaker Testing

Circuit breakers that interrupt faults approaching their ratings shall be inspected and tested in accordance with the manufacturer's instructions.

Circuit breakers that do not operate within their prescribed ratings and ranges can result in catastrophic effects. One effect might be a significant increase in danger to personnel. For example, if the circuit breaker is cracked or has a higher current induced through a fault condition, it could explode, striking the worker with shrapnel traveling in excess of 700 miles per hour.

If the circuit breaker does not trip within its prescribed range, it can result in an increase in incident energy, exposing the worker to increased risk. For example, a typical arc flash situation where the worker is 18 in. from a 20-kA short-circuit and 5-cycle tripping time results in an incident energy exposure of 6.44 cal/cm$^2$. If the tripping time is increased to 30 cycles, due to the circuit breaker being out of calibration or improper maintenance, the incident energy is 38.64 cal/cm$^2$.

Circuit breakers should have an acceptance test and maintenance testing at recommended intervals. NFPA 70B and the NETA *Maintenance Testing Specifications* are documents that can assist a company in understanding the specific tests and testing intervals that are required to ensure reliability.

# ARTICLE 230
# Rotating Equipment

Movement of rotating equipment and motors presents safety hazards to workers, so adequate maintenance of their guards is extremely important.

### 230.1 Terminal Boxes

Terminal chambers, enclosures, and terminal boxes shall be maintained to guard against accidental contact with live parts and other electrical hazards.

### 230.2 Guards, Barriers, and Access Plates

Guards, barriers, and access plates shall be maintained to prevent employees from contacting moving or energized parts.

# ARTICLE 235
# Hazardous (Classified) Locations

Maintenance of equipment in hazardous areas in a facility presents a special safety problem. Personnel who are assigned to such maintenance must be trained to understand the explosive nature of the materials within the areas and how the equipment maintenance is important to a safe environment.

### 235.1 Scope

This article covers maintenance requirements in those areas identified as hazardous (classified) locations in accordance with Article 440 of this standard.

FPN: These locations require special types of equipment and installation that ensure safe performance under conditions of proper use and maintenance. It is important that inspection authorities and users exercise more than ordinary care with regard to installation and maintenance. The maintenance requirements for specific equipment and materials covered elsewhere in Chapter 2 are applicable to hazardous (classified) locations. Other maintenance is required to ensure that the form of construction and of installation that makes the equipment and materials suitable for the particular location are not nullified.

The maintenance required for specific hazardous (classified) locations requires that the classification of the specific location be known. The design principles and equipment characteristics—for example, use of positive pressure ventilation, explosionproof, nonincendive, intrinsically safe, and purged and pressurized equipment—that were applied in the installation to meet the requirements of the area classification must also be known. With this information, the employer and the inspection authority are able to determine whether the installation as maintained has retained the condition necessary for a safe workplace.

### 235.2 Maintenance Requirements for Hazardous (Classified) Locations

Equipment and installations in these locations shall be maintained such that the following apply:

(1) No energized parts are exposed.

*Exception to (1): Intrinsically safe and nonincendive circuits.*

(2) There are no breaks in conduit systems, fittings, or enclosures from damage, corrosion, or other causes.
(3) All bonding jumpers are securely fastened and intact.
(4) All fittings, boxes, and enclosures with bolted covers have all bolts installed and bolted tight.
(5) All threaded conduit shall be wrenchtight and enclosure covers shall be tightened in accordance with the manufacturer's instructions.
(6) There are no open entries into fittings, boxes, or enclosures that would compromise the protection characteristics.
(7) All close-up plugs, breathers, seals, and drains are securely in place.
(8) Marking of luminaires (lighting fixtures) for maximum lamp wattage and temperature rating is legible and not exceeded.
(9) Required markings are secure and legible.

Equipment maintenance should be performed only by qualified personnel trained in safe maintenance practices and the special considerations necessary to maintain electrical equipment for use in hazardous (classified) locations. These individuals should be familiar with requirements for safe electrical installations. They should be trained to identify and eliminate ignition sources such as high surface temperatures, stored electrical energy, and the buildup of static charges and to identify the need for special tools, equipment, tests, and protective clothing.

## ARTICLE 240
## Batteries and Battery Rooms

Because of the special safety hazards of explosive gases present in battery rooms, maintenance of these areas requires adequate ventilation and special safety equipment.

### 240.1 Ventilation

Ventilation systems, forced or natural, shall be maintained to prevent buildup of explosive mixtures. This maintenance shall include a functional test of any associated detection and alarm systems.

Ventilation systems must be tested periodically to ensure a minimum of two air changes per hour to remove gases generated by vented batteries during charging or caused by equipment malfunction. NFPA 111, *Standard on Stored Electrical Energy Emergency and Standby Power Systems,* provides guidance on the installation requirements for battery rooms.

### 240.2 Eye and Body Wash Apparatus

Eye and body wash apparatus shall be maintained in operable condition.

### 240.3 Cell Flame Arresters and Cell Ventilation

Battery cell ventilation openings shall be unobstructed, and cell flame arresters shall be maintained.

## ARTICLE 245
## Portable Electric Tools and Equipment

Most hazardous conditions associated with portable electric tools and equipment result from improper handling or storage. Maintenance and inspection of such equipment must be included in the facility's electrical safety program as a matter of course.

### 245.1 Maintenance Requirements for Portable Electric Tools and Equipment

Attachment plugs, receptacles, cover plates, and cord connectors shall be maintained such that the following apply:

(1) There are no breaks, damage, or cracks exposing live parts.
(2) There are no missing cover plates.
(3) Terminations have no stray strands or loose terminals.
(4) There are no missing, loose, altered, or damaged blades, pins, or contacts.
(5) Polarity is correct.

Periodic electrical testing of tools can uncover operating defects. Immediate correction of these defects ensures continued safe operation and prevents breakdown and more costly repairs. A visual inspection is recommended when a tool is issued as well as after each use just before the tool is returned to the storage area.

Employees should be trained to recognize visible defects such as cut, frayed, spliced, or broken cords; cracked or broken attachment plugs; and missing or deformed grounding prongs. Such defects should be reported immediately and the tool removed from service until it is repaired.

Employees should be instructed to report all shocks immediately, no matter how minor, and to cease using the tool. Tools causing shocks should be examined and repaired before further use.

## ARTICLE 250
## Personal Safety and Protective Equipment

Personal protective equipment is a worker's final chance to avoid injury in the event of an incident. Therefore, workers should invest a great deal of interest and energy in maintaining this special equipment.

### 250.1 Maintenance Requirements for Personal Safety and Protective Equipment

Personal safety and protective equipment such as the following shall be maintained in a safe working condition:

(1) Grounding equipment
(2) Hot sticks

(3) Rubber gloves, sleeves, and leather protectors
(4) Voltage test indicators
(5) Blanket and similar insulating equipment
(6) Insulating mats and similar insulating equipment
(7) Protective barriers
(8) External circuit breaker rack-out devices
(9) Portable lighting units
(10) Safety grounding equipment
(11) Dielectric footwear
(12) Protective clothing

To ensure reliability, all equipment must be maintained in accordance with the manufacturers' recommendations or listings.

### 250.2 Inspection and Testing of Protective Equipment and Protective Tools

**(A) Visual.** Safety and protective equipment and protective tools shall be visually inspected for damage and defects before initial use and at intervals thereafter as service conditions require, but in no case shall the interval exceed 1 year.

**(B) Testing.** The insulation of protective equipment and protective tools, such as items (1) through (12) of 250.1, shall be verified by the appropriate test and visual inspection to ascertain that insulating capability has been retained before initial use, and at intervals thereafter as service conditions and applicable standards and instructions require, but in no case shall the interval exceed 3 years.

### 250.3 Safety Grounding Equipment

**(A) Visual.** Personal protective ground cable sets shall be inspected for cuts in the protective sheath and damage to the conductors. Clamps and connector strain relief devices shall be checked for tightness. These inspections shall be made at intervals thereafter as service conditions require, but in no case shall the interval exceed 1 year.

**(B) Testing.** Prior to being returned to service, safety grounds that have been repaired or modified shall be tested to ascertain that 30- and 15-cycle maximum voltage drop values are not exceeded for the rating of the ground set. These tests shall be conducted at intervals as service conditions and applicable required standards and instructions require, but in no case shall the interval exceed 3 years.

## REFERENCES CITED IN COMMENTARY

*NETA Maintenance Testing Specifications,* International Electrical Testing Association, Morrison, CO, 2001.

NFPA 70, *National Electrical Code,®* 2002 edition, National Fire Protection Association, Quincy, MA.

NFPA 70B, *Recommended Practice for Electrical Equipment Maintenance,* 2002 edition, National Fire Protection Association, Quincy, MA.

NFPA 111, *Standard on Stored Electrical Energy Emergency and Standby Power Systems,* 2001 edition, National Fire Protection Association, Quincy, MA.

# CHAPTER 3

# Safety Requirements for Special Equipment

Some facilities that use electrical energy have special circumstances. The electrical energy in these cases is used in ways that differ from most general industries. In some cases, the electrical energy is an integral part of the manufacturing process. In others the electrical energy is converted to a form that exposes workers to unique hazards. When electrical energy is used as a process variable, the safe work practices defined in Chapter 1 might become unsafe. Chapter 3 is intended to identify work practices for use in these situations.

## ARTICLE 300 Introduction

Article 300 names the special equipment discussed in Chapter 3 and identifies the purpose of the chapter as supplementing and modifying the safety-related work practices in Chapters 1 and 4 for work with special equipment. Article 300 also points out the responsibility of the employer to provide safety-related work practices and training.

### 300.1 Scope

Chapter 3 covers electrical safety installation requirements and safety-related work practices and procedures for employees who work on or near special electrical equipment in the workplace. Chapter 3 supplements or modifies the general requirements of Chapter 1 and Chapter 4.

### 300.2 Responsibility

The employer shall provide safety-related work practices and employee training. The employee shall follow those work practices.

### 300.3 Organization

Chapter 3 of this standard is divided into articles. Article 300 applies generally. Article 310 applies to electrolytic cells as described in 430.8. Article 320 applies to batteries and battery rooms. Article 330 applies to lasers. Article 340 applies to power electronic equipment.

FPN: The NFPA 70E Technical Committee might develop additional chapters for other types of special equipment in the future.

# ARTICLE 310
# Safety-Related Work Practices for Electrolytic Cells

Article 310 identifies the supplementary or replacement safe work practices workers should use in electrolytic cell line working zones and the special hazards of working with ungrounded dc voltage (see Exhibit 310.1).

***EXHIBIT 310.1.*** *An electrolytic cell line. (Photo by David Pace and Michael Petry, courtesy of Olin Corporation.)*

### 310.1 Scope

The requirements of this chapter shall apply to the electrical safety-related work practices used in the types of electrolytic cell areas set forth in 430.8.

FPN No. 1: See Annex L for a typical application of safeguards in the cell line working zone.

FPN No. 2: For further information, see IEEE *Standard for Electrical Safety Practices in Electrolytic Cell Line Working Zones*, IEEE Std. 463-1993.

### 310.2 Definitions

For the purposes of this chapter, the following definitions shall apply.

**Battery Effect.** A voltage that exists on the cell line after the power supply is disconnected.

FPN: Electrolytic cells could exhibit characteristics similar to an electrical storage battery, and thus a hazardous voltage could exist after the power supply is disconnected from the cell line.

**Safeguarding.** Safeguards for personnel include the consistent administrative enforcement of safe work practices. Safeguards include training in safe work practices, cell line design, safety equipment, personal protective equipment, operating procedures, and work checklists.

### 310.3 Safety Training

**(A) General.** The training requirements of this chapter shall apply to employees who are exposed to the risk of electrical hazard in the cell line working zone defined in 110.6 and shall supplement or modify the requirements of 110.8, 120.1, 130.1, and 130.5.

**(B) Training Requirements.** Employees shall be trained to understand the specific hazards associated with electrical energy on the cell line. Employees shall be trained in safety-related work practices and procedural requirements to provide protection from the electrical hazards associated with their respective job or task assignment.

All safety-training requirements defined in Chapter 1 apply to Article 310. However, the safety training requirements of Chapter 1 must be supplemented by training to consider the unique exposures associated with electrolytic cell lines.

### 310.4 Employee Training

**(A) Qualified Persons.**

**(1) Training.** Qualified persons shall be trained and knowledgeable in the operation of cell line working zone equipment and specific work methods and shall be trained to avoid the electrical hazards that are present. Such persons shall be familiar with the proper use of precautionary techniques and personal protective equipment. Training for a qualified person shall include the following:

(1) The skills and techniques to avoid dangerous contact with hazardous voltages between energized surfaces and between energized surfaces and ground. Skills and techniques might include temporarily insulating or guarding parts to permit the employee to work on energized parts.
(2) The method of determining the cell line working zone area boundaries.

**(2) Qualified Persons.** Qualified persons shall be permitted to work within the cell line working zone.

In electrolytic cell working zones, individual cells normally act as batteries. Direct current voltage is supplied by the cells as well as by rectifying equipment. Because the dc voltage normally is ungrounded, hand tools that might contact the dc bus work must not be grounded. Employees who work within the area of the dc bus must be trained to understand the unique hazards associated with ungrounded dc voltage.

Hazard/risk analyses used for ac circuits might not be appropriate for use with dc circuits. Each employee must be trained to understand how he or she might be exposed to a thermal hazard associated with an arc flash. Employees must understand how to select PPE for use when they are exposed to an arcing fault.

**(B) Unqualified Persons.**

**(1) Training.** Unqualified persons shall be trained to recognize electrical hazards to which they may be exposed and the proper methods of avoiding the hazards.

**(2) In Cell Line Working Zone.** When there is a need for an unqualified person to enter the cell line working zone to perform a specific task, that person shall be advised by the designated qualified person-in-charge of the possible hazards to ensure the unqualified person is safeguarded.

### 310.5 Safeguarding of Employees in the Cell Line Working Zone.

**(A) General.** Operation and maintenance of electrolytic cell lines may require contact by employees with exposed energized surfaces such as buses, electrolytic cells, and their attachments. The approach distances referred to in Table 130.2(C) shall not apply to work performed by qualified persons in the cell line working zone. Safeguards such as safety-related work practices and other safeguards shall be used to protect employees from injury while working in the cell line working zone. These safeguards shall be consistent with the nature and extent of the related electrical hazards. Safeguards might be different for energized cell lines and deenergized cell lines. Hazardous battery effect voltages shall be dissipated to consider a cell line deenergized.

FPN No. 1: Exposed energized surfaces might not establish a hazardous condition. A hazardous electrical condition is related to current flow through the body causing shock and flash burns and arc blasts. Shock is a function of many factors, including resistance through the body and through skin, of return paths, of paths in parallel with the body, and of system voltages. Arc flash burns and arc blasts are a function of the current available at the point involved and the time of arc exposure.

FPN No. 2: A cell line or group of cell lines operated as a unit for the production of a particular metal, gas, or chemical compound might differ from other cell lines producing the same product because of variations in the particular raw materials used, output capacity, use of proprietary methods or process practices, or other modifying factors. Detailed standard electrical safety-related work practice requirements could become overly restrictive without accomplishing the stated purpose of Chapter 1 of this standard.

Employers must define an electrical safety program that addresses the issues identified in Chapter 1. However, the work practices can be modified, as necessary, to recognize the different types of exposure to electrical hazards. For instance, because each cell probably acts like a battery, the employer must define actions that are necessary if a worker must contact the dc bus structure. Those procedures must be consistent with the risk associated with the work task.

**(B) Signs.** Permanent signs shall clearly designate electrolytic cell areas.

**(C) Electrical Flash Hazard Analysis.** The requirements of 130.3, Flash Hazard Analysis, shall not apply to electrolytic cell line work zones.

**(1) Flash Hazard Analysis Procedure.** Each task performed in the electrolytic cell line working zone shall be analyzed for the risk of flash hazard injury. If there is risk of personal injury, appropriate measures shall be taken to protect persons exposed to the flash hazards. These measures shall include one or more of the following:

(1) Provide appropriate personal protective equipment *[see 310.5(D)(2)]* to prevent injury from the flash hazard.
(2) Alter work procedures to eliminate the possibility of the flash hazard.
(3) Schedule the task so that work can be performed when the cell line is deenergized.

**(2) Routine Tasks.** Flash hazard analysis shall be done for all routine tasks performed in the cell line work zone. The results of the flash hazard analysis shall be used in training employees in job procedures that minimize the possibility of flash hazards. The training shall be included in the requirements of 310.3.

**(3) Nonroutine Tasks.** Before a nonroutine task is performed in the cell line working zone, a flash hazard analysis shall be done. If flash hazard is a possibility during nonroutine work, appropriate instructions shall be given to employees involved on how to minimize the possibility of a hazardous flash.

**(4) Flash Hazards.** If the possibility of a flash hazard exists for either routine or nonroutine tasks, employees shall use appropriate safeguards.

**(D) Safeguards.** Safeguards shall include one or a combination of the following means.

**(1) Insulation.** Insulation shall be suitable for the specific conditions, and its components shall be permitted to include glass, porcelain, epoxy coating, rubber, fiberglass, plastic, and when dry, such materials as concrete, tile, brick, and wood. Insulation shall be permitted to be applied to energized or grounded surfaces.

**(2) Personal Protective Equipment.** Personal protective equipment shall provide protection from hazardous electrical conditions. Personal protective equipment shall include one or more of the following as determined by authorized management:

(1) Shoes, boots, or overshoes for wet service
(2) Gloves for wet service
(3) Sleeves for wet service
(4) Shoes for dry service
(5) Gloves for dry service
(6) Sleeves for dry service
(7) Electrically insulated head protection
(8) Protective clothing
(9) Eye protection

a. Standards for Personal Protective Equipment. Personal and other protective equipment shall be appropriate for conditions, as determined by authorized management, and shall not be required to meet the equipment standards in 130.7(C)(8) through 130.7(F) and in Table 130.7(C)(8) and Table 130.7(F).

b. Testing of Personal Protective Equipment. Personal protective equipment shall be verified with regularity and by methods that are consistent with the exposure of employees to hazardous electrical conditions.

**(3) Barriers.** Barriers shall be devices that prevent contact with energized or grounded surfaces that could present a hazardous electrical condition.

**(4) Voltage Equalization.** Voltage equalization shall exist where conductive surfaces are bonded to an energized surface, either directly or through a resistance, so that there is insufficient voltage between the surfaces to result in a hazardous electrical condition.

**(5) Isolation.** Isolation shall be the placement of equipment or items in locations such that employees are unable to simultaneously contact exposed conductive surfaces that could present a hazardous electrical condition.

**(6) Safe Work Practices.** Employees shall be trained in safe work practices. The training shall include why the work practices in a cell line working zone are different from similar work situations in other areas of the plant. Employees shall comply with established safe work practices and the safe use of protective equipment.

(a) Attitude Awareness. Safe work practice training shall include attitude awareness instruction. Simultaneous contact with energized parts and ground can cause serious electrical shock. Of special importance is the need to be aware of body position where contact may be made with energized parts of the electrolytic cell line and grounded surfaces.

(b) Bypassing of Safety Equipment. Safe work practice training shall include techniques to prevent bypassing the protection of safety equipment. Clothing may bypass protective equipment if the clothing is wet. Trouser legs should be kept at appropriate length, and shirt sleeves should be a good fit so as not to drape while reaching. Jewelry and other metal accessories that may bypass protective equipment shall not be worn while working in the cell line working zone.

**(7) Tools.** Tools and other devices used in the energized cell line work zone shall be selected to prevent bridging between surfaces at hazardous potential difference.

FPN: Tools and other devices of magnetic material could be difficult to handle in energized cells' areas due to their strong dc magnetic fields.

Significant magnetic forces normally exist in areas in which electrolytic cells are present. Tools that contain magnetic materials or materials that are affected by magnetic fields should not be used in the cell area.

**(8) Portable Cutout Type Switches.** Portable cell cutout switches that are connected shall be considered as energized and as an extension of the cell line working zone. Appropriate procedures shall be used to ensure proper cutout switch connection and operation.

**(9) Cranes and Hoists.** Cranes and hoists shall meet the requirements of 430.8(I). Insulation required for safeguarding employees, such as insulated crane hooks, shall be periodically tested.

**(10) Attachments.** Attachments that extend the cell line electrical hazards beyond the cell line working zone shall utilize one or more of the following:

(1) Temporary or permanent extension of the cell line working zone
(2) Barriers
(3) Insulating breaks
(4) Isolation

**(11) Pacemakers and Metallic Implants.** Employees with implanted pacemakers, ferromagnetic medical devices, or other electronic devices vital to life shall not be permitted in cell areas unless written permission is obtained from the employee's physician.

FPN: The American Conference of Government Industrial Hygienists (ACGIH) recommends that persons with implanted pacemakers should not be exposed to magnetic flux densities above 10 gauss.

Employers must take steps to ensure that workers who wear pacemakers and similar medical devices are not exposed to the magnetic fields that normally exist in the cell area.

**(12) Testing.** Equipment safeguards for employee protection shall be tested to ensure they are in a safe working condition.

### 310.6 Portable Tools and Equipment

**(A) Portable Electrical Equipment.** The grounding requirements of 110.9(B)(2) shall not be permitted within an energized cell line working zone. Portable electrical equipment shall meet the requirements of 430.8(E). Power supplies for portable electric equipment shall meet the requirements of 430.8(F).

**(B) Auxiliary Nonelectric Connections.** Auxiliary nonelectric connections such as air, water, and gas hoses shall meet the requirements of 430.8(H). Pneumatic-powered tools and equipment shall be supplied with nonconductive air hoses in the cell line working zone.

**(C) Welding Machines.** Welding machine frames shall be considered at cell potential when within the cell line working zone. Safety-related work practices shall require that the cell line not be grounded through the welding machine or its power supply. Welding machines located outside the cell line working zone shall be barricaded to prevent employees from touching the welding machine and ground simultaneously where the welding cables are in the cell line working zone.

**(D) Portable Test Equipment.** Test equipment in the cell line working zone shall be suitable for use in areas of large magnetic fields and orientation.

FPN: Test equipment that is not suitable for use in such magnetic fields could result in an incorrect response. When such test equipment is removed from the cell line working zone, its performance might return to normal, giving the false impression that the results were correct.

Portable tools must not be grounded when used in the cell area. Although a grounding conductor normally decreases exposure to an electrical hazard, in cell areas any grounded conductor increases exposure to an electrical hazard. All equipment and tools (including pneumatic tools) must be free from any grounding circuit. Pneumatic tools must be fitted with nonconductive hoses.

# ARTICLE 320 Safety Requirements Related to Batteries and Battery Rooms

Article 320 identifies work practices associated with installation and maintenance of batteries containing many cells, such as those used with uninterruptible power supplies (UPS) and unit substation dc power supplies.

### 320.1 Scope

The requirements of this article shall apply to the safety requirements related to installations of batteries and battery rooms with a stored capacity exceeding 1 kWh or a floating voltage that exceeds 115 volts but does not exceed 650 volts.

FPN: For further information, refer to the following documents:

(1) NFPA 70-2002, *National Electrical Code*, Article 480, Storage Batteries
(2) IEEE Std. 484-2002, *Recommended Practice for Installation Design and Installation of Vented Lead-Acid Batteries for Stationary Applications*
(3) IEEE Std. 937-1987 (R1993), *Recommended Practice for Installation and Maintenance of Lead-Acid Batteries for Photovoltaic Systems*
(4) IEEE Std. 1187-1996, *Recommended Practice for Installation Design and Installation of Valve-Regulated Lead-Acid Storage Batteries for Stationary Applications*
(5) OSHA 1926.403, *Battery Rooms and Battery Charging*
(6) OSHA 1910.178(g), *Changing and Charging Batteries*
(7) OSHA 1910.305(j)(7), *Storage Batteries*

Working with batteries exposes a worker to both shock and arc flash. A person's body might react to contact with dc voltage differently from contact with ac voltage. However, this standard and OSHA take a conservative position and consider the risk of shock or electrocution to be the same for both ac and dc exposure.

In addition to the same electrical hazards, batteries also expose a worker to hazards associated with the chemical electrolyte used in the battery. The worker also must understand that battery charging might generate significant quantities of hydrogen and other flammable gases. When selecting work practices and personal protective equipment, the worker must consider exposure to these hazards as well.

### 320.2 Definitions

These definitions are intended to apply to Article 320 only. However, the definitions are consistent with the *Authoritative Dictionary of IEEE Standards Terms.*

For the purposes of this chapter, the following definitions shall apply.

**Accessories.** Items supplied with the battery to facilitate the continued operation of the battery.

**Authorized Person.** The person in charge of the premises, or other person appointed or selected by the person in charge of the premises, to perform certain duties associated with the battery installation on the premises.

**Battery.** An electrochemical system capable of storing under chemical form the electric energy received and which can give it back by reconversion.

**Battery Enclosure.** An enclosure containing batteries that is suitable for use in an area other than a battery room or an area restricted to authorized personnel.

**Battery Room.** Room specifically intended for the installation of batteries that have no other protective enclosure.

**Capacity.** The quantity of electricity (electric charge) usually expressed in ampere-hour (A h) that a fully charged battery can deliver under specified conditions.

**Cell.** An assembly of electrodes and electrolyte that constitutes the basic unit of the battery.

**Charging.** An operation during which a battery receives electric energy that is converted to chemical energy from an external circuit. The quantity of electric energy then is known as the charge and is usually measured in ampere-hour.

**Constant Current Charge.** A charge during which the current is maintained at a constant value.

**Constant Voltage Charge.** A charge during which the voltage across the battery terminals is maintained at a constant value.

**Container.** A container for the plate pack and electrolyte of a cell of a material impervious to attack by the electrolyte.

**Discharging.** An operation during which a battery delivers current to an external circuit by the conversion of chemical energy to electric energy.

**Electrolyte.** A solid, liquid, or aqueous salt solution that permits ionic conduction between positive and negative electrodes of a cell.

**Electrolyte Density.** Density of the electrolyte, measured in kilograms per cubic meter at a specific temperature (density of pure water = 1000 kilograms per cubic meter at 4°Celsius).

FPN: The density of an electrolyte was formerly indicated by its specific gravity. Specific gravity is the ratio of the density of the electrolyte to the density of pure water. S.G. = (electrolyte density in kilograms per cubic meter)/1000.

**Flame-Arrested Vent Plug.** A vent plug design that provides protection against internal explosion when the cell or battery is exposed to a naked flame or external spark.

**Gassing.** The formation of gas produced by electrolyte.

**Intercell and Interrow Connection.** Connections made between rows of cells or at the positive and negative terminals of the battery that might include lead-plated terminal plates, cables with lead, plated lugs, and lead-plated rigid copper connectors, and for nickel-cadmium cells, nickel-plated copper intercell connections.

**Intercell Connector Safety Cover.** Insulated cover to shroud the terminals and intercell connectors from inadvertent contact by personnel or accidental short circuiting.

**Nominal Voltage.** An approximate value of voltage used to identify a type of battery.

**Pilot Cell.** A selected cell of a battery that is considered to be representative of the average state of the battery or part thereof.

**Prospective Fault Current.** The highest level of fault current that can occur at a point on a circuit. This is the fault current that can flow in the event of a zero impedance short-circuit and if no protection devices operate.

**Rate.** The current expressed in amperes at which a battery is discharged.

**Sealed Battery.** A battery that has no provision for the addition of water or electrolyte or for external measurement of electrolyte specific gravity.

**Secondary Battery.** Two or more cells electrically connected and used as a source of energy.

**Secondary Cell.** An assembly of electrodes and electrolytes that constitutes the basic unit of a battery.

**Stepped Stand.** Containers placed in rows and these rows are placed at different levels to form a stepped arrangement.

**Terminal Post.** A part provided for the connection of a cell or a battery to external conductors.

**Tiered Stand.** Where rows of containers are placed above containers of the same or another battery.

**Valve Regulated Battery.** A battery in which the venting of the products of electrolysis is controlled by a reclosing pressure-sensitive valve.

**Vented Battery.** A battery in which the products of electrolysis and evaporation are allowed to escape freely to the atmosphere.

**Vent Plug.** A part closing the filling hole that is also employed to permit the escape of gas.

**VRLA.** Valve-regulated lead-acid storage battery.

### 320.3 Battery Connections

Batteries are sources of energy. Therefore, it is not possible to isolate the source of voltage from a cell. However, it is possible to limit the number of cells that are connected together. Workers are exposed to shock and arc flash when performing tasks associated with batteries.

Exhibit 320.1 illustrates a method of performing maintenance that decreases the hazard to the worker. The degree of arc flash hazard increases as the capacity of the battery increases. The hazard/risk analysis must consider the number of cells associated with the work task.

***EXHIBIT 320.1.*** *A large battery installation.*

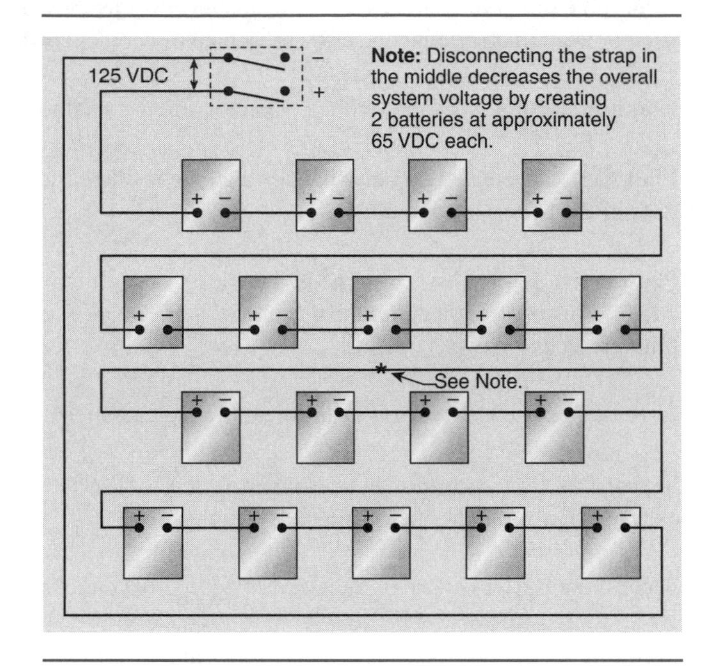


## (A) Method of Connection.

FPN No. 1: Batteries usually consist of a number of identical cells connected in series. The voltage of a series connection of cells is the voltage of a single cell multiplied by the number of cells. If cells of sufficiently large capacity are available, then two or more series-connected strings of equal numbers of cells could be connected in parallel to achieve the desired rated capacity. The rated capacity of such a battery is the sum of the capacities of a group of cells comprising a single cell from each of the parallel branches.

FPN No. 2: Cells of unequal capacity should not be connected in series.

FPN No. 3: Parallel connections of batteries should be limited to 4 strings.

FPN No. 4: Parallel connections of batteries are not recommended for constant current-charging applications.

FPN No. 5: Cells connected in series have high voltages that could produce a shock hazard.

**(B) Battery Short-Circuit Current.** The battery manufacturer shall be consulted regarding the sizing of the battery short-circuit protection.

*Exception: If information regarding the short-circuit protection of a battery is not available from the manufacturer, the prospective fault level at the battery terminals shall be considered to be twenty times the nominal battery capacity at the 3-hour rate.*

FPN: Battery short-circuit current = (battery voltage)/(internal resistance).

**(C) Connection Between Battery and DC Switching Equipment.**

**(1) General.** Any cable, busbar, or busway forming the connection between the battery terminal and the dc switching equipment shall be rated to withstand the prospective short-circuit current.

FPN: The available short-circuit current should be assumed for a time period of at least 1 second.

Outside busbars and cables should be both of the following:

(1) Insulated from the battery terminals to a height of 3.75 m (12 ft 4 in.), or to the battery room ceiling, whichever is lower
(2) Be clearly identified and segregated from any other supply circuits

**(2) Cable.** Cables shall be effectively clamped and sufficient support shall be provided throughout the length of cables to minimize sag and prevent undue strain from being imposed on the cable.

**(3) Busbars.**

FPN: Busbars should be insulated throughout their length by an insulating material not affected by the acid fumes that are present in a battery room. The steelwork supporting the busbar system should be installed so as not to restrict access to the battery for the purpose of maintenance.

**(4) Busways.**

FPN: Busways should be fully enclosed and able to withstand high levels of fault current without danger.

**(D) DC Switching Equipment.** Switching equipment shall comply with the *NEC*.

**(E) Terminals and Connectors.** Intercell and battery terminal connections shall be constructed of materials, either intrinsically resistant to corrosion or suitably protected by surface finish against corrosion. The joining of materials that are incompatible in a corrosive atmosphere shall be avoided.

FPN No. 1: To prevent mechanical stress on the battery terminal posts, the connection between the battery and any busbar system or large cable should be by insulated flexible cable of suitable rating.

FPN No. 2: The takeoff battery terminals and busbar connections should be shrouded or protected by physical barriers to prevent accidental contact.

**(F) DC Systems Grounding and Ground-Fault Detection.** One of the four types of available dc grounding systems, described as Type 1 through Type 4, shall be used.

FPN: Stationary battery systems should not be grounded.

(1) Type 1. The ungrounded dc system in which neither pole of the battery is connected to ground

FPN: Work on such a system should be carried out with the battery isolated from the battery charger. If an intentional ground is placed at one end of the battery, an increased shock hazard would exist between the opposite end of the battery and ground. Also, if another ground develops within the system (e.g., dirt and acid touching the battery rack), it creates a short-circuit that could cause a fire. An ungrounded dc system should be equipped with an alarm to indicate the presence of a ground-fault.

(2) Type 2. The solidly grounded dc system where either the positive or negative pole of the battery is connected directly to ground
(3) Type 3. The resistance grounded dc system, where the battery is connected to ground through a resistance

FPN: The resistance is used to permit operation of a current relay, which in turn initiates an alarm.

(4) Type 4. A tapped solid ground, either at the center point or at another point to suit the load system

**(G) Protection of DC Circuits.** DC circuits shall be protected in accordance with the *NEC.*

**(H) Alarms.**

**(1) Abnormal Battery Conditions.** Alarms shall be provided for early warning of the following abnormal conditions of battery operation:

(1) For vented batteries:
   a. Overvoltage
   b. Undervoltage
   c. Overcurrent
   d. Ground-fault
(2) For VRLA batteries, items (1)(a) though (1)(d) plus overtemperature, as measured at the pilot cell

**(2) Warning Signal.** The alarm system shall provide an audible alarm and visual indication at the battery location, and where applicable, at a remote manned control point.

### 320.4 Installations of Batteries

Installations using secondary batteries vary considerable in size, form large uninterruptible power supply systems, telecommunication systems, and demand load leveling installations to small emergency lighting installations. Secondary batteries permanently installed in or on buildings, structures, or premises, having a nominal voltage exceeding 24 volts and a capacity exceeding 10 ampere-hours at the 1-hour rate, shall be installed in a battery room or battery enclosure.

**(A) Location.** Batteries shall be installed in one of the following:

(1) Dedicated battery rooms
(2) An area accessible only to authorized personnel
(3) An enclosure with lockable doors or a suitable housing that shall be lockable and provide protection against electrical contact and damage to the battery

NFPA 111, *Standard on Stored Electrical Energy Emergency and Standby Power Systems,* provides valuable information for the installation, operation, and maintenance of battery rooms and systems. Compliance with these requirements can enhance worker safety in these environments.

**(B) Arrangement of Cells.** The space between adjacent containers shall be at least 12.5 mm (½ in.) and meet the following requirements:

(1) All cells shall be readily accessible for examination of the electrolyte level, refilling, cleaning, or removal as applicable.

(2) Each cell shall be readily accessible without having to reach over another cell or alternatively all exposed live surfaces shall be shrouded.

**(C) Ventilation for Batteries of the Vented Type.**

**(1) Installation.** Batteries shall be located in rooms or enclosures with outside vents or in well-ventilated rooms, so arranged to prevent the escape of fumes, gases, or electrolyte spray into other areas.

**(2) Ventilation.** Ventilation shall be provided so as to prevent liberated hydrogen gas from exceeding 1 percent concentration.

(a) Adequacy. Room ventilation shall be adequate to assure that pockets of trapped hydrogen gas do not occur, particularly at the ceiling, to prevent the accumulation of an explosive mixture.

(b) Equipment Considerations. Exhaust air shall not pass over electrical equipment unless the equipment is listed for the use.

(c) Location of Inlets. Inlets shall be no higher than the tops of the battery cells and outlets at the highest level in the room.

FPN: Ventilation rates should be based on the maximum hydrogen evolution rate for the applicable batteries. The maximum hydrogen evolution rate for lead antimony batteries should be considered as 0.000440 $m^3$/min (0.000269 $ft^3$/min) per charging ampere per cell at 25°C (77°F), with the maximum charging current available from the battery charger applied into a fully charged battery. The maximum hydrogen evolution rate for other types of batteries (e.g., lead calcium and nickel cadmium) should be obtained for the condition when the maximum charging current available from the battery charger is applied into a fully charged battery.

**(3) Mechanical Ventilation.** Where mechanical ventilation is installed, the following shall be required:

(1) Airflow sensors shall be installed to initiate an alarm if the ventilation fan becomes inoperative.
(2) Control equipment for the exhaust fan shall be located more than 1800 mm (6 ft) from the battery and a minimum of 100 mm (4 in.) below the lowest point of the highest ventilation opening.
(3) Where mechanical ventilation is used in a dedicated battery room, all exhaust air shall be discharged outside the building.
(4) Fans used to remove air from a battery room shall not be located in the duct unless the fan is listed for the use.

**(D) Ventilation for VRLA Type.**

**(1) Ventilation Requirements.** Ventilation shall be provided so as to prevent liberated hydrogen gas from exceeding a 1 percent concentration.

(a) Adequacy. Room ventilation shall be adequate to ensure that pockets of trapped hydrogen gas do not occur, particularly at the ceiling, to prevent the accumulation of an explosive mixture.

(b) Exhaust. Exhaust air shall not pass over electrical equipment unless the equipment is listed for the use.

(c) Inlets. Inlets shall be no higher than the tops of the battery cells and outlets at the highest level in the room.

**(2) Mechanical Ventilation.** Where mechanical ventilation is installed, the following shall be required:

(1) Airflow sensors shall be installed to initiate an alarm if the ventilation fan becomes inoperative.
(2) Control equipment for the exhaust fan shall be located more than 1800 mm (6 ft) from the battery and a minimum of 100 mm (4 in.) below the lowest point of the highest ventilation opening.
(3) Where mechanical ventilation is used in a dedicated battery room, all exhaust air shall be discharged outside the building.
(4) Fans used to remove air from a battery room shall not be located in the duct unless the fan is listed for the use.

**(3) Temperature Requirements.** Ventilation shall be provided to maintain design temperature to prevent thermal runaway that can cause cell meltdown leading to a fire or explosion.

**(E) Ventilation for Sealed Gelled Electrolyte Type.**

**(1) Temperature Requirements.** Ventilation shall be provided to maintain design temperature to prevent thermal runaway that can cause cell meltdown leading to a fire or explosion.

**(2) Mechanical Ventilation.** Where mechanical ventilation is installed, airflow sensors shall be installed to initiate an alarm if the ventilation fan becomes inoperative.

**320.5 Battery Room Requirements**

**(A) General.** The battery room shall be accessible only to authorized personnel and shall be locked when unoccupied.

**(1) Battery Rooms or Areas Restricted to Authorized Personnel.**

(a) Doors. The battery room and enclosure doors shall open outward. The doors shall be equipped with quick-release, quick-opening hardware.

(b) Location. The battery room shall be located so that access to the batteries is unobstructed. Direct-current switching equipment, rotating machinery other than exhaust fans, and other equipment not directly part of the battery and charging facilities shall be external to the battery room. Alternatively, dc switching equipment shall be separated from the battery by a partition of a height no less than 2 m (6 ft 6 in.) and of sufficient length to prevent accidental contact with live surfaces.

(c) Foreign Piping. Foreign piping shall not pass through the battery room.

(d) Passageways. Passageways shall be of sufficient width to allow the replacement of all battery room equipment.

(e) Emergency Exits. Emergency exits shall be provided as required.

(f) Access. Access and entrance to working space about the battery shall be provided as required by 400.15.

FPN: Provision to include emergency services personnel and their equipment should be made.

**(2) Battery Enclosures.** All cells shall be readily accessible for examination of the electrolyte level, refilling, cleaning, and removal.

**(3) Battery Room Floor Loading.** Floor loading shall take into account the seismic activity.

**(4) Battery Room Floor Construction and Finish.** Where the grading of the floor is not practicable, suitable drip trays or sumps shall be installed to restrict the spread of spilled electrolyte.

FPN No. 1: The battery room floor should be of concrete construction. The floor should be graded so any spillage of electrolyte will drain to an area where the electrolyte could be neutralized before disposal. (The battery manufacturer should be consulted on the appropriate floor grading so as to reduce connection alignment problems.)

FPN No. 2: The floor should be covered with an electrolyte-resistant, durable, antistatic, and slip-resistant surface overall, to a height 100 mm (4 in.) on each wall. Where batteries are mounted against a wall, the wall behind and at each end of the battery should be coated to a distance of 500 mm (20 in.) around the battery with an electrolyte-resistant paint.

**(B) Battery Layout and Floor Area.** The battery layout and floor area shall meet the following requirement:

**(1) Battery Layout.** The installation shall be so designed that, unless there is a physical barrier, potential differences exceeding 120 volts shall be separated by a distance of not less than 900 mm (36 in.) measured in a straight line in any direction.

**(2) Floor Area.** The floor area shall allow for the following clearances.

(a) Aisle Width. The minimum aisle width shall be 900 mm (36 in.).

(b) Single Row Batteries. In addition to the minimum aisle width, there shall be a minimum clearance of 25 mm (1 in.) between a cell and any wall or structure on the side not requiring access for maintenance. This required clearance does not preclude battery stands touching adjacent walls or structures, provided that the battery shelf has a free air space for no less than 90 percent of its length.

(c) Double-Row Batteries. The minimum aisle width shall be maintained on one end and both sides of the battery. The remaining end shall have a minimum clearance of 100 mm (4 in.) between any wall or structure and a cell.

(d) Tiered Batteries. Tiered batteries shall meet the requirements of 320.5(B)(2)(a), 320.5(B)(2)(b), and 320.5(B)(2)(c). In addition, there shall be a minimum clearance of 300 mm (12 in.) between the highest point of the battery located on the bottom tier and the lowest point of the underside of the upper runner bearers.

(e) Where a charger, or other associated electrical equipment, is located in a battery room, the aisle width between any battery and any part of the battery-charging equipment (including the doors when fully open) shall be at least 900 mm (36 in.).

**(C) Takeoff Battery Terminals and Outgoing Busbars and Cables.**

**(1) Takeoff Battery Terminals.** Outgoing busbars and cables shall meet the following requirements:

(1) Be insulated from the battery terminals to a height of 3.75 m (12 ft 4 in.) or the battery room ceiling, whichever is lower
(2) Be clearly identified and segregated from any other supply circuits

**(2) Outgoing Busbars and Cables.** The takeoff battery terminals and busbar connections shall comply with either of the following:

(1) Be shrouded
(2) Be protected by physical barriers to prevent accidental contact

**(D) Intertier and Interrow Connections.** The battery terminals and busbar and cable interconnections between rows shall comply with either of the following:

(1) Be shrouded
(2) Be protected by insulating barriers to prevent accidental contact

**(E) Barriers.** To avoid accidental contact with intercell connections, the following insulating barriers shall be installed.

**(1) Double-Row Batteries.** Insulating barriers between double-row batteries shall be installed for the entire length of the battery extending 100 mm (4 in.) past the end terminal unless those terminals are shrouded. The barrier shall extend vertically a minimum of 400 mm (16 in.) above the exposed portion of the intercell connections and a minimum of 25 mm (1 in.) below the top of the battery container.

**(2) Batteries Above 120 Volts.** Where the nominal voltage of the battery exceeds 120 volts, interblock barriers shall be installed to sectionalize the battery into voltage blocks not exceeding 120 volts. Barriers shall extend a minimum of 50 mm (2 in.) out from the exposed side of the battery and a minimum of 400 mm (16 in.) above the top of the container.

**(F) Illumination.**

**(1) Battery Room Lighting.** Battery room lighting shall be installed to provide a minimum level of illumination of 300 lux (30 ft-candles).

**(2) Emergency Lighting.** Emergency illumination shall be provided for safe egress from the battery room.

**(G) Location of Luminaires (Lighting Fixtures) and Switches.** Luminaires (lighting fixtures) shall not be installed directly over cells or exposed live parts. Switches for the control of the luminaires (lighting fixtures) shall be readily accessible.

**(H) Power.** General-purpose outlets shall be installed for the maintenance of the battery.

**(I) Location of General-Purpose Outlets.** General-purpose outlets shall be installed at least 1800 mm (6 ft) from the battery and a minimum of 100 mm (4 in.) below the lowest point of the highest ventilation opening.

**320.6 Battery Enclosure Requirements.**

**(A) Enclosure Construction.**

**(1) General.** Where enclosures are designed to accommodate the battery, the battery charger, and other equipment, separate compartments shall be provided for each.

**(2) Ventilation.** The ventilation openings for the compartments shall be spaced as far apart as practicable.

**(B) Battery Takeoff Terminals and Outgoing Busbars and Cables.** Outgoing busbars and cables shall be fully insulated, and the battery takeoff terminals shall comply with either of the following:

(1) Be fully shrouded
(2) Have physical barriers installed between them

**(C) Battery Compartment Circuits.** Only circuits associated with the battery shall be installed within a battery compartment of the enclosure.

**320.7 Protection.**

**(A) General.**

**(1) Marking.** When the battery capacity exceeds 100 ampere-hours or where the nominal battery voltage is in excess of 50 volts, suitable warning notices indicating the battery voltage and the prospective short-circuit current of the installation shall be displayed.

**(2) Overcurrent Protection.** Each output conductor shall be individually protected by a fuse or circuit breaker positioned as close as practicable to the battery terminals.

**(3) Protective Equipment.** Protective equipment shall not be located in the battery compartment of the enclosure.

**(B) Switching and Control Equipment.** Switching and control equipment shall comply with NFPA 70, *National Electrical Code,* and shall be listed for the application.

**(C) Ground-Fault Protection.** For an ungrounded battery of nominal voltage in excess of 120 volts, a ground-fault detector shall be provided to initiate a ground-fault alarm.

**(D) Main Isolating Switch.** The battery installation shall have an isolating switch installed as close as practicable to the main terminals of the battery. Where a busway system is installed, the isolating switch may be incorporated into the end of the busway.

**(E) Section Isolating Equipment.** Where the battery section exceeds 120 volts, the installation shall include an isolating switch, plugs, or links, as required, to isolate sections of the battery, or part of the battery for maintenance.

**(F) Warning Signs.** The following signs shall be posted in appropriate locations:

(1) Electrical hazard warning signs indicating the shock hazard due to the battery voltage and the arc hazard due to the prospective short-circuit current
(2) Chemical hazard warning signs indicating the danger of hydrogen explosion from open flame and smoking and the danger of chemical burns from the electrolyte
(3) Notice for personnel to use and wear protective equipment and apparel
(4) Notice prohibiting access to unauthorized personnel

### 320.8 Personnel Protective Equipment

The following protective equipment shall be available to employees performing battery maintenance:

(1) Goggle and face shields
(2) Chemical-resistant gloves
(3) Protective aprons
(4) Protective overshoes
(5) Portable or stationary water facilities for rinsing eyes and skin in case of electrolyte spillage

### 320.9 Tools and Equipment

Tools and equipment for work on batteries shall comply with the following:

(1) Be of the nonsparking type
(2) Be equipped with handles listed as insulated for the maximum working voltage

# ARTICLE 330
## Safety-Related Work Practices for Use of Lasers

Article 330 is limited to tasks performed in the laboratory or in the shop. It is not intended to cover the application and use of lasers in the workplace. However, employees who work with lasers might be exposed to hazards associated with the laser output in addition to the electrical hazards associated with the equipment.

### 330.1 Scope

The requirements of this article shall apply to the use of lasers in the laboratory and the workshop.

### 330.2 Definitions

For the purposes of this article, the following definitions shall apply.

**Fail Safe.** The design consideration in which failure of a component does not increase the hazard. In the failure mode, the system is rendered inoperative or nonhazardous.

**Fail Safe Safety Interlock.** An interlock that in the failure mode does not defeat the purpose of the interlock, for example, an interlock that is positively driven into the off position as soon as a hinged cover begins to open, or before a detachable cover is removed, and that is positively held in the off position until the hinged cover is closed or the detachable cover is locked in the closed position.

**Laser.** Any device that can be made to produce or amplify electromagnetic radiation in the wavelength range from 100 nm to 1 mm primarily by the process of controlled stimulated emission.

**Laser Controlled Area.** An area where the occupancy and activity of those within are subject to control and supervision for the purpose of protection from radiation hazards.

**Laser Energy Source.** Any device intended for use in conjunction with a laser to supply energy for the excitation of electrons, ions, or molecules. General energy sources, such as electrical supply services or batteries, shall not be considered to constitute laser energy sources.

**Laser Fiber Optic Transmission System.** A system consisting of one or more laser transmitters and associated fiber optic cable.

**Laser Hazard Area.** The area within which the beam irradiance or radiant exposure exceeds the appropriate corneal maximum permissible exposure (MPE), including the possibility of accidental misdirection of the beam.

**Laser Product.** Any product or assembly of components that constitutes, incorporates, or is intended to incorporate a laser or laser system.

**Laser Radiation.** All electromagnetic radiation emitted by a laser product between 100 nm and 1 mm that is produced as a result of a controlled stimulated emission.

**Laser System.** A laser in combination with an appropriate laser energy source with or without additional incorporated components.

### 330.3 Safety Training

**(A) Personnel to Be Trained.** Employers shall provide training for all operator and maintenance personnel.

**(B) Scope of Training.** The training shall include, but is not limited to, the following:

(1) Familiarization with laser principles of operation, laser types, and laser emissions
(2) Laser safety, including the following:
   a. System operating procedures
   b. Hazard control procedures
   c. The need for personnel protection
   d. Accident reporting procedures
   e. Biological effects of the laser upon the eye and the skin
   f. Electrical and other hazards associated with the laser equipment, including the following:
      i. High voltages (> 1 kV) and stored energy in the capacitor banks
      ii. Circuit components, such as electron tubes, with anode voltages greater than 5 kV emitting X-rays
      iii. Capacitor bank explosions
      iv. Production of ionizing radiation
      v. Poisoning from the solvent or dye switching liquids or laser media
      vi. High sound intensity levels from pulsed lasers

**(C) Proof of Qualification.** Proof of qualification of the laser equipment operator shall be available and in possession of the operator at all times.

Employees who work on lasers in the laboratory or in the shop must demonstrate an understanding of both electrical hazards and hazards associated with the laser output. Employees who have demonstrated this understanding should be issued a certificate indicating successful completion of safety training for work on or with the specific lasers available on site.

### 330.4 Safeguarding of Employees in the Laser Operating Area

**(A) Eye Protection.** Employees shall be provided with eye protection as required by federal regulation.

**(B) Warning Signs.** Warning signs shall be posted at the entrances to areas or protective enclosures containing laser products.

**(C) Master Control.** High power laser equipment shall include a key-operated master control.

**(D)** High-power laser equipment shall include a fail-safe laser radiation emission audible and visible warning when it is switched on or if the capacitor banks are charged.

**(E)** Beam shutters or caps shall be utilized, or the laser switched off, when laser transmission is not required. The laser shall be switched off when unattended for 30 minutes or more.

**(F)** Laser beams shall not be aimed at employees.

**(G)** Laser equipment shall bear a label indicating its maximum output.

**(H)** Personnel protective equipment shall be provided for users and operators of high-power laser equipment.

### 330.5 Employee Responsibility

Employees shall be responsible for the following:

(1) Obtaining authorization for laser use
(2) Obtaining authorization for being in a laser operating area
(3) Observing safety rules
(4) Reporting laser equipment failures and accidents to the employer

# ARTICLE 340
# Safety-Related Work Practices: Power Electronic Equipment

Chapters 1, 2, and 4 apply to electrical equipment that operates at frequencies normally supplied for consumer use. The reaction of the human body changes as the frequency increases. When an increase in frequency reaches the microwave band, joule heating can result in internal burns. Employees who work on or with equipment within the scope of Article 340 must be qualified to perform tasks on specific electronic equipment.

Employees who are qualified to work on or with this type of equipment should be trained to understand the unique hazards associated with the specific equipment on which he or she will perform work tasks. The worker should demonstrate understanding of the specific hazards and how to avoid exposure to them. The worker should be issued a certificate indicating successful completion of a safe work-practice training program.

## 340.1 Scope

This article shall apply to safety-related work practices around power electronic equipment, including the following:

(1) Electric arc welding equipment
(2) High-power radio, radar, and television transmitting towers and antenna
(3) Industrial dielectric and RF induction heaters
(4) Shortwave or radio frequency diathermy devices
(5) Process equipment that includes rectifiers and inverters such as the following:
   a. Motor drives
   b. Uninterruptible power supply systems
   c. Lighting controllers

## 340.2 Definition

For the purposes of this article, the following definition shall apply.

**Radiation Worker.** A person who is required to work in electromagnetic fields, the radiation levels of which exceed those specified for nonoccupational exposure.

## 340.3 Application

This purpose of this article is to provide guidance for safety personnel in preparing specific safety-related work practices within their industry.

## 340.4 Reference Standards

The following are reference standards for use in the preparation of specific guidance to employees:

(1) International Electrotechnical Commission [IEC]: 479 Effects of current passing through the human body:
   a. 479-1 Part 1 General aspects
   b. 479-1-1 Chapter 1: Electrical impedance of the human body
   c. 479-1-2 Chapter 2: Effects of ac in the range of 15 Hz to 100 Hz
   d. 479-2 Part 2: Special aspects

e. 479-2-4: Chapter 4: Effects of ac with frequencies above 100 Hz
f. 479-2-5 Chapter 5: Effects of special waveforms of current
g. 479-2-6 Chapter 6: Effects of unidirectional single impulse currents of short duration

(2) International Commission on Radiological Protection [IRCP]:
International Commission for Radiological Protection Publication 15: Protection against ionizing radiation from external sources

### 340.5 Hazards Associated with Power Electronic Equipment

Employer and employees shall be aware of the following hazards associated with power electronic equipment.

(1) Results of Power Frequency Current.
a. At 5 mA, shock is perceptible.
b. At 10 mA, a person may not be able to voluntarily let go of the hazard.
c. At about 40 mA, the shock, if lasting for 1 second or longer, may be fatal due to ventricular fibrillation.
d. Further increasing current leads to burns and cardiac arrest.

(2) Results of Direct Current.
a. A dc current of 2 mA is perceptible.
b. A dc current of 10 mA is considered the threshold of the let-go current.

(3) Results of Voltage. A voltage of 30 V rms, or 60 V dc, is considered safe except when the skin is broken the internal body resistance can be as low as 500 ohms so fatalities can occur.

(4) Results of Short Contact.
a. For contact less than 0.1 second and with currents just greater than 0.5 mA, ventricular fibrillation may occur only if the shock is in a vulnerable part of the cardiac cycle.
b. For contact of less than 0.1 second and with currents of several amperes, ventricular fibrillation may occur if the shock is in a vulnerable part of the cardiac cycle.
c. For contact of greater than 0.8 second and with currents just greater than 0.5 A, cardiac arrest (reversible) may occur.
d. For contact greater than 0.8 second and with currents of several amperes, burns and death are probable.

(5) Results of ac at Frequencies Above 100 Hz. When the threshold of perception increases from 10 kHz to 100 kHz, the threshold of let-go current increases from 10 mA to 100 mA.

(6) Effects of Waveshape. Contact with voltages from phase controls usually causes effects between those of ac and dc sources.

(7) Effects of Capacitive Discharge.
a. A circuit of capacitance of 1 microfarad having a 10 kV capacitor charge may cause ventricular fibrillation.
b. A circuit of capacitance of 20 microfarad having a 10 kV capacitor charge may be dangerous and probably cause ventricular fibrillation.

### 340.6 Hazards Associated with Power Electronic Equipment

Employer and employees shall be aware of the hazards associated with the following:

(1) High voltages within the power supplies
(2) Radio frequency energy–induced high voltages
(3) Effects of radio frequency, RF, fields in the vicinity of antennas and antenna transmission lines, which can introduce electrical shock and burns

(4) Ionizing (X-radiation) hazards from magnetrons, klystrons, thyratrons, cathode-ray tubes, and similar devices
(5) Non-ionizing RF radiation hazards from the following:
   a. Radar equipment
   b. Radio communication equipment, including broadcast transmitters
   c. Satellite earth-transmitters
   d. Industrial scientific and medical equipment
   e. RF induction heaters and dielectric heaters
   f. Industrial microwave heaters and diathermy radiators

**340.7 Specific Measures for Personnel Safety**

**(A) Employer Responsibility** The employer shall be responsible for the following:

(1) Proper training and supervision by properly qualified personnel including the following:
   a. The nature of the associated hazard
   b. Strategies to minimize the hazard
   c. Methods of avoiding or protecting against the hazard
   d. The necessity of reporting any hazardous incident
(2) Properly installed equipment.
(3) Proper access to the equipment.
(4) Availability of the correct tools for operation and maintenance.
(5) Proper identification and guarding of dangerous equipment.
(6) Provision of complete and accurate circuit diagrams and other published information to the employee prior to the employee starting work. The circuit diagrams should be marked to indicate the hazardous components.
(7) Maintenance of clear and clean work areas around the equipment to be worked.
(8) Provision of adequate and proper illumination of the work area.

**(B) Employee Responsibility.** The employee is responsible for the following:

(1) Being continuously alert and aware of the possible hazards
(2) Using the proper tools and procedures for the work
(3) Informing the employer of malfunctioning protective measures, such as faulty or inoperable enclosures and locking schemes
(4) Examining all documents provided by the employer relevant to the work, especially those documents indicating the hazardous components location
(5) Maintaining good housekeeping around the equipment and work space
(6) Reporting any hazardous incident

## REFERENCES CITED IN COMMENTARY

*Authoritative Dictionary of IEEE Standards Terms,* 7th edition, IEEE, Piscataway, NJ, 2000.
NFPA 111, *Standard on Stored Electrical Energy Emergency and Standby Power Systems,* 2002 edition, National Fire Protection Association, Quincy, MA.

CHAPTER 4

# Installation Safety Requirements

The installation requirements of NFPA 70 (the *National Electrical Code*®, *NEC*®, or *Code*) define the first step toward minimizing exposure of the general public or other unqualified persons to fire and shock. The requirements in Chapter 4 are extracted from the 2002 edition of the *NEC* and contain the installation requirements that pertain to issues regarding personnel safety.

Chapter 4 is not intended to be applied as a design, installation, modification, or construction standard for an electrical installation or system. Its content has been intentionally limited in comparison to the content of the *NEC* to apply to an electrical installation or system as part of an employee's workplace. It is compatible with corresponding provisions of the *NEC*, but it cannot be used in lieu of that *Code*.

## ARTICLE 400 General Requirements for Electrical Installations

Article 400 contains the basic requirements necessary for an installation to be considered safe for the public when an electrical system is operating normally. When parts of the system require service and maintenance, the requirements contained in NFPA 70E, Chapter 1, apply to the equipment and to the worker.

### I. General

#### 400.1 Scope

**(A) Introduction.** The requirements contained in Chapter 4 shall be based on the provisions of NFPA 70, *National Electrical Code.* Where installations of electric conductors and equipment have been found to conform with the safety requirements of the *National Electrical Code* in use at the time of installation by governmental bodies or agencies having legal jurisdiction for enforcement of the *National Electrical Code,* this conformance shall be prima facie evidence that such installations were adequately designed and installed.

**(B) Arrangement of the Chapter.** Chapter 4 of this standard is divided into six articles. Article 400, 410, and 420 apply generally. Article 430 applies to specific-purpose equipment installations. Articles 440 and 450 apply to hazardous (classified) locations and special systems. Articles 430, 440, and 450 supplement or modify the general rules, and 450.5 covers communications systems and is independent of the other paragraphs and chapters except where specifically referenced. Articles 400, 410, and 420 apply except as amended by Articles 430, 440, and 450 for the particular condition.

### 400.2 Approval

The conductors and equipment required or permitted by this standard shall be acceptable only if approved.

FPN: See the definitions of *Approved, Identified, Labeled,* and *Listed* in Article 100.

All electrical equipment is required by 400.2 to be approved as defined in *NEC,* Article 100, and, as such, to be acceptable to the authority having jurisdiction (also defined in *NEC,* Article 100). Section 400.3 provides guidance for evaluating equipment and recognizes listing or labeling as a means of establishing equipment suitability.

Approval of equipment is the responsibility of the electrical inspection authority. Many such approvals are based on tests and listings of third-party testing laboratories.

### 400.3 Examination, Identification, Installation, and Use of Equipment

**(A) Examination.** In judging equipment, considerations such as the following shall be evaluated:

For wire-bending and connection space in cabinets and cutout boxes, see 312.6, Table 312.6(A) and Table 312.6(B), and 312.7, 312.9, and 312.11 of the *NEC.* For wire-bending and connection space in other equipment, see the appropriate *NEC* article and section. For example, see 314.16 and 314.28 of the *NEC* for outlet, device, pull, and junction boxes, as well as conduit bodies; *NEC* 404.3 and 404.18 for switches; *NEC* 408.3(F) for switchboards and panelboards; and *NEC* 430.10 for motors and motor controllers.

(1) Suitability for installation and use in conformity with the provisions of this standard

FPN: Suitability of equipment use can be identified by a description marked on or provided with a product to identify the suitability of the product for a specific purpose, environment, or application. Suitability of equipment can be evidenced by listing or labeling.

(2) Mechanical strength and durability, including, for parts designed to enclose and protect other equipment, the adequacy of the protection thus provided
(3) Wire-bending and connection space
(4) Electrical insulation
(5) Heating effects under normal conditions of use and also under abnormal conditions likely to arise in service
(6) Arcing effects
(7) Classification by type, size, voltage, current capacity, and specific use
(8) Other factors that contribute to the practical safeguarding of persons using or likely to come in contact with the equipment

**(B) Installation and Use.** Listed or labeled equipment shall be installed and used in accordance with any instructions included in the listing or labeling.

Manufacturers usually supply installation instructions with equipment, and users should follow those instructions. For example, the second paragraph of *NEC* 210.52 permits permanently installed electric baseboard heaters to be equipped with receptacle outlets that meet the requirements for the wall space utilized by such heaters. The installation instructions for such permanent baseboard heaters indicate that these heaters should not be mounted beneath a receptacle. Therefore, to meet the provisions of both *NEC* 210.52(A) and the installation instructions, a receptacle must be either part of the heating unit or installed in the floor close to the wall but not above the heating unit. (See *NEC* 210.52, FPN, for more specific details.)

*NEC* 400.3 does not require listing or labeling of equipment. It does, however, require considerable evaluation of equipment. Section 400.2 requires that equipment be acceptable only if approved. The term *approved* is defined as "acceptable to the authority having jurisdiction." Before issuing approval, the authority having jurisdiction may require evidence of compliance with 400.3(A). The most common form of this evidence is a listing or labeling by a third party.

Some *NEC* sections require listed or labeled equipment. For instance, *NEC* 250.8 includes the phrase "listed pressure connectors, listed clamps, or other listed means."

### 400.4 Insulation Integrity

Completed wiring installations shall be free from short circuits and from grounds other than as required or permitted in conformity with this standard.

Insulation is the nonconductive material that prevents current flow between points in the system that are at a different voltage. In both high-voltage and low-voltage systems, failure of the insulation system is one of the most common causes of problems in electrical installations. Insulation tests are performed to determine the quality or condition of the insulation of conductors and equipment. The principal causes of insulation failure are heat, moisture, dirt, and physical damage (abrasion or nicks) that occur during and after installation. Insulation also can degrade due to chemical exposure, sunlight (UV) exposure, and excessive voltage stresses.

The electrical insulation must not fail in the event of an overcurrent condition. Overcurrent protective devices must be selected and coordinated using tables of insulation thermal-withstand ability to ensure that the damage point of an insulated conductor is never reached. These tables, entitled Allowable Short-Circuit Currents for Insulated Copper (or Aluminum) Conductors, are contained in the Insulated Cable Engineers Association's publication ICEA P-32–382, *Short-Circuit Characteristics of Insulated Cable*. (See 400.6 for other circuit components.)

In an insulation resistance test, constant voltage, ranging from 100 to 5000, is applied across the insulation. A megohmmeter is usually the voltage source. The instrument indicates directly the insulation resistance on a scale calibrated in megohms. The quality of the insulation is evaluated on the basis of the level of the insulation resistance.

The resistance of many types of insulation varies with temperature. Therefore, field data must be temperature corrected for the class of equipment being tested. The value of insulation resistance in megohms is inversely proportional to the volume of insulation tested. For example, a cable 1000 ft long would be expected to have one-tenth the insulation resistance of a cable 100 ft long, if all other conditions are identical.

The insulation resistance test is relatively easy to perform and is useful on all types and classes of electrical equipment. Its main value lies in the charting of data from periodic tests, corrected for temperature, over a long period, so that trends can be detected.

Manuals on this subject are available from instrument manufacturers. Thorough knowledge in the use of insulation testers is essential if the test results are to be meaningful. Exhibit 400.1 shows a typical megohmmeter insulation tester.

### 400.5 Interrupting Rating

Equipment intended to interrupt current at fault levels shall have an interrupting rating sufficient for the nominal circuit voltage and the current that is available at the line terminals of the equipment. Equipment intended to interrupt current at other than fault levels shall have an interrupting rating at nominal circuit voltage sufficient for the current that must be interrupted.

In the 1999 *NEC,* 110.9 was revised by substituting the word *interrupt* for the word *break* in two places.

The interrupting rating of overcurrent protective devices is determined under standard test conditions that must match the actual installation needs. Section 110.9 states that all fuses and circuit breakers intended to interrupt the circuit at fault levels must have adequate interrupting

***EXHIBIT 400.1** A manual multivoltage, multirange insulation tester.*

ratings wherever they are used in the electrical system. Fuses or circuit breakers that do not have adequate interrupting ratings could rupture while attempting to clear a short circuit.

Interrupting ratings should not be confused with short-circuit current ratings.

### 400.6 Circuit Impedance and Other Characteristics

The overcurrent protective devices, the total impedance, the component short-circuit current ratings, and other characteristics of the circuit to be protected shall be selected and coordinated to permit the circuit-protective devices used to clear a fault to do so without extensive damage to the electrical components of the circuit. This fault shall be assumed to be either between two or more of the circuit conductors or between any circuit conductor and the grounding conductor or enclosing metal raceway. Listed products applied in accordance with their listing shall be considered to meet the requirements of this section.

In the 1999 *NEC,* 110.10 was revised by substituting the word *current* for *withstand.* That change helped to correlate the *NEC* language with the standard marking language used on equipment. Withstand ratings are *not* marked on equipment; short-circuit current ratings *are* marked on equipment. This marking appears on many pieces of equipment, such as panelboards, switchboards, busways, contactors, and starters. A new final sentence was added to 110.10 of the 1999 *NEC* to address concerns about what exactly constitutes "extensive damage." Because electrical equipment is evaluated for indications of extensive damage, listed products used within their ratings are considered to have met the requirements of this section.

The basic purpose of overcurrent protection is to open the circuit before conductors or conductor insulation is damaged. An overcurrent condition can be the result of an overload, a ground fault, or a short circuit. The overcurrent condition must be removed before the conductor insulation damage point is reached. Overcurrent protective devices (such as fuses and circuit breakers) should be selected to ensure that the short-circuit current ratings of system components are not exceeded.

System components include wire, bus structures, switching devices, protection and disconnecting devices, and distribution equipment. These system components have limited short-circuit ratings and would be damaged or destroyed if short-circuit ratings were exceeded. Overcurrent devices with sufficient interrupting rating might not ensure adequate short-circuit protection for the system components. When the available short-circuit current exceeds the short-circuit (fault) current rating of a system component, the overcurrent protective device must limit the let-through energy to the rating of the system component.

Utility companies determine and usually provide information about available short-circuit (fault) current levels at the service equipment. Instructions on how to determine available short-circuit current at each system component can be obtained by contacting the manufacturers of overcurrent protective devices or by referring to IEEE 141–1993, *IEEE Recommended Practice for Electric Power Distribution for Industrial Plants* ("Red Book"). Computer software to perform a system analysis is widely available in the marketplace.

For example: A typical single-family dwelling with a 100-ampere service uses 2 AWG aluminum conductors supplied by a 37½ kVA transformer with 1.72 percent impedance located at a distance of 25 ft. In this instance, the available short-circuit current is approximately 6000 amperes.

Available fault (short circuit) current to a multifamily structure, where a pad-mounted transformer is located close to the multimetering location, can be relatively high. For example, the line-to-line fault-current values close to a low-impedance transformer could exceed 22,000 amperes. At the secondary of a single-phase, center-tapped transformer, the line-to-neutral fault current is approximately 1½ times that of the line-to-line fault current. The fault-current rating of utilization equipment located and connected near the service equipment should be known. For example, HVAC equipment is tested at 3500 amperes through a 40-ampere load rating and at 5000 amperes for loads rated more than 40 amperes.

Adequate short-circuit protection can be provided by fuses, molded-case circuit breakers, and low-voltage power circuit breakers, depending on specific circuit and installation requirements.

### 400.7 Deteriorating Agents

**(A) Location.** Unless identified for use in the operating environment, no conductors or equipment shall be located in damp or wet locations; where exposed to gases, fumes, vapors, liquids, or other agents that have a deteriorating effect on the conductors or equipment; or where exposed to excessive temperatures.

FPN No. 1: In general, areas where acids and alkali chemicals are handled and stored could present such corrosive conditions, particularly when wet or damp. Severe corrosive conditions could also be present in portions of meat-packing plants, tanneries, glue houses, and some stables; in installations immediately adjacent to a seashore and swimming pool areas; in areas where chemical deicers are used; and in storage cellars or rooms for hides, casings, fertilizer, salt, and bulk chemicals.

FPN No. 2: Some cleaning and lubricating compounds can cause severe deterioration of many plastic materials used for insulating and structural applications in equipment.

**(B) Protection of Type 1 Equipment.** Equipment identified only as "dry locations," "Type 1," or "indoor use only" shall be protected against permanent damage from the weather during building construction.

### 400.8 Mechanical Execution of Work

Electric equipment shall be installed in a neat and workmanlike manner.

The requirement for "neat and workmanlike" installations has appeared in the *NEC* as currently worded for more than half a century. The phrase stands as a basis for pride in one's work and has been emphasized by persons involved in the training of apprentice electricians for many years. Many *Code* conflicts or violations have been cited by the authority having jurisdiction on the basis of the authority's interpretation of "neat and workmanlike" manner. Many electrical inspection authorities use their own experience or precedents in their local areas as the basis for their judgments.

Examples of installations that do not qualify as "neat and workmanlike" include exposed runs of cables or raceways that are improperly supported (e.g., sagging between supports or using improper support methods); field-bent and kinked, flattened, or poorly measured raceways; or cabinets, cutout boxes, and enclosures that are not plumb or not properly secured.

**(A) Unused Openings.** Unused cable or raceway openings in boxes, raceways, auxiliary gutters, cabinets, cutout boxes, meter socket enclosures, equipment cases, or housings shall be effectively closed to afford protection substantially equivalent to the wall of the equipment. Where metallic plugs or plates are used with nonmetallic enclosures, they shall be recessed at least 6 mm (¼ in.) from the outer surface of the enclosure.

In the 2002 *NEC,* 110.12(A) has been revised to clarify that openings such as weep holes are not required to be closed up.

**(B) Subsurface Enclosures.** Conductors shall be racked to provide ready and safe access in underground and subsurface enclosures into which persons enter for installation and maintenance.

**(C) Integrity of Electric Equipment and Connections.** Internal parts of electric equipment, including busbars, wiring terminals, insulators, and other surfaces, shall not be damaged or contaminated by foreign materials such as paint, plaster, cleaners, abrasives, or corrosive residues. There shall be no damaged parts, such as parts that are broken; bent; cut; or deteriorated by corrosion, chemical action, or overheating, that could adversely affect safe operation or mechanical strength of the equipment.

### 400.9 Mounting and Cooling of Equipment

**(A) Mounting.** Electric equipment shall be firmly secured to the surface on which it is mounted. Wooden plugs driven into holes in masonry, concrete, plaster, or similar materials shall not be used.

**(B) Cooling.** Electric equipment that depends on the natural circulation of air and convection principles for cooling of exposed surfaces shall be installed so that room airflow over such surfaces is not prevented by walls or by adjacent installed equipment. For equipment designed for floor mounting, clearance between top surfaces and adjacent surfaces shall be provided to dissipate rising warm air. Electric equipment provided with ventilating openings shall be installed so that walls or other obstructions do not prevent the free circulation of air through the equipment.

Ventilation for motor locations is covered in 430.14(A) and 430.16 of the *NEC.* Ventilation for transformer locations is covered in *NEC* 450.9 and 450.45.

For example: A ventilated busway must be located where no walls or other objects exist that might interfere with the natural circulation of air for cooling. (See the definition of *ventilated* in Article 100.)

Panelboards, transformers, and other types of equipment might be adversely affected if surfaces normally exposed to room air restrict normal convection cooling. Ventilating openings in equipment are provided to allow the circulation of room air around internal components. Circulation of air by convection must not be restricted.

### 400.10 Electrical Connections

Because of different characteristics of dissimilar metals, devices such as pressure terminals or pressure splicing connectors and soldering lugs shall be identified for the material of the conductor and shall be properly installed and used. Conductors of dissimilar metals shall not be intermixed in a terminal or splicing connector where physical contact occurs between dissimilar conductors (such as copper and aluminum, copper and copper-clad aluminum, or aluminum and copper-clad aluminum), unless the device is identified for the purpose and conditions of use. Materials such as solder, fluxes, inhibitors, and compounds, where employed, shall be suitable for the use and shall be of a type that will not adversely affect the conductors, installation, or equipment.

FPN: Many terminations and equipment are marked with a tightening torque.

Section 400.3(B) applies where terminations and equipment are marked with tightening torques. Commentary Tables 400.1 through 400.4 provide torque data used by Underwriters Laboratories® (UL®) for wire connectors that have torque value recommended by the manufacturer. These tables should be used only for guidance if no tightening information on the specific wire connector is available. The data in these tables must not be used in lieu of the manufacturer's instructions.

Information in Commentary Tables 400.1 through 400.4 was extracted from UL Standard 486B, *Wire Connections for Use with Aluminum Conductors,* in effect at the time of the print-

***COMMENTARY TABLE 400.1*** *Tightening Torques for Screws* in Pound-Inches*

| | *Slotted Head No. 10 and Larger* | | *Hexagonal Head-External Drive Socket Wrench* | |
|---|---|---|---|---|
| *Wire Size (AWG or kcmil)* | *Slot Width to 3/64 in. or Slot Length to 1/4 in.†* | *Slot Width over 3/64 in. or Slot Length over 1/4 in.†* | *Split-Bolt Connectors* | *Other Connectors* |
| 30–10 | 20 | 35 | 80 | 75 |
| 8 | 25 | 40 | 80 | 75 |
| 6 | 35 | 45 | 165 | 110 |
| 4 | 35 | 45 | 165 | 110 |
| 3 | 35 | 50 | 275 | 150 |
| 2 | 40 | 50 | 275 | 150 |
| 1 | — | 50 | 275 | 150 |
| 1/00 | — | 50 | 385 | 180 |
| 2/00 | — | 50 | 385 | 180 |
| 3/00 | — | 50 | 500 | 250 |
| 4/00 | — | 50 | 500 | 250 |
| 250 | — | 50 | 500 | 325 |
| 300 | — | 50 | 650 | 325 |
| 350 | — | 50 | 650 | 325 |
| 400 | — | 50 | 825 | 325 |
| 500 | — | 50 | 825 | 375 |
| 600 | — | 50 | 1000 | 375 |
| 700 | — | 50 | 1000 | 375 |
| 750 | — | 50 | 1000 | 375 |
| 800 | — | 50 | 1100 | 500 |
| 900 | — | 50 | 1100 | 500 |
| 1000 | — | 50 | 1100 | 500 |
| 1250 | — | — | 1100 | 600 |
| 1500 | — | — | 1100 | 600 |
| 1750 | — | — | 1100 | 600 |
| 2000 | — | — | 1100 | 600 |

*Clamping screws with multiple tightening means. For example, for a slotted hexagonal head screw, use the torque value associated with the tool used in the installation. UL uses both values when testing.

†For values of slot width or length other than those specified, select the largest torque value associated with conductor size.

***COMMENTARY TABLE 400.2*** *Torques in Pound-Inches for Slotted Head Screws* Smaller Than No. 10, for Use with 8 AWG and Smaller Conductors*

| Screw-Slot Length (in.)† | Screw-Slot Width Less Than 3⁄64 in. | Screw-Slot Width 3⁄64 in. and Larger |
|---|---|---|
| To 5⁄32 | 7 | 9 |
| 5⁄32 | 7 | 12 |
| 3⁄16 | 7 | 12 |
| 7⁄32 | 7 | 12 |
| 1⁄4 | 9 | 12 |
| 9⁄32 | — | 15 |
| Above 9⁄32 | — | 20 |

*Clamping screws with multiple tightening means. For example, for a slotted hexagonal head screw, use the torque value associated with the tool used in the installation. UL uses both values when testing.

†For slot lengths of intermediate values, select torques pertaining to next-shorter slot length.

***COMMENTARY TABLE 400.3*** *Torques for Recessed Allen Head Screws*

| Socket Size Across Flats (in.) | Torque (lb-in.) |
|---|---|
| 1⁄8 | 45 |
| 5⁄32 | 100 |
| 3⁄16 | 120 |
| 7⁄32 | 150 |
| 1⁄4 | 200 |
| 5⁄16 | 275 |
| 3⁄8 | 375 |
| 1⁄2 | 500 |
| 9⁄16 | 600 |

***COMMENTARY TABLE 400.4*** *Lug-Bolting Torques for Connection of Wire Connectors to Busbars*

| Bolt Diameter | Tightening Torque (lb-ft) |
|---|---|
| No. 8 or smaller | 1.5 |
| No. 10 | 2 |
| 1⁄4 in. or less | 6 |
| 5⁄16 in. | 11 |
| 3⁄8 in. | 19 |
| 7⁄16 in. | 30.0 |
| 1⁄2 in. | 40 |
| 9⁄16 in. or larger | 55 |

ing of the 2002 edition of the *NEC Handbook.* Similar information can be found in UL 486A, *Wire Connections and Solder Lugs for Use with Copper Conductors.*

**(A) Terminals.** Connection of conductors to terminal parts shall ensure a thoroughly good connection without damaging the conductors and shall be made by means of pressure connectors (including set-screw type), solder lugs, or splices to flexible leads. Connection by means of wire-bending screws or studs and nuts having upturned lugs or equivalent shall be permitted for 10 AWG or smaller conductors. Terminals for more than one conductor and terminals used to connect aluminum shall be so identified.

**(B) Splices.** Conductors shall be spliced or joined with splicing devices identified for the use or by brazing, welding, or soldering with a fusible metal or alloy. Soldered splices shall first be spliced or joined so as to be mechanically and electrically secure without solder and then be soldered. All splices and joints and the free ends of conductors shall be covered with an insulation equivalent to that of the conductors or with an insulating device identified for the purpose. Wire connectors or splicing means installed on conductors for direct burial shall be listed for such use.

Reports of fire investigations suggest that failures of electrical connections are the cause of fires. Other anecdotal information suggests that connection failures result in equipment damage. Connection failures are normally the result of improper terminations, poor workmanship, the differing characteristics of dissimilar metals, and improper binding screws or splicing devices.

UL's listing requirements for solid aluminum conductors in 12 AWG and 10 AWG gauges and for snap switches and receptacles for use on 15- and 20-ampere branch circuits incorporate stringent tests that consider normal connection failures. For further information regarding receptacles and switches using CO/ALR-rated terminals, refer to *NEC* 404.14(C) and *NEC* 406.2(C).

Screwless pressure terminal connectors of the conductor push-in type are for use with solid copper and copper-clad aluminum conductors only. Each coil of No. 12 AWG aluminum and No. 10 AWG aluminum wire must contain instructions that describe acceptable installation techniques. The instructions emphasize good workmanship. (See also the commentary on tightening torque that follows 400.10, FPN.)

Manufacturers have developed product and material designs that improve aluminum wire terminations. The following information is provided to help the user understand how to use aluminum conductors accurately. This commentary is based on a report prepared by the Ad Hoc Committee on Aluminum Terminations prior to the 1975 *Code.* This information is still pertinent today and is necessary to comply with 400.10(A) when aluminum wire is used in installations.

**Marking.** For direct connection, only 15- and 20-ampere receptacles and switches marked "CO/ALR" should be used, and they must be connected as indicated by the manufacturer. "CO/ALR" is marked on the device mounting yoke or strap. That marking means the devices have been tested according to stringent heat-cycling requirements to determine their suitability for use with UL-labeled aluminum, copper, or copper-clad aluminum wire.

If using aluminum wire, only listed solid aluminum wire, 12 AWG or 10 AWG, should be used, marked with the label of the aluminum insulated wire. Installation must be completed according to the installation instructions that are packaged with the wire.

**Installation Method.** Exhibit 400.2 illustrates the correct method of connection to be used, as follows:

1. The freshly stripped end of the wire should be wrapped two-thirds to three-quarters of the distance around the wire-binding screw post, as shown in Step A of Exhibit 400.2. The loop is made so that rotation of the screw during tightening will tend to wrap the wire around the post rather than unwrap it.
2. The screw should be tightened until the wire is snugly in contact with the underside of the screw head and with the contact plate on the wiring device, as shown in Step B of Exhibit 400.2.

*EXHIBIT 400.2 Correct method of terminating aluminum wire at wire-binding screw terminals of receptacles and snap switches. (Redrawn courtesy of Underwriters Laboratories Inc.)*

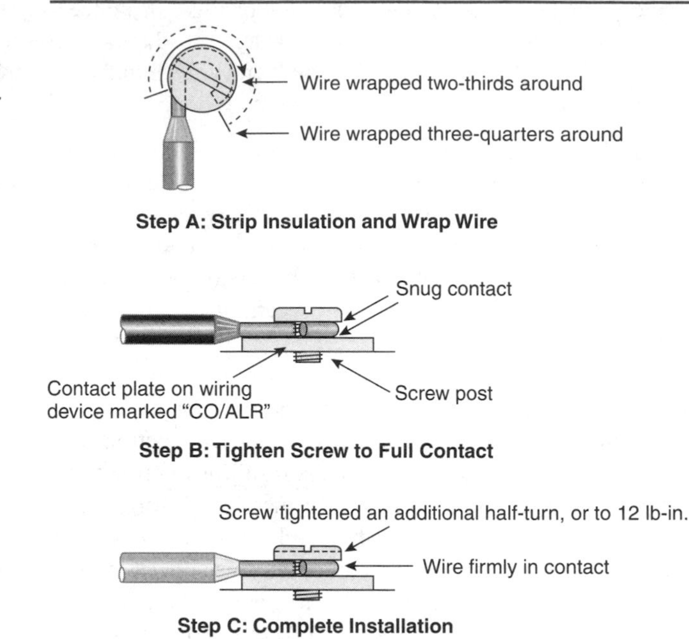


3. The screw should be tightened an additional half-turn, thereby providing a firm connection, as shown in Step C of Exhibit 400.2. Where torque screwdrivers are used, the screw should be tightened to 12 in.-lb, as shown in Step C of Exhibit 400.2.
4. The wires should be positioned behind the wiring device to decrease the likelihood of the terminal screws loosening when the device is positioned into the outlet box.

Exhibit 400.3 illustrates incorrect methods for connection. These methods should not be used.

*EXHIBIT 400.3 Incorrect methods of terminating aluminum wire at wire-binding screw terminals of receptacles and snap switches. (Redrawn courtesy of Underwriters Laboratories Inc.)*

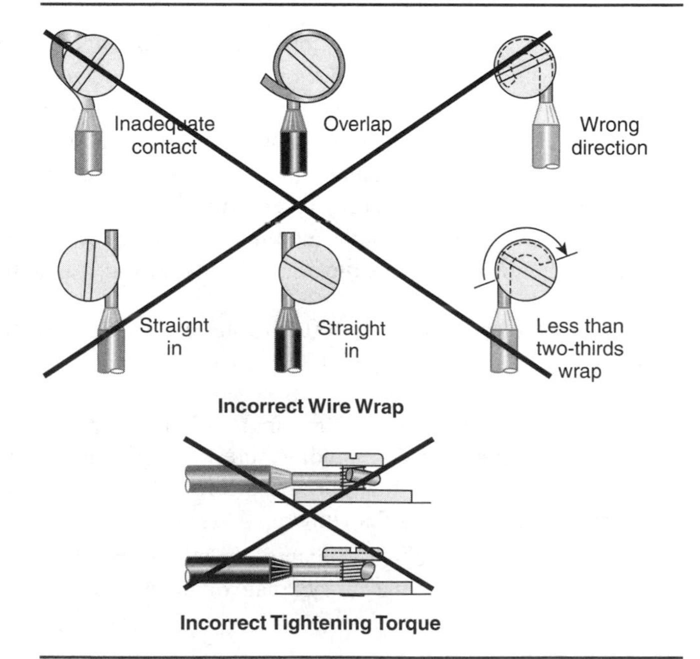

**Existing Inventory Only.** Labeled 12 AWG or 10 AWG solid aluminum wire that does not bear the UL label should be used with wiring devices marked "CO/ALR" and connected as described in the preceding Installation Method. This is the preferred and recommended method for using such wire.

In the following types of devices, the terminals should not be directly connected to aluminum conductors but may be used with labeled copper or copper-clad conductors:

1. Receptacles and snap switches marked "AL-CU"
2. Receptacles and snap switches having no conductor marking
3. Receptacles and snap switches that have back-wired terminals or screwless terminals of the push-in type

**For Existing Installations.** If examination discloses overheating or loose connections, the recommendations described under Existing Inventory Only should be followed.

**Twist-On Wire Connectors.** Because 400.10(B) requires conductors to be spliced with "splicing devices identified for the use," wire connectors are required to be marked for conductor suitability. Twist-on wire connectors are not suitable for splicing aluminum conductors or copper-clad aluminum to copper conductors unless it is so stated and marked as such on the shipping carton. The marking is typically "AL-CU (dry locations)." Currently, one style of wire nut and one style of crimp-type connector are listed as having met these requirements.

On February 2, 1995, Underwriters Laboratories announced the listing of a twist-on wire connector suitable for use with aluminum-to-copper conductors, in accordance with UL 486C, *Standard for Splicing Wire Connectors.* That was the first listing of a twist-on type connector for aluminum-to-copper conductors since 1987. The UL listing does not cover aluminum-to-aluminum combinations. However, more than one aluminum or copper conductor is allowed when used in combination.

These listed wire-connecting devices are available for pigtailing short lengths of copper conductors to the original aluminum branch-circuit conductors, as shown in Exhibit 400.4. Primarily, these pigtailed conductors supply 15- and 20-ampere wiring devices. Pigtailing is permitted, provided suitable space is available within the enclosure.

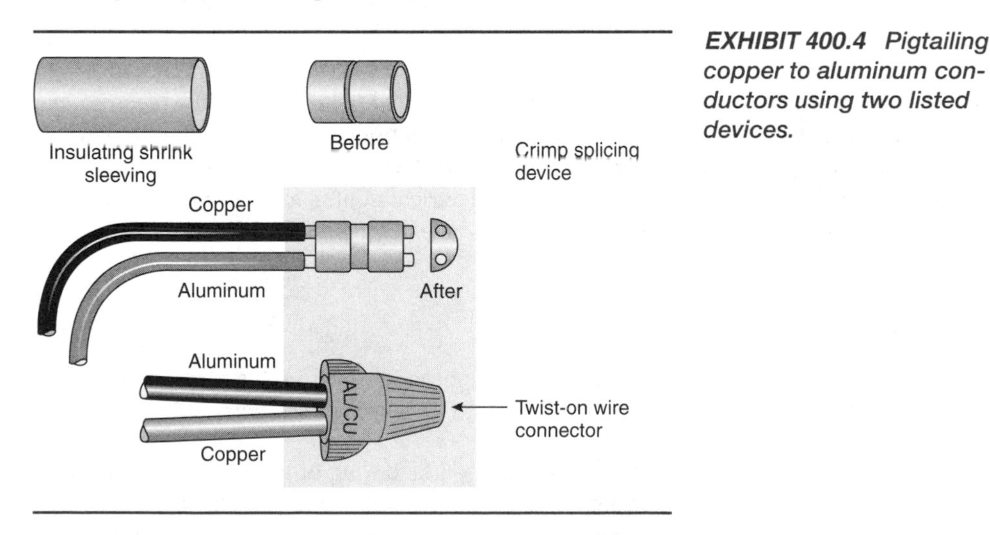


***EXHIBIT 400.4** Pigtailing copper to aluminum conductors using two listed devices.*

### 400.11 Flash Protection

Switchboards, panelboards, industrial control panels, and motor control centers that are in other than dwelling occupancies and are likely to require examination, adjustment, servicing, or maintenance while energized shall be field marked to warn qualified persons of potential electric arc flash hazards. The marking shall be located so as to be clearly visible to qualified persons before examination, adjustment, servicing, or maintenance of the equipment.

Most injuries to electrical workers result from exposure to an electrical arc. The temperature of an electrical arc frequently reaches 15,000°F to 20,000°F, even in electrical circuits that are protected from short circuit in accordance with other articles in Chapter 4. The field-installed label warns workers that removing a cover or opening a door exposes them to a risk of thermal injury. The field-installed label provides workers with a warning that, should an arcing fault be initiated, they are likely to receive a significant burn injury unless adequately protected from the thermal hazard. Personal protective equipment (PPE) that mitigates exposure to the thermal hazard is available. Chapter 1 of this standard describes how to select PPE that will minimize potential for thermal injury.

Exhibit 400.5 depicts an electrical employee working inside the Flash Protection Boundary and in front of a large-capacity service-type switchboard where an electrically safe work condition has not been established (see Chapter 1 for safety-related work practices and the applicable commentary). The worker is wearing PPE considered to be appropriate flash protection clothing for the flash hazard involved with this specific exposure.

***EXHIBIT 400.5*** *Electrical worker wearing flame-resistant PPE.*

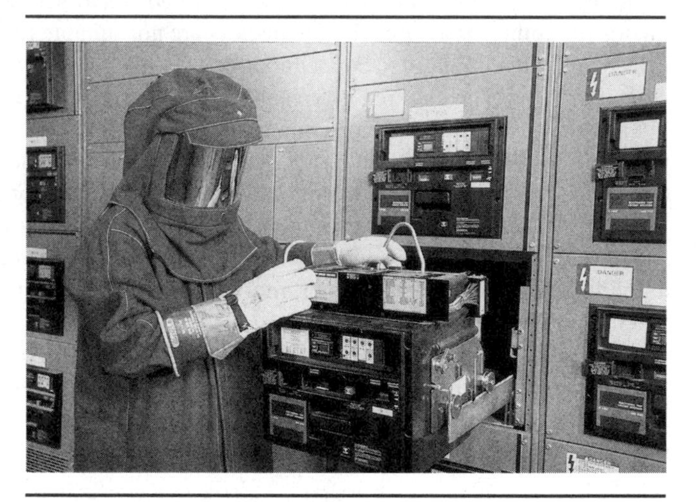

Electrocutions and burn injuries recorded by industry and reported as public records continue to confirm that workers responsible for the installation or maintenance of electrical equipment often accept the increased risk associated with working on or near live parts. The underlying purpose of this requirement is to alert electrical contractors, electricians, facility owners and managers, and other interested parties to some of the hazards of working on or near live parts.

### 400.12 Arcing Parts

Parts of electric equipment that in ordinary operation produce arcs, sparks, flames, or molten metal shall be enclosed or separated and isolated from all combustible material.

Examples of electrical equipment that can produce sparks during ordinary operation include open motors having a centrifugal starting switch, open motors with commutators, and collector rings. Adequate separation from combustible material is essential if open motors with these features are used.

### 400.13 Marking

The manufacturer's name, trademark, or other descriptive marking by which the organization responsible for the product can be identified shall be placed on all electric equipment. Other markings that indicate voltage, current, wattage, or other ratings shall be provided as specified elsewhere in this standard. The marking shall be of sufficient durability to withstand the environment involved.

Section 400.13 requires that equipment be marked with identifying information. The markings must be visible or easily accessible during or after installation.

### 400.14 Identification of Disconnecting Means

**(A) General.** Each disconnecting means shall be legibly marked to indicate its purpose unless located and arranged so the purpose is evident. The marking shall be of sufficient durability to withstand the environment involved.

Identification must be legible and specific. For example, the label should indicate not merely "motor" but rather "motor, water pump" and not merely "lights" but rather "lights, front lobby." The label should be maintained in a readable condition. Equipment can be assigned unique names or numbers. Including the unique name or number on the record drawings facilitates locating the disconnecting means.

**(B) Series Combination Ratings.** Where circuit breakers or fuses are applied in compliance with the series combination ratings marked on the equipment by the manufacturer, the equipment enclosure(s) shall be legibly marked in the field to indicate the equipment has been applied with a series combination rating. The additional series combination interrupting rating shall be marked on the end use equipment, such as switchboards and panelboards. The marking shall be readily visible and state the following:

CAUTION
SERIES COMBINATION SYSTEM RATED __ AMPERES.
IDENTIFIED REPLACEMENT COMPONENTS REQUIRED

Series-rated overcurrent devices should be legibly marked as indicated. The equipment manufacturer may rate and label equipment to be used at a series combination rating. If the equipment is applied at its marked series combination rating, an additional label indicating that the series combination rating applies must be installed on the equipment enclosure that reads as indicated.

## II. 600 Volts, Nominal, or Less

### 400.15 Spaces About Electric Equipment

Sufficient access and working space shall be provided and maintained about all electric equipment to permit ready and safe operation and maintenance of such equipment. Enclosures that house electric apparatus and are controlled by lock and key shall be considered accessible to qualified persons.

Key to understanding 400.15 is to divide the space around electrical equipment in two separate and distinct categories: working space and dedicated equipment space. Working space is intended to provide the worker with sufficient space to perform his or her duties. Dedicated space is intended to apply to the space reserved for access to the electrical equipment. Equipment that is not associated with the electrical equipment must not impinge into the dedicated space. Material storage that impinges on either working space or dedicated space must be avoided.

**(A) Working Space.** Working space for equipment operating at 600 volts, nominal, or less to ground and likely to require examination, adjustment, servicing, or maintenance while energized shall comply with the dimensions of 400.15(A)(1), 400.15(A)(2), and 400.15(A)(3) or as required or permitted elsewhere in this standard.

The intent of 400.15(A) is to provide enough space for personnel to perform any necessary operations, such as examination, adjustment, servicing, and maintenance of equipment.

Examples of such equipment likely to require examination, adjustment, servicing, or maintenance while energized include panelboards, switches, circuit breakers, controllers, and controls on heating and air-conditioning equipment. The word *examination* can also include such tasks as checking for the presence of voltage.

Minimum working clearances are not required if the equipment is not likely to require examination, adjustment, servicing, or maintenance while energized. However, "sufficient" access and working space still are required.

**(1) Depth of Working Space.** The depth of the working space in the direction of live parts shall be not less than that indicated in Table 400.15(A)(1) unless the requirements of 400.15(A)(1)(a), 400.15(A)(1)(b), or 400.15(A)(1)(c) are met. Distances shall be measured from the exposed live parts if such are exposed or from the enclosure or opening if the live parts are enclosed.

***TABLE 400.15(A)(1)*** *Working Spaces*

| Nominal Voltage to Ground | Minimum Clear Distance | | | | | |
|---|---|---|---|---|---|---|
| | Condition 1 | | Condition 2 | | Condition 3 | |
| 0–150 | 900 mm | (3 ft) | 900 mm | (3 ft) | 900 mm | (3 ft) |
| 151–600 | 900 mm | (3 ft) | 1 m | (3½ ft) | 1.2 m | (4 ft) |

Note: Where the conditions are as follows:

*Condition 1*—Exposed live parts on one side and no live or grounded parts on the other side of the working space, or exposed live parts on both sides effectively guarded by suitable wood or other insulating materials. Insulated wire or insulated busbars operating at not over 300 volts to ground shall not be considered live parts.

*Condition 2*—Exposed live parts on one side and grounded parts on the other side. Concrete, brick, or tile walls shall be considered as grounded surfaces.

*Condition 3*—Exposed live parts on both sides of the work space (not guarded as provided in Condition 1) with the operator between.

Included in these clearance requirements is the step-back distance from the face of the equipment. Table 400.15(A)(1) provides requirements for clearances away from the equipment, based on the circuit voltage to ground and whether grounded or ungrounded objects are present in the step-back space or exposed live parts across from each other. The voltages to ground consist of two groups: 0 to 150 and 151 to 600, inclusive. In an ungrounded system, the voltage to ground is the greatest voltage between the given conductor and any other conductor of the circuit. For example, the voltage to ground for a 480-volt ungrounded delta system is 480 volts.

See Exhibit 400.6 for general working clearance requirements for each of the three conditions expressed in Table 400.15(A)(1). If any assemblies, such as switchboards or motor-control centers, are accessible from the back and expose live parts, the working clearance dimensions would be required at the rear of the equipment, as illustrated in Exhibit 400.6. Note that for Condition 3, where there is an enclosure on opposite sides of the working space, the clearance for only one working space is required.

(a) Dead-Front Assemblies. Working space shall not be required in the back or sides of assemblies, such as dead-front switchboards or motor control centers, where all connections and all renewable or adjustable parts, such as fuses or switches, are accessible from locations other than the back or sides. Where rear access is required to work on nonelectrical parts on the back of enclosed equipment, a minimum horizontal working space of 762 mm (30 in.) shall be provided.

The intent of 400.15(A)(1)(a) is to point out that working space is required only from the side(s) of the enclosure that requires access. The general rule still applies: Equipment that re-

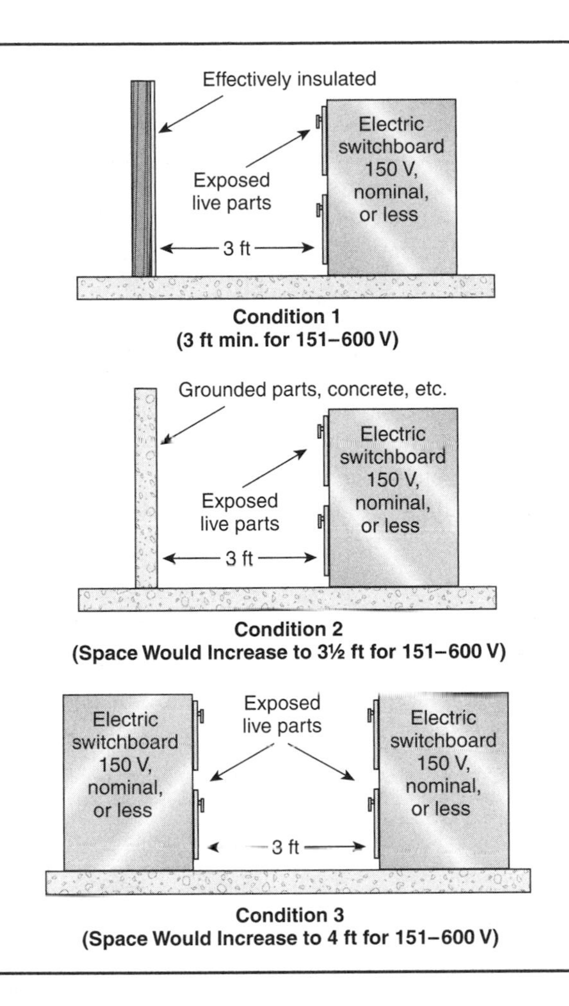


***EXHIBIT 400.6*** *Distances measured from the live parts if the live parts are exposed or from the enclosure front if the live parts are enclosed.*

quires front, rear, or side access for electrical activities described in 400.15(A) must meet the requirements of Table 400.15(A)(1). In many cases, equipment of "dead-front" assemblies requires front access only. Where equipment requires rear access for nonelectrical activity, a reduced working space of at least 30 in. must be provided. Exhibit 400.7 illustrates a reduced working space of 30 in. at the rear of equipment to allow work on nonelectrical parts.

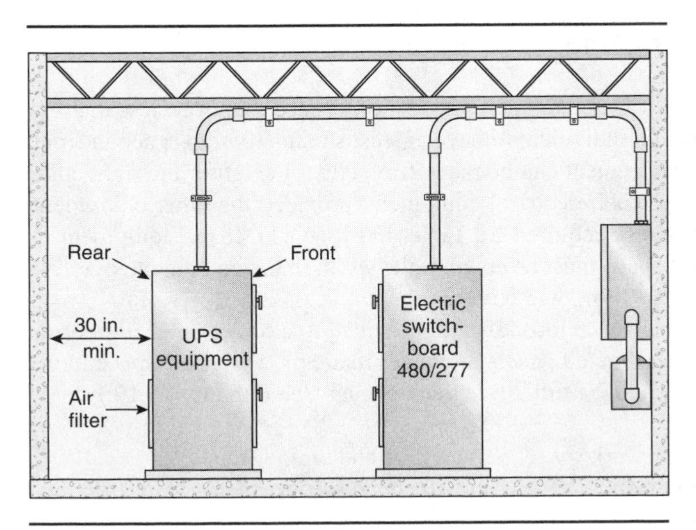


***EXHIBIT 400.7*** *Example of the 30-in. minimum working space at the rear of equipment to allow work on nonelectrical parts, such as the replacement of an air filter.*

(b) Low Voltage. By special permission, smaller working spaces shall be permitted where all uninsulated parts operate at not greater than 30 volts rms, 42 volts peak, or 60 volts dc.

(c) Existing Buildings. In existing buildings where electric equipment is being replaced, Condition 2 working clearance shall be permitted between dead-front switchboards, panelboards, or motor control centers located across the aisle from each other where conditions of maintenance and supervision ensure that written procedures have been adopted to prohibit equipment on both sides of the aisle from being open at the same time. Qualified persons who are authorized will service the installation.

Section 400.15(A)(1)(c) permits some relief for installations being upgraded. When dead-front switchboards, panelboards, or motor-control centers are replaced in an existing building, the working clearance allowed is that required by Table 400.15(A)(1), Condition 2. The reduction from a Condition 3 to a Condition 2 clearance is allowed only where a written procedure prohibits facing doors of equipment from being open at the same time and where only authorized and qualified persons service the installation. Exhibit 400.8 illustrates this relief for existing buildings.

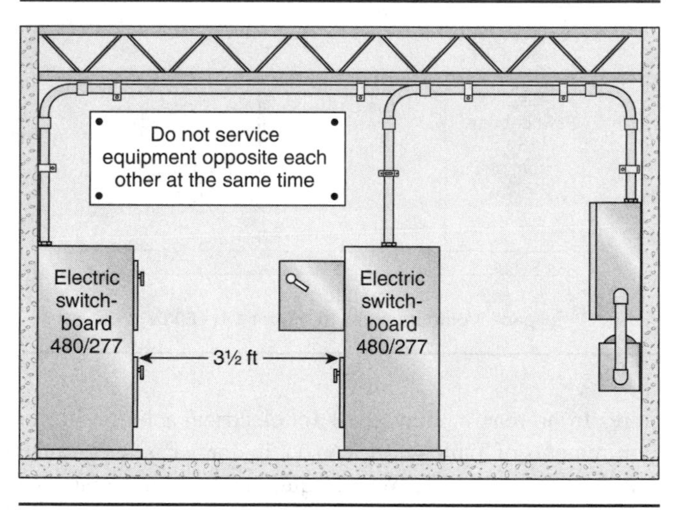


***EXHIBIT 400.8* Permitted reduction from a Condition 3 to a Condition 2 clearance according to 400.15(A)(1)(c).**

**(2) Width of Working Space.** The width of the working space in front of the electric equipment shall be the width of the equipment or 750 mm (30 in.), whichever is greater. In all cases, the work space shall permit at least a 90 degree opening of equipment doors or hinged panels.

Regardless of the width of the electrical equipment, the working space cannot be less than 30 in. wide. This requirement allows an individual to have at least shoulder-width space in front of the equipment. This 30-in. measurement can be made from either the left or the right edge of the equipment and can overlap other electrical equipment, provided the other equipment does not extend beyond the clearance required by Table 400.15(A)(1). If the equipment is wider than 30 in., the left-to-right space must be equal to the width of the equipment. See Exhibit 400.9 for an explanation of the 30-in. width requirement.

Sufficient depth in the working space must also be provided to allow a panel or door to open at least 90 degrees. If doors or hinged panels are wider than 3 ft, a working space more than 3 ft deep must be provided to allow a full 90-degree opening (see Exhibit 400.10.)

**(3) Height of Working Space.** The work space shall be clear and extend from the grade, floor, or platform to the height required by 400.15(E). Within the height requirements of this section,

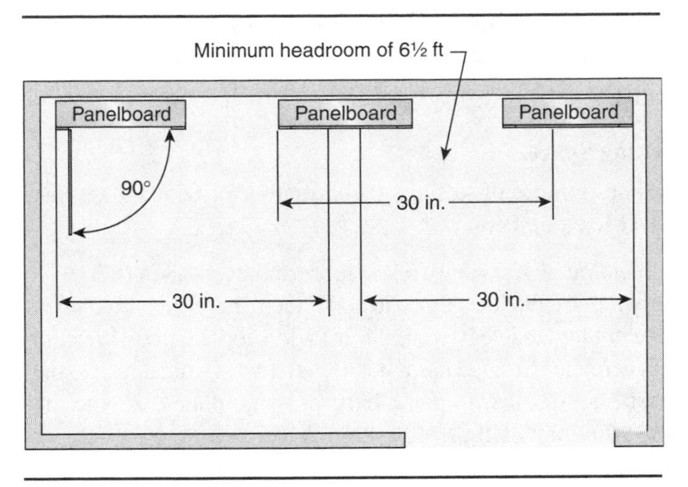


***EXHIBIT 400.9*** *The 30-in. wide front working space not required to be directly centered on the electrical equipment if space is sufficient for safe operation and maintenance of such equipment.*

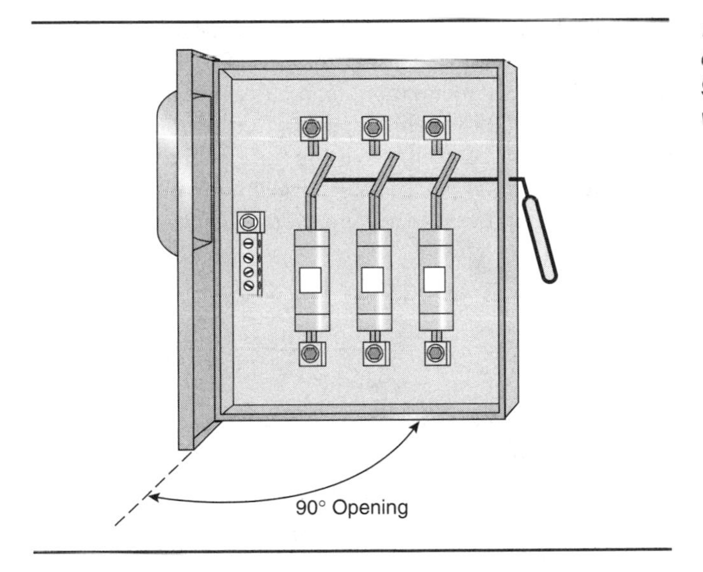


***EXHIBIT 400.10*** *Equipment doors required to open a full 90 degrees to ensure a safe working space.*

other equipment that is associated with the electrical installation and is located above or below the electrical equipment shall be permitted to extend not more than 150 mm (6 in.) beyond the front of the electrical equipment.

In addition to requiring a working space to be clear from the floor up to a height of 6½ ft or to the height of the equipment, whichever is greater, 400.15(A)(3) permits electrical equipment located above or below other electrical equipment to extend into the "working space" not more than 6 in. This requirement allows the placement of a 12-in. × 12-in. wireway on the wall directly above or below a 6-in.-deep panelboard without impinging into the working space or compromising practical working clearances. The requirement continues to prohibit large differences in depth of equipment below or above other equipment that specifically requires working space. Electrical equipment that produces heat or that otherwise requires ventilation also must comply with 400.3(B) and 400.9.

**(B) Clear Spaces.** Working space required by this standard shall not be used for storage. When normally enclosed live parts operating at 50 volts or more are exposed for inspection or servicing, the working space, if in a passageway or general open space, shall be suitably guarded.

Panelboards may be placed in corridors or passageways. If it is necessary to remove a cover of such equipment, measures must be taken to protect any unqualified person who might approach exposed live parts.

**(C) Access and Entrance to Working Space.**

**(1) Minimum Required.** At least one entrance of sufficient area shall be provided to give access to the working space about electric equipment.

**(2) Large Equipment.** For equipment rated 1200 amperes or more and over 1.8 m (6 ft) wide that contains overcurrent devices, switching devices, or control devices, there shall be one entrance to the required working space not less than 610 mm (24 in.) wide and 2.0 m (6½ ft) high at each end of the working space. Where the entrance has a personnel door(s), the door(s) shall open in the direction of egress and be equipped with panic bars, pressure plates, or other devices that are normally latched but open under simple pressures. A single entrance to the required working space shall be permitted where either of the conditions in 400.15(C)(2)(a) or 400.15(C)(2)(b) is met.

Paragraph 400.15(C)(2) requires that panic hardware be installed on doors that are likely to be used for egress from electrical rooms that house large electrical equipment. Large equipment includes switchboards, panelboards, and the like, that are more than 6 ft wide and rated 1200 amperes or more. The panic hardware is intended to enable a person to exit from a room in which exists an unexpected condition such as an arcing fault, a fire, or a similar emergency condition. Panic hardware enables an injured worker to exit an electrical room quickly.

For a graphical explanation of access and entrance requirements to a working space, see Exhibits 400.11, 400.12, and 400.13.

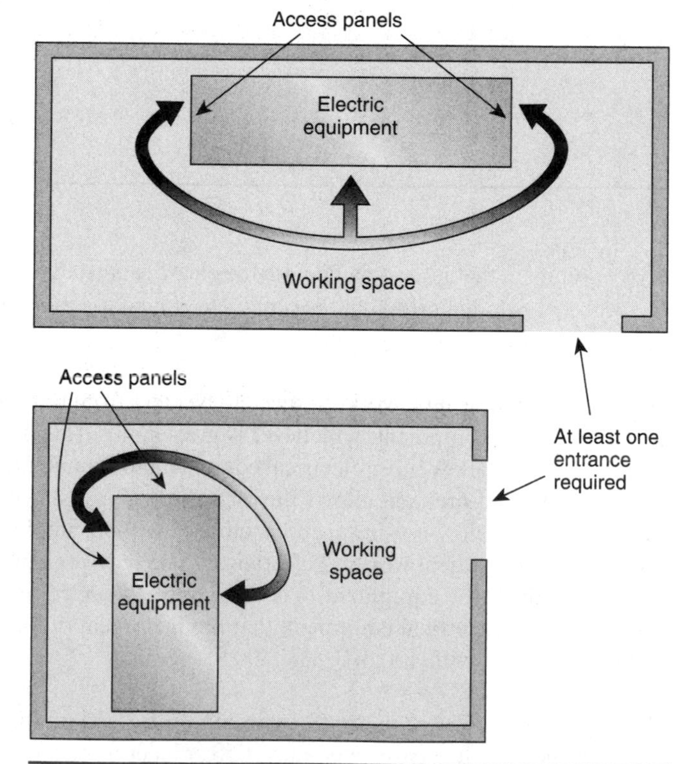


***EXHIBIT 400.11*** ***Basic Rule, first paragraph. At least one entrance is required to provide access to the working space around electrical equipment [400.15(C)(1)]. The lower installation would not be acceptable for a switchboard more than 6 ft wide and rated 1200 amperes or more.***

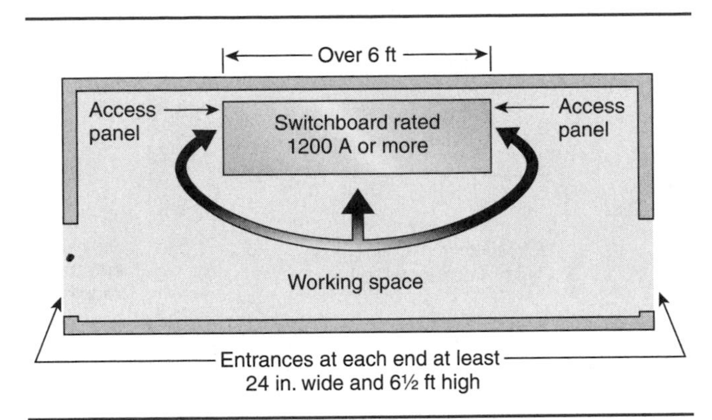


***EXHIBIT 400.12** Basic Rule, second paragraph. For equipment rated 1200 amperes or more and more than 6 ft wide, one entrance no less than 24 in. wide and 6½ ft high is required at each end [400.15(C)(2)].*

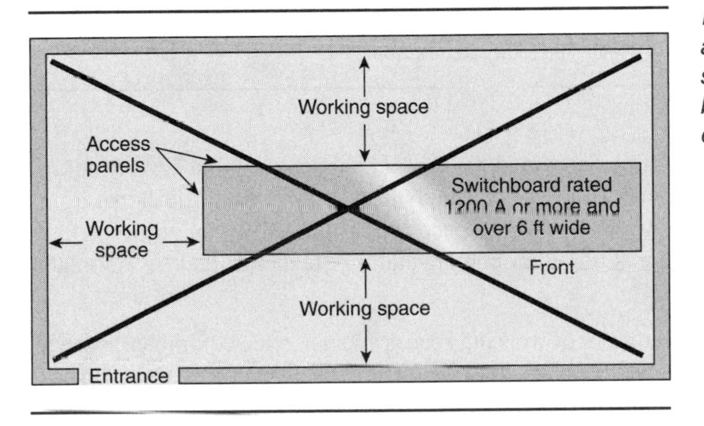


***EXHIBIT 400.13** Unacceptable arrangement of a large switchboard. A person could be trapped behind arcing electrical equipment.*

(a) Unobstructed Exit. Where the location permits a continuous and unobstructed way of exit travel, a single entrance to the working space shall be permitted.

(b) Extra Working Space. Where the depth of the working space is twice that required by 400.15(A)(1), a single entrance shall be permitted. It shall be located so that the distance from the equipment to the nearest edge of the entrance is not less than the minimum clear distance specified in Table 400.15(A)(1) for equipment operating at that voltage and in that condition.

For an explanation of 400.15(C)(2)(a) and (b), see Exhibits 400.14 and 400.15.

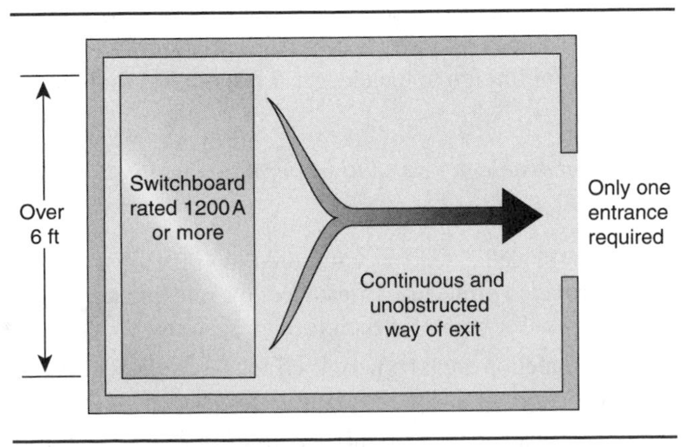


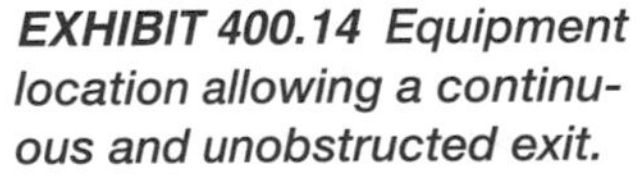

***EXHIBIT 400.14** Equipment location allowing a continuous and unobstructed exit.*

*EXHIBIT 400.15 Working space with one entrance. Only one entrance is required if the working space required by 400.15(A) is doubled. [See Table 400.15(A)(1) for permitted dimensions for X.]*

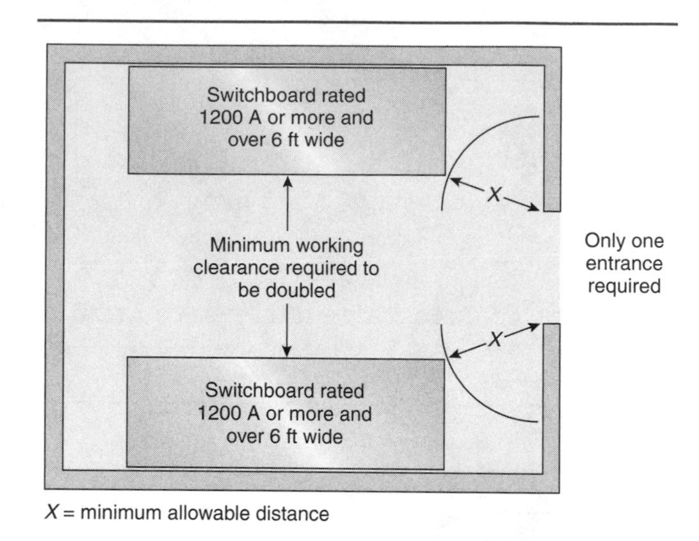


**(D) Illumination.** Illumination shall be provided for all working spaces about service equipment, switchboards, panelboards, or motor control centers installed indoors. Additional lighting outlets shall not be required where the work space is illuminated by an adjacent light source. In electrical equipment rooms, the illumination shall not be controlled by automatic means only.

**(E) Headroom.** The minimum headroom of working spaces about service equipment, switchboards, panelboards, or motor control centers shall be 2.0 m (6½ ft). Where the electrical equipment exceeds 2.0 m (6½ ft) in height, the minimum headroom shall not be less than the height of the equipment.

**(F) Dedicated Equipment Space.** All switchboards, panelboards, distribution boards, and motor control centers shall be located in dedicated spaces and protected from damage.

*Exception: Control equipment that by its very nature or because of other rules of the standard must be adjacent to or within sight of its operating machinery shall be permitted in those locations.*

**(1) Indoor.** Indoor installations shall comply with 400.15(F)(1)(a) through 400.15(F)(1)(d).

(a) Dedicated Electrical Space. The space equal to the width and depth of the equipment and extending from the floor to a height of 1.8 m (6 ft) above the equipment or to the structural ceiling, whichever is lower, shall be dedicated to the electrical installation. No piping, ducts, leak protection apparatus, or equipment foreign to the electrical installation shall be located in this zone.

*Exception: Suspended ceilings with removable panels shall be permitted within the 1.8 m (6 ft) zone.*

(b) Foreign Systems. The area above the dedicated space required in 400.15(F)(1) shall be permitted to contain foreign systems, provided protection is installed to avoid damage to the electrical equipment from condensation, leaks, or breaks in such foreign systems.

(c) Sprinkler Protection. Sprinkler protection shall be permitted for the dedicated space where the piping complies with this section.

(d) Suspended Ceilings. A dropped, suspended, or similar ceiling that does not add strength to the building structure shall not be considered a structural ceiling.

Dedicated space includes the three-dimensional space defined by extending the footprint of the switchboard or panelboard from the floor to a height of 6 ft above the height of the equipment or to the structural ceiling, whichever is lower. This reserved space permits busways, conduits, raceways, and cables to enter the equipment. The dedicated electrical space must be clear of any piping, ducts, leak protection apparatus, or equipment foreign to the electrical installation. Plumbing, heating, ventilation, and air-conditioning piping, ducts, and equipment must be installed outside the width and depth of the zone.

Foreign systems installed directly above the dedicated space reserved for electrical equipment must include protective equipment that provides assurance that leaks, condensation, and breaks are not capable of damaging the electrical equipment located below.

Sprinklers are permitted, provided the sprinkler system complies with 400.15(F). A dropped, suspended, or similar ceiling is permitted to be located directly in the dedicated space, as are building structural members. The electrical equipment also must be protected from physical damage. Damage can be caused by activities occurring near this equipment, such as material handling by personnel or the operation of a forklift or other mobile equipment. See 400.16(B) for other provisions relating to the protection of electrical equipment.

Exhibits 400.16, 400.17, and 400.18 illustrate the two distinct indoor installation spaces required in 400.15(A) and 400.15(F), that is, the working space and the dedicated electrical space.

In Exhibit 400.16, the dedicated electrical space required by 400.15(F) is the space outlined by the width and depth of the equipment (the footprint) and extending from the floor to 6 ft above the equipment or to the structural ceiling (whichever is lower). The dedicated electrical space is reserved for electrical equipment and for conduits, cable trays, and so on, entering or exiting that equipment. The outlined area in front of the electrical equipment in Exhibit 400.16 is working space required by 400.15(A). Note that sprinkler protection is afforded the entire dedicated electrical space and working space without actually entering either space. Also, note that the exhaust duct is not located in or directly above the dedicated electrical space. Although not specifically required to be located here, this duct location can be a cost-effective solution that avoids the substantial physical protection requirements of 400.15(F)(1)(b).

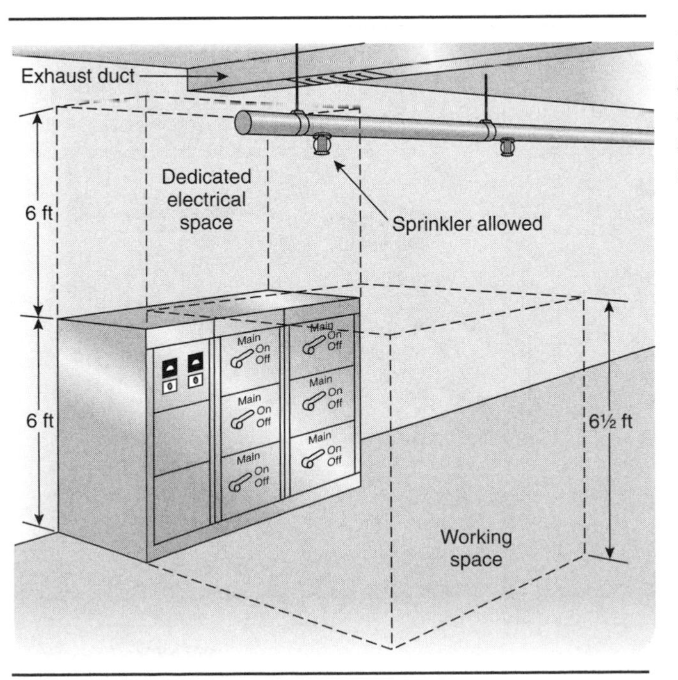


***EXHIBIT 400.16*** *The two distinct indoor installation spaces required by 400.15(A) and 400.15(F): the working space and the dedicated electrical space.*

Exhibit 400.17 illustrates the working space required in front of the panelboard by 400.15(A). No equipment, electrical or otherwise, is allowed in this working space.

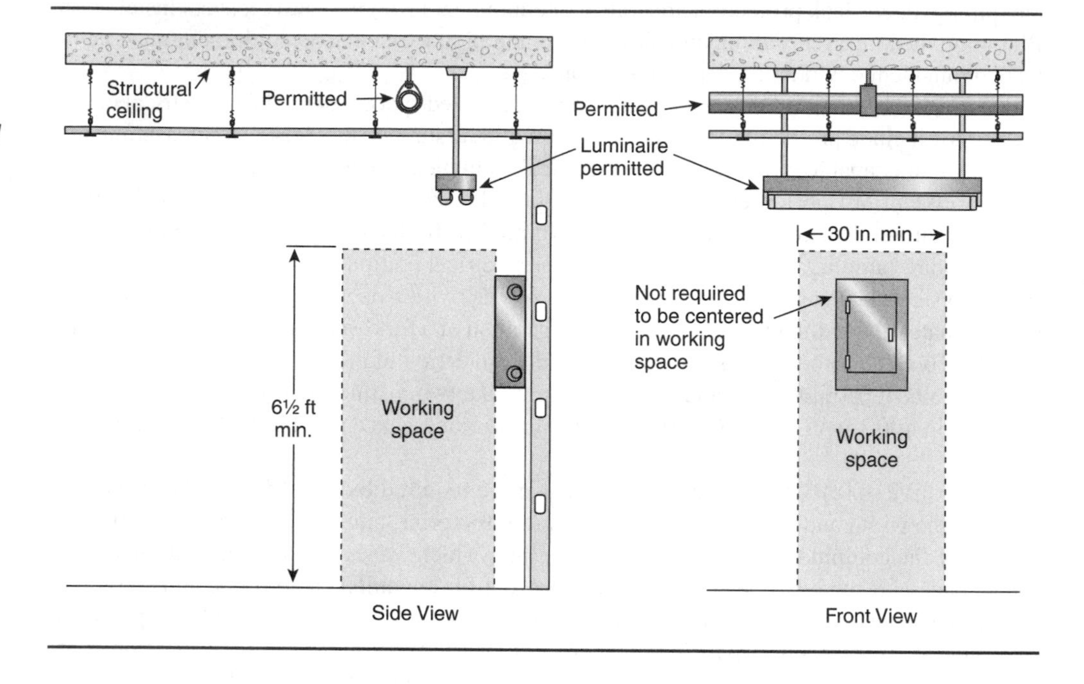


***EXHIBIT 400.17*** *The working space in front of a panelboard as required by 400.15(A). This illustration supplements the dedicated electrical space shown in Exhibit 400.16.*

Exhibit 400.18 illustrates the dedicated electrical space required over and under the panelboard by 400.15(F)(1). This space is for the cables, raceways, and so on that run to and from the panelboard.

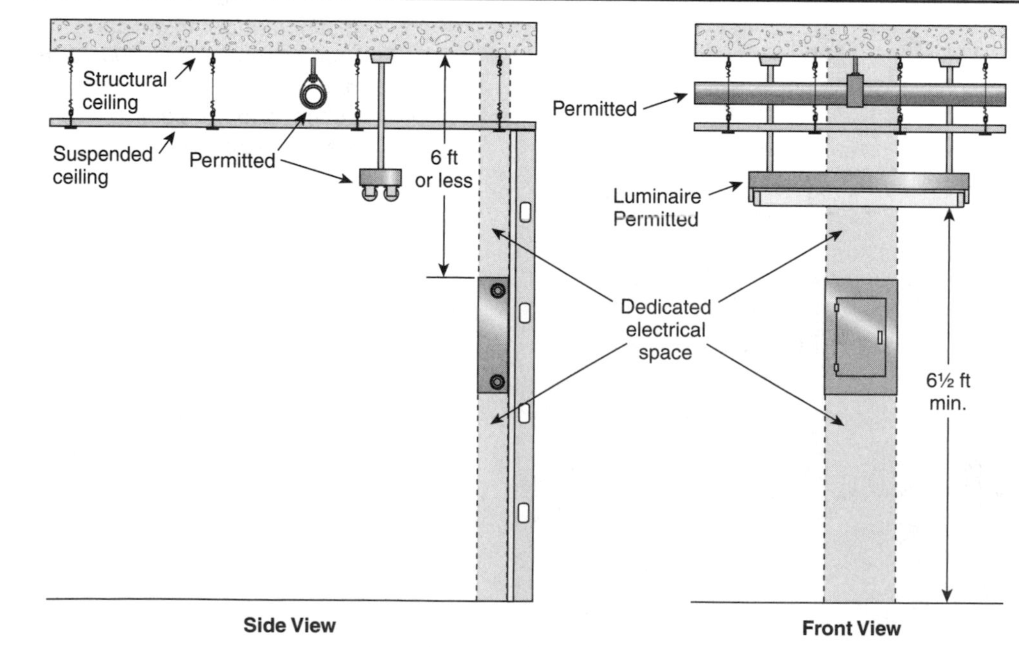


***EXHIBIT 400.18*** *Dedicated electrical space above and below a panelboard as required by 400.15(F)(1).*

**(2) Outdoor.** Outdoor electric equipment shall be installed in suitable enclosures and shall be protected from accidental contact by unauthorized personnel, or by vehicular traffic, or by accidental spillage or leakage from piping systems. The working clearance space shall include the zone described in 400.15(A). No architectural appurtenance or other equipment shall be located in this zone.

Caution must be exercised when protection required by 400.15(F)(2) is added to an existing installation. Excavating or driving fence posts into the ground should be done only after thoroughly investigating for below-grade obstructions.

### 400.16 Guarding of Live Parts

**(A) Live Parts Guarded Against Accidental Contact.** Except as elsewhere required or permitted by this standard, live parts of electric equipment operating at 50 volts or more shall be guarded against accidental contact by approved enclosures or by any of the following means:

(1) By location in a room, vault, or similar enclosure that is accessible only to qualified persons.
(2) By suitable permanent, substantial partitions or screens arranged so that only qualified persons have access to the space within reach of the live parts. Any openings in such partitions or screens shall be sized and located so that persons are not likely to come into accidental contact with the live parts or to bring conducting objects into contact with them.
(3) By location on a suitable balcony, gallery, or platform elevated and arranged so as to exclude unqualified persons.
(4) By elevation of 2.5 m (8 ft) or more above the floor or other working surface.

Contact conductors used for traveling cranes are permitted to be bare by 610.13(B) and 610.21(A) of the *NEC.* Although contact conductors obviously have to be bare for contact shoes on the moving member to make contact with the conductor, guards can be placed near the conductor, to prevent accidental contact, that still have slots or spaces through which the moving contacts can operate. Note that the *NEC* recognizes the guarding of live parts by elevation.

**(B) Prevention of Physical Damage.** In locations where electric equipment is likely to be exposed to physical damage, enclosures or guards shall be so arranged and of such strength as to prevent such damage.

**(C) Warning Signs.** Entrances to rooms and other guarded locations that contain exposed live parts operating at 50 volts or more shall be marked with conspicuous warning signs forbidding unqualified persons to enter.

Live parts of electrical equipment should be covered, shielded, enclosed, or otherwise protected by covers, barriers, mats, or platforms to minimize the chance of contact by persons or objects. See the definitions of *dead front* and *isolated (as applied to location)* in Article 100.

## III. Over 600 Volts, Nominal

### 400.17 General

Conductors and equipment used on circuits over 600 volts, nominal, shall comply with 400.1(A) of this standard and with the following sections, which supplement or modify 400.1(A). In no case shall the provisions of 400.18, 400.19, and 400.20 apply to equipment on the supply side of the service point.

Equipment on the supply side of the service point is outside the scope of the *NEC.* ANSI C2, *National Electrical Safety Code,* published by the Institute of Electrical and Electronic Engineers (IEEE), covers such equipment.

### 400.18 Enclosure for Electrical Installations

**(A) Indoor or Controlled or Locked Installations.** Electrical installations in a vault, room, or closet or in an area surrounded by a wall, screen, or fence, access to which is controlled by lock and key or other approved means, shall be considered to be accessible to qualified persons only. The type of enclosure used in a given case shall be designed and constructed according to the nature and degree of the hazard(s) associated with the installation.

**(B) Outdoor Installations.** For installations other than equipment as described in 400.18(E), a wall, screen, or fence shall be used to enclose an outdoor electrical installation to deter access by unqualified persons. A fence shall not be less than 2.1 m (7 ft) in height or a combination of 1.8 m (6 ft) or more of fence fabric and a 300 mm (1 ft) or more extension utilizing three or more strands of barbed wire or equivalent. The distance from the fence to live parts shall be not less than that given in Table 400.18(B).

FPN: For clearances of conductors for specific system voltages and typical BIL ratings, see ANSI C2–2002, *National Electrical Safety Code.*

***TABLE 400.18(B)*** *Minimum Distance from Fence to Live Parts*

| | Minimum Distance to Live Parts | |
|---|---|---|
| **Nominal Voltage** | **m** | **ft** |
| 601–13,799 | 3.05 | 10 |
| 13,800–230,000 | 4.57 | 15 |
| Over 230,000 | 5.49 | 18 |

**(C) Fire Resistivity of Electrical Vaults.** The walls, roof, floors, and doorways of vaults containing conductors and equipment over 600 volts, nominal, shall be constructed of material with structural strength adequate for the conditions, with a minimum fire rating of 3 hours. The floors of vaults in contact with the earth shall be of concrete that is not less than 4 in. (102 mm) thick, but where the vault is constructed with a vacant space or other stories below it, the floor shall have adequate structural strength for the load imposed on it and a minimum fire resistance of 3 hours. For the purpose of 400.18(A), studs and wallboards shall not be considered acceptable.

**(D) Indoor Installations.**

**(1) In Places Accessible to Unqualified Persons.** Indoor electrical installations that are accessible to unqualified persons shall be made with metal-enclosed equipment. Metal-enclosed switchgear, unit substations, transformers, pull boxes, connection boxes, and other similar associated equipment shall be marked with appropriate caution signs. Openings in ventilated dry-type transformers and similar openings in other equipment shall be designed so that foreign objects inserted through these openings are deflected from energized parts.

**(2) In Places Accessible to Qualified Persons Only.** Indoor electrical installations considered accessible only to qualified persons in accordance with this section shall comply with 400.19.

**(E) Outdoor Installations.**

**(1) In Places Accessible to Unqualified Persons.** Outdoor electrical installations that are open to unqualified persons shall comply with 410.7.

FPN: For clearances of conductors for system voltages over 600 volts, nominal, see ANSI C2–2002, *National Electrical Safety Code.*

**(2) In Places Accessible to Qualified Persons Only.** Outdoor electrical installations that have exposed live parts operating at 50 volts or more shall be accessible only to qualified persons in accordance with 400.18(A) and shall comply with 400.19.

**(F) Enclosed Equipment Accessible to Unqualified Persons.** Ventilating or similar openings in equipment shall be designed such that foreign objects inserted through these openings are deflected from energized parts. Where exposed to physical damage from vehicular traffic, suitable guards shall be provided. Nonmetallic or metal-enclosed equipment located outdoors and accessible to the general public shall be designed so that exposed nuts or bolts cannot be readily removed, permitting access to live parts. Where nonmetallic or metal-enclosed equipment is accessible to the general public and the bottom of the enclosure is less than 2.5 m (8 ft) above the floor or grade level, the enclosure door or hinged cover shall be kept locked. Doors and covers of enclosures used solely as pull boxes, splice boxes, or junction boxes shall be locked, bolted, or screwed on. Underground box covers that weigh over 45.4 kg (100 lb) shall be considered as meeting this requirement.

### 400.19 Work Space About Equipment

Sufficient space shall be provided and maintained about electric equipment to permit ready and safe operation and maintenance of such equipment. Where energized parts are exposed, the minimum clear work space shall not be less than 2.0 m (6½ ft) high (measured vertically from the floor or platform), or less than 900 mm (3 ft) wide (measured parallel to the equipment). The depth shall be as required in 400.21. In all cases, the work space shall permit at least a 90-degree opening of doors or hinged panels.

### 400.20 Entrance and Access to Work Space

**(A) Entrance.** At least one entrance not less than 610 mm (24 in.) wide and 2.0 m (6½ ft) high shall be provided to give access to the working space about electric equipment. Where the entrance has a personnel door(s), the door(s) shall open in the direction of egress and be equipped with panic bars, pressure plates, or other devices that are normally latched but open under simple pressure.

**(1) Large Equipment.** On switchboard and control panels exceeding 1.8 m (6 ft) in width, there shall be one entrance at each end of the equipment. A single entrance to the required working space shall be permitted where either of the conditions in 400.20(A)(1)(a) or 400.20(A)(1)(b) is met.

(a) Unobstructed Exit. Where the location permits a continuous and unobstructed way of exit travel, a single entrance to the working space shall be permitted.

(b) Extra Working Space. Where the depth of the working space is twice that required by 400.21, a single entrance shall be permitted. It shall be located so that the distance from the equipment to the nearest edge is not less than the minimum clear distance specified in Table 400.21 for equipment operating at that voltage and in that condition.

**(2) Guarding.** Where bare energized parts at any voltage or insulated energized parts above 600 volts, nominal, to ground are located adjacent to such entrance, they shall be suitably guarded.

Section 400.20(A) contains requirements similar to those of 400.15(C). Because of the higher voltages covered, 400.20(A) differs slightly in that a minimum working space entrance dimension is required, even for switchboards of 6 ft or less in width.

**(B) Access.** Permanent ladders or stairways shall be provided to give safe access to the working space around electric equipment installed on platforms, balconies, mezzanine floors, or in attic or roof rooms or spaces.

### 400.21 Work Space and Guarding

**(A) Working Space.** Except as elsewhere required or permitted in this standard, the minimum clear working space in the direction of access to live parts of electric equipment shall be not less than specified in Table 400.21. Distances shall be measured from the live parts, if such are exposed, or from the enclosure front or opening if such are enclosed.

*Exception: Working space shall not be required in back of equipment such as dead-front switchboards or control assemblies where there are no renewable or adjustable parts (such as fuses or switches) on the back and where all connections are accessible from locations other than the back. Where rear access is required to work on the deenergized parts on the back of enclosed equipment, a minimum working space of 750 mm (30 in.) horizontally shall be provided.*

***TABLE 400.21*** *Minimum Depth of Clear Working Space at Electric Equipment*

| | Minimum Clear Distance | | | | | |
|---|---|---|---|---|---|---|
| | Condition 1 | | Condition 2 | | Condition 3 | |
| Nominal Voltage to Ground | m | ft | m | ft | m | ft |
| 601–2500 V | 0.9 | 3 | 1.2 | 4 | 1.5 | 5 |
| 2501–9000 V | 1.2 | 4 | 1.5 | 5 | 1.8 | 6 |
| 9001–25,000 V | 1.5 | 5 | 1.8 | 6 | 2.8 | 9 |
| 25,001–75 kV | 1.8 | 6 | 2.5 | 8 | 3.0 | 10 |
| Above 75 kV | 2.5 | 8 | 3.0 | 10 | 3.7 | 12 |

Note: Where the conditions are as follows:

*Condition 1*—Exposed live parts on one side and no live or grounded parts on the other side of the working space, or exposed live parts on both sides effectively guarded by suitable wood or other insulating materials. Insulated wire or insulated busbars operating at not over 300 volts shall not be considered live parts.

*Condition 2*—Exposed live parts on one side and grounded parts on the other side. Concrete, brick, or tile walls will be considered as grounded surfaces.

*Condition 3*—Exposed live parts on both sides of the work space (not guarded as provided in Condition 1) with the operator between.

**(B) Separation from Low-Voltage Equipment.** Where switches, cutouts, or other equipment operating at 600 volts, nominal, or less are installed in a room or enclosure where there are exposed live parts or exposed wiring operating at over 600 volts, nominal, the high-voltage equipment shall be effectively separated from the space occupied by the low-voltage equipment by a suitable partition, fence, or screen.

*Exception: Switches or other equipment operating at 600 volts, nominal, or less and serving only equipment within the high-voltage vault, room, or enclosure shall be permitted to be installed in the high-voltage enclosure, room, or vault if accessible to qualified persons only.*

**(C) Locked Rooms or Enclosures.**

**(1) General.** The entrances to all buildings, rooms, or enclosures containing exposed live parts or exposed conductors operating at over 600 volts, nominal, shall be kept locked unless such entrances are under the observation of a qualified person at all times.

Equipment used on circuits over 600 volts, nominal, and having exposed live parts or exposed conductors, is required by 400.21(C)(1) to be located in a locked room or an enclosure. Locking is not required if the location is under observation at all times, as in the case of some engine rooms.

**(2) Warning Signs.** Where the voltage exceeds 600 volts, nominal, permanent and conspicuous warning signs shall be provided, reading as follows:

DANGER—HIGH VOLTAGE—KEEP OUT

FPN: For further information on hazard signs and labels, see ANSI Z535.4, *Product Signs and Safety Labels.*

**(D) Illumination.** Illumination shall be provided for all working spaces about electric equipment. The lighting outlets shall be so arranged that persons changing lamps or making repairs on the lighting system are not endangered by live parts operating at 50 volts or more or by other equipment. The points of control shall be so located that persons are not likely to come in contact with any live part operating at 50 volts or more or moving part of the equipment while turning on the lights.

**(E) Elevation of Unguarded Live Parts.** Unguarded live parts above working space shall be maintained at elevations not less than required in Table 400.21(E).

**(F) Protection of Service Equipment, Metal-Enclosed Power Switchgear, and Industrial Control Assemblies.** Pipes or ducts foreign to the electrical installation and requiring periodic maintenance or whose malfunction would endanger the operation of the electrical system shall not be located in the vicinity of the service equipment, metal-enclosed power switchgear, or industrial control assemblies. Protection shall be provided where necessary to avoid damage from condensation leaks and breaks in such foreign systems. Piping and other facilities shall not be considered foreign if provided for fire protection of the electrical installation.

***TABLE 400.21(E)*** *Elevation of Unguarded Live Parts Above Working Space*

| | Elevation | |
|---|---|---|
| **Nominal Voltage Between Spaces** | **m** | **ft** |
| 601–7500 V | 2.8 | 9 |
| 7501–35,000 V | 2.9 | 9½ |
| Over 35 kV | 2.9 + 9.5 mm/kV above 35 | 9½ ft + 0.37 in./kV above 35 |

# ARTICLE 410
## Wiring Design and Protection

Many injuries occur as a result of misapplication of the requirements in Article 410 or a failure to maintain wiring installations adequately. The primary hazard associated with wiring design is shock from touch potential, as discussed in Chapter 1 of this standard. Article 410 identifies requirements that are intended to minimize the opportunity for exposure to electrocution or shock. The article defines requirements necessary to establish an effective path for ground-fault current to return safely to the source of energy.

### 410.1 Use and Identification of Grounded and Grounding Conductors

**(A) Identification of Conductors.** A conductor used as a grounded conductor shall be identifiable and distinguishable from all other conductors.

**(B) Polarity of Connections.** No grounded conductor shall be attached to any terminal or lead so as to reverse designated polarity.

### 410.2 Branch Circuits

**(A) Identification of Ungrounded Circuits.** Where more than one nominal voltage system exists in a building, each ungrounded conductor of a multiwire branch circuit, where accessible, shall be identified by phase and system. The means of identification shall be permitted to be by separate color coding, marking tape, tagging, or other approved means and shall be permanently posted at each branch circuit panelboard.

Exhibit 410.1 illustrates two different nominal voltage systems in a building. Each ungrounded system conductor is identified by color-coded marking tape. A notice indicating the means of the identification is permanently located at each panelboard; however, it should be noted that this requirement applies only to multiwire branch circuits.

Although color-coding systems provide one indication that a specific conductor is at ground potential (or not), workers must rely only on testing for voltage as defined in Chapter 1 of this standard.

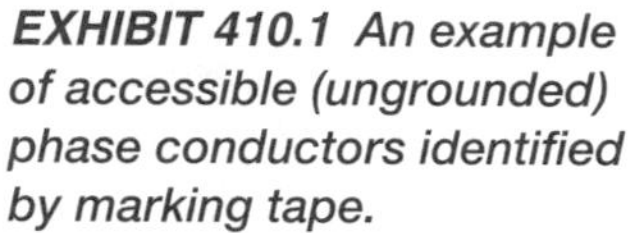

***EXHIBIT 410.1*** *An example of accessible (ungrounded) phase conductors identified by marking tape.*

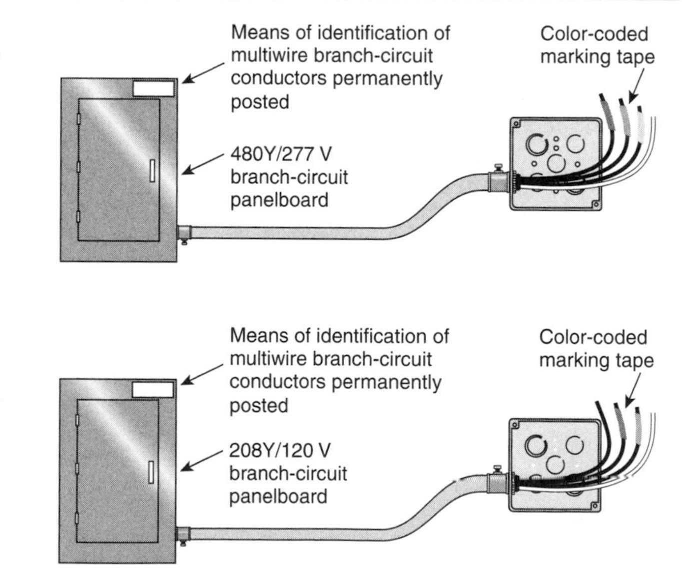


**(B) Receptacles and Cord Connectors.**

**(1) Grounding Type.** Receptacles installed on 15- and 20-ampere branch circuits shall be of the grounding type. Grounding-type receptacles shall be installed only on circuits of the voltage class and current for which they are rated, except as provided in Table 410.5(B)(2) and Table 410.5(B)(3).

*Exception: Nongrounding-type receptacles installed in accordance with 410.2(B)(4)(c).*

**(2) To Be Grounded.** Receptacles and cord connectors that have grounding contacts shall have those contacts effectively grounded.

*Exception No. 1: Receptacles mounted on portable and vehicle-mounted generators in accordance with this standard.*

*Exception No. 2: Replacement receptacles as permitted by 410.2(B)(4).*

**(3) Methods of Grounding.** The grounding contacts of receptacles and cord connectors shall be grounded by connection to the equipment grounding conductor of the circuit supplying the receptacle or cord connector. The branch circuit wiring method shall include or provide an equipment grounding conductor to which the grounding contacts of the receptacle or cord connector shall be connected.

**(4) Replacements.** Replacement of receptacles shall comply with 410.2(B)(4)(a), 410.2(B)(4)(b), and 410.2(C) as applicable.

(a) Grounding-Type Receptacles. Where a grounding means exists in the receptacle enclosure or a grounding conductor is installed, grounding-type receptacles shall be used and shall be connected to the grounding conductor.

(b) Ground-Fault Circuit Interrupters. Ground-fault circuit-interrupter—protected receptacles shall be provided where replacements are made at receptacle outlets that are required to be so protected elsewhere in this standard.

**(C) Non–Grounding-Type Receptacles.** Where a grounding means does not exist in the receptacle enclosure, the installation shall comply with (1), (2), or (3):

(1) A non–grounding-type receptacle(s) shall be permitted to be replaced with another non–grounding-type receptacle(s).
(2) A non–grounding-type receptacle(s) shall be permitted to be replaced with a ground-fault circuit-interrupter–type of receptacle(s). Such receptacle shall be marked "No Equipment Ground." An equipment grounding conductor shall not be connected from the ground-fault circuit-interrupter–type receptacle to any outlet supplied from the ground-fault circuit–interrupter receptacle.
(3) A non–grounding-type receptacle(s) shall be permitted to be replaced with a grounding-type receptacle(s) where supplied through a ground-fault circuit interrupter. Grounding-type receptacles supplied through the ground-fault circuit-interrupter shall be marked "GFCI Protected" and "No Equipment Ground." An equipment grounding conductor shall not be connected between the grounding-type receptacles.

**(5) Cord-and-Plug-Connected Equipment.** The installation of grounding-type receptacles shall not be used as a requirement that all cord-and-plug-connected equipment be of the grounded type.

**(6) Noninterchangeable Types.** Receptacles connected to circuits that have different voltages, frequencies, or types of current (ac or dc) on the same premises shall be of such design that the attachment plugs used on these circuits are not interchangeable.

### 410.3 Identification of Ungrounded Conductors

Where more than one nominal voltage system exists in a building, each ungrounded conductor of a multiwire branch circuit, where accessible, shall be identified by phase and system. This means of identification shall be permitted to be by separate color coding, marking tape, tagging, or other approved means and shall be permanently posted at each branch-circuit panelboard.

### 410.4 Ground-Fault Circuit-Interrupter Protection for Personnel

Section 410.4 is the main rule for using ground-fault circuit interrupters (GFCIs). Since GFCIs were introduced as a requirement in the 1971 *Code,* these devices have prevented thousands of electrocutions. After NFPA 70 (the *National Electrical Code®*, *NEC®*, or *Code*) adopted a requirement for GFCIs, a similar requirement was added to OSHA Subpart S for both general industry and construction.

Such application of technology through requirements offers satisfying and visible progress in safety of people as they use electricity. According to David Wallis, Director of Electrical and Electronic Standards for OSHA, GFCIs have saved between 100 and 150 lives each year between 1990 and 1996. Wallis offered that observation at the IEEE Electrical Safety Workshop in February 2004.

Exhibit 410.2 illustrates a typical GFCI circuit. The hot and neutral circuit conductors are passed through a sensor and are connected to a shunt-trip device. If the current in each conductor is equal, the contact stays closed. If the hot conductor leaks current to ground, either directly or through a person's body, and some current returns by an alternative path, the current is unbalanced. The coil senses the unbalanced current, and the shunt-trip mechanism opens the circuit. Note that the circuit design does not require the presence of an equipment grounding conductor, which is the reason 410.2(C) permits the use of GFCIs as replacements for receptacles where a grounding means does not exist.

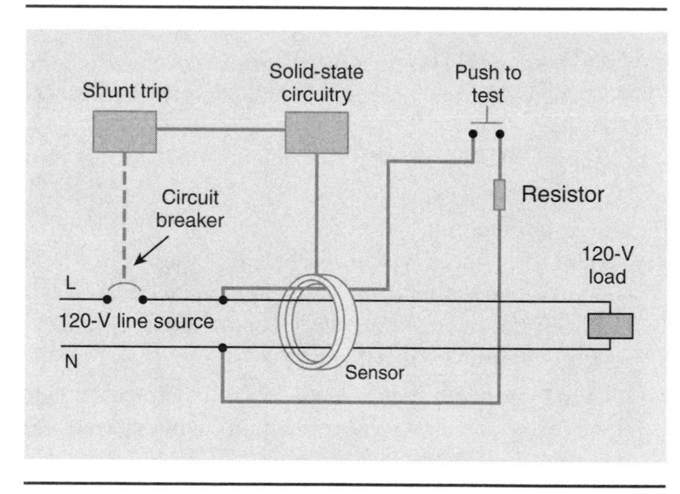


***EXHIBIT 410.2** The circuitry and components of a typical GFCI.*

GFCIs operate on currents of 5 mA. Listing standards permit a differential of 4 mA to 6 mA. At trip levels of 5 mA (the instantaneous current could be much higher), a shock can be felt during the time of the fault. The shock can lead to involuntary reactions that could cause secondary accidents, such as falls. GFCIs will not protect persons from shock hazards where contact is between phase and neutral or between phase conductors.

A variety of GFCIs are available, including portable, plug-in, circuit-breaker, those built into attachment plug caps, and receptacle types. Each type has a test switch so that units can be checked periodically to ensure proper operation. See Exhibits 410.3 and 410.4.

Although 210.8 of the *NEC* is the main rule for GFCIs, other specific applications require the use of GFCIs. These additional applications are listed in Commentary Table 410.1.

**(A) Other Than Dwelling Units.** All 125-volt, single-phase, 15- and 20-ampere receptacles installed in the locations specified in (1), (2), and (3) shall have ground-fault circuit-interrupter protection for personnel:

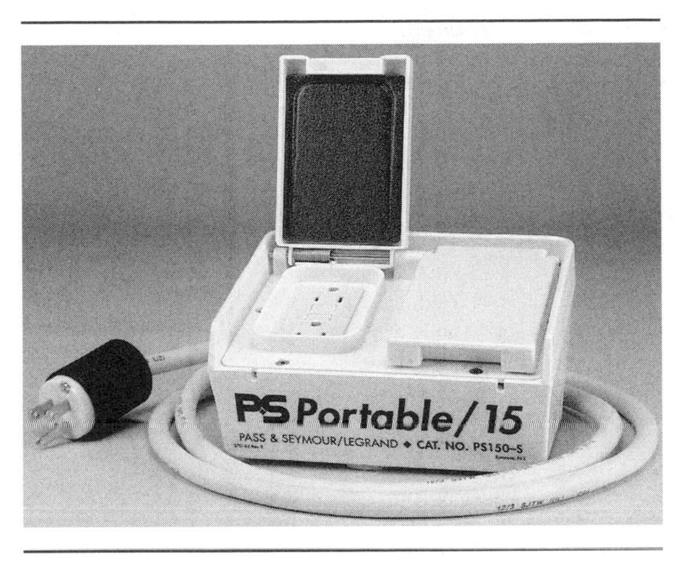


*EXHIBIT 410.3 A portable plug-in type of GFCI. (Courtesy of Pass & Seymour/Legrand®)*

*EXHIBIT 410.4 A 15-ampere duplex receptacle with integral GFCI that also protects downstream loads. (Courtesy of Pass & Seymour/Legrand®)*

(1) Bathrooms

If receptacles are provided in bathroom areas of hotels and motels, GFCI-protected receptacles are required. Lavatories in airports, commercial buildings, industrial facilities, and other nondwelling occupancies are required to have all their receptacles protected with GFCIs. The only exception to this rule is found in 517.21 of the *NEC,* which permits receptacles in hospital critical-care areas to be non-GFCI if the toilet and basin are installed in the patient room

***COMMENTARY TABLE 410.1*** *Additional NEC Requirements for the Application of Ground-Fault Circuit-Interrupter Protection*

| *Location* | *Applicable NEC Section(s)* |
|---|---|
| Audio system equipment | 640.10(A) |
| Boathouses | 555.19(B)(1) |
| Carnivals, circuses, fairs, and similar events | 525.23 |
| Commercial garages | 511.12 |
| Electric vehicle charging systems | 625.22 |
| Electronic equipment, sensitive | 647.7(A) |
| Elevators, escalators, and moving walkways | 620.85 |
| Feeders | 215.9 |
| Fountains | 680.51(A) |
| Health care facilities | 517.20(A), 517.21 |
| High-pressure spray washers | 422.49 |
| Hydromassage bathtubs | 680.71 |
| Marinas | 555.19(B)(1) |
| Mobile and manufactured homes | 550.13(B) and (E); 550.32(E) |
| Park trailers | 552.41(C) |
| Pools, permanently installed | 680.22(A)(1), (A)(5), and (B)(4); 680.23(A)(3) |
| Pools, storable | 680.32 |
| Sensitive electronic equipment | 647.7(A) |
| Signs with fountains | 680.57(B) |
| Signs, mobile or portable | 600.10(C)(2) |
| Recreational vehicles | 551.40(C), 551.41(C) |
| Recreational vehicle parks | 551.71 |
| Replacement receptacles | 406.3(D)(2) |
| Temporary installations | 527.6 |

rather than in a separate bathroom. Some motel and hotel bathrooms, like the one shown in Exhibit 410.5, have the basin located outside the door to the room containing the tub, toilet, or another basin. The definition of *bathroom,* as found in Article 100, applies to motel and hotel bathrooms, as does the GFCI requirement of 410.4(A)(1).

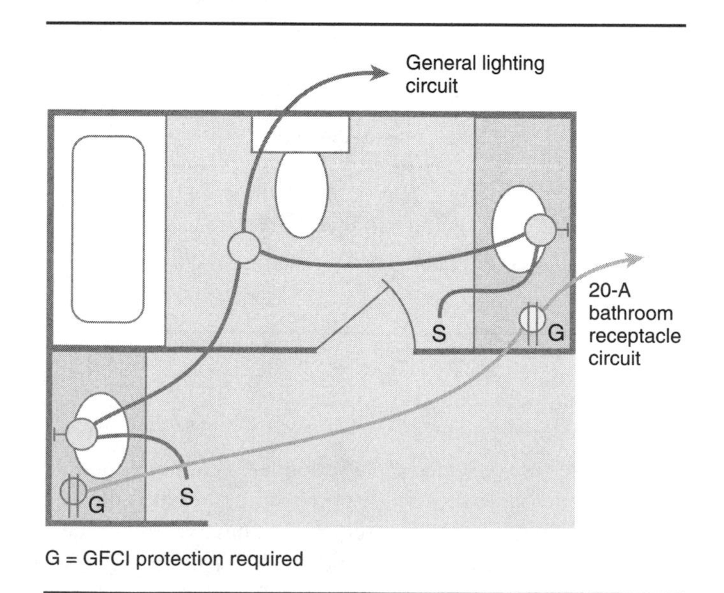


***EXHIBIT 410.5*** ***GFCI-protected receptacles shown in accordance with 410.4(A)(1) in a motel/hotel bathroom where one basin is located outside the door to the rest of the bathroom area.***

(2) Rooftops

Section 410.4(A)(2) requires all rooftop 15- and 20-ampere receptacles in nondwelling occupancies to be protected with GFCIs. For rooftops that also have heating, air-conditioning, and refrigeration equipment, see 210.63 of the *NEC.*

(3) Kitchens

Section 410.4(A)(3) is new to the 2004 edition of NFPA 70E and requires all 15- and 20-ampere, 125-volt receptacles in nondwelling-type kitchens to be GFCI protected. This requirement applies to each and every 15- and 20-ampere, 125-volt kitchen receptacle, whether or not the receptacle serves countertop appliances.

Accident data related to electrical incidents in nondwelling kitchens reveal the presence of many hazards, including poorly maintained electrical apparatus, damaged electrical cords, wet floors, and employees without proper electrical safety training. Mandating some limited form of GFCI protection for high-hazard areas such as nondwelling kitchens should help prevent electrical accidents.

**(B) Ground-Fault Protection for Personnel.** Ground-fault protection for personnel for all temporary wiring installations shall be provided to comply with 410.4(B)(1) or 410.4(B)(2). This section shall apply only to temporary wiring installations used to supply temporary power to equipment used by personnel during construction, remodeling, maintenance, repair, or demolition of buildings, structures, equipment or similar activities.

**(1) Receptacle Outlets.** All 125-volt, single-phase, 15-, 20-, and 30-ampere receptacle outlets that are not a part of the permanent wiring of the building or structure and that are in use by personnel shall have ground-fault circuit-interrupter protection for personnel. If a receptacle(s) is installed or exists as part of the permanent wiring of the building or structure and is used for temporary electric power, ground-fault circuit-interrupter protection for personnel shall be provided. For the purposes of this section, cord sets or devices incorporating listed ground-fault circuit-interrupter protection for personnel identified for portable use shall be permitted.

*Exception: In industrial establishments only, where conditions of maintenance and supervision ensure that only qualified personnel are involved, an assured equipment grounding conductor program as specified in 410.4(B)(2) shall be permitted only for those receptacle outlets used to supply equipment that would create a greater hazard if power was interrupted or having a design that is not compatible with GFCI protection.*

**(2) Use of Other Outlets.** Receptacles other than 125-volt, single-phase, 15-, 20-, and 30-ampere receptacles shall have protection in accordance with 410.4(B)(2)(a) or the assured equipment grounding conductor program in accordance with 410.4(B)(2)(b).

(a) GFCI Protection. Ground-fault circuit-interrupter protection for personnel.

(b) Assured Equipment Grounding Conductor Program. A written assured equipment grounding conductor program continuously enforced at the site by one or more designated persons to ensure that equipment grounding conductors for all cord sets, receptacles not part of the permanent wiring of the building or structure, and equipment connected by cord and plug are installed and maintained. The following tests shall be performed on all cord sets, receptacles not part of the permanent wiring of the building or structure, and cord-and-plug-connected equipment required to be grounded:

(1) All equipment grounding conductors shall be tested for continuity and shall be electrically continuous.

(2) Each receptacle and attachment plug shall be tested for correct attachment of the equipment grounding conductor. The equipment grounding conductor shall be connected to its proper terminal.
(3) All required tests shall be performed as follows:
   a. Before first use on site
   b. When there is evidence of damage
   c. Before equipment is returned to service following any repairs
   d. At intervals not exceeding 3 months
(4) The test required by 410.4(B)(2)(b) shall be recorded and made available to the authority having jurisdiction.

Severe environmental conditions exist during construction, remodeling, maintenance, repair, and demolition. Personnel using temporary wiring are exposed to an elevated risk of electrical shock or electrocution. All temporarily installed 125-volt, single-phase, 15-, 20-, and 30-ampere receptacles must be protected by GFCIs.

The exception to 410.4(A) is limited in scope and application. It applies only to those industrial occupancies in which only qualified persons will be using 125-volt, single-phase, 15-, 20-, and 30-ampere receptacles. Any equipment being supplied by these receptacles must be demonstrated to be incompatible with the proper operation of GFCI protective devices. Some electrically operated testing equipment has proven to be incompatible with GFCI protection. Where the conditions exist for the exception to apply, the assured equipment grounding conductor program specified in 410.4(B)(2)(b) is permitted.

Receptacle configurations other than the 125-volt, single-phase, 15-, 20- and 30-ampere types must be GFCI protected or installed and maintained in accordance with the assured equipment grounding conductor program of 410.4(B)(2)(b).

According to OSHA 29 CFR 1926.404(b)(1)(iii):

> The employer shall establish and implement an assured equipment grounding conductor program on construction sites covering all cord sets, receptacles which are not a part of the building or structure, and equipment connected by cord and plug which are available for use or used by employees. This program shall comply with the following minimum requirements:
>
> (A) A written description of the program, including the specific procedures adopted by the employer, shall be available at the jobsite for inspection and copying by the Assistant Secretary and any affected employee.
>
> (B) The employer shall designate one or more competent persons. . . .

These OSHA requirements are very similar to the present *NEC* requirements for an assured grounding program.

GFCI protection for construction or maintenance personnel using receptacles that are part of the permanent wiring and are not GFCI protected may be provided by using cord sets or listed portable GFCIs identified for portable use. An example of a GFCI cord set that is identified for portable use is shown in Exhibit 410.6

Exhibits 410.6 through 410.9 show some examples of ways to implement the temporary wiring requirements of 410.6.

### 410.5 Outlet Devices

Outlet devices shall have an ampere rating that is not less than the load to be served and shall comply with 410.5(A) and 410.5(B).

**(A) Lampholders.** Where connected to a branch circuit having a rating in excess of 20 amperes, lampholders shall be of the heavy-duty type. A heavy-duty lampholder shall have a rat-

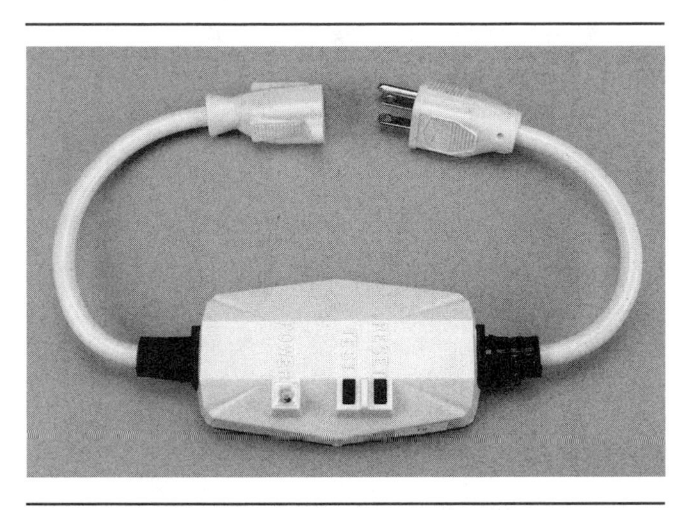

***EXHIBIT 410.6*** *A raintight GFCI with open neutral protection that is designed for use on the line end of a flexible cord. (Courtesy of Pass & Seymour/Legrand®)*

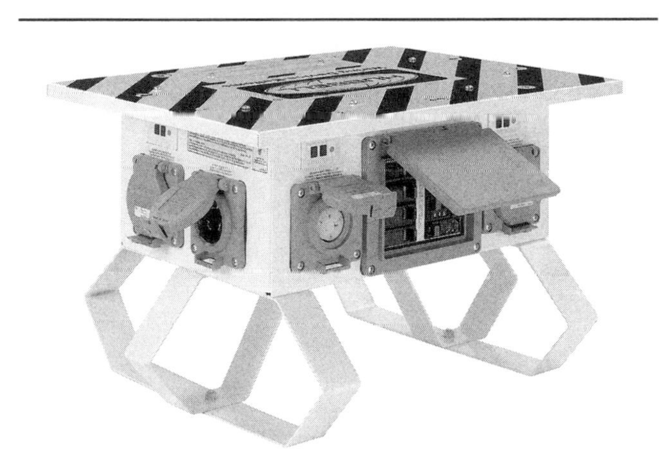

***EXHIBIT 410.7*** *A temporary power outlet unit commonly used on construction sites with a variety of configurations, including GFCI protection. (Courtesy of Hubbell, Inc.)*

***EXHIBIT 410.8*** *A watertight plug and connector used to prevent tripping of GFCI protective devices in wet or damp weather. (Courtesy of Hubbell, Inc.)*

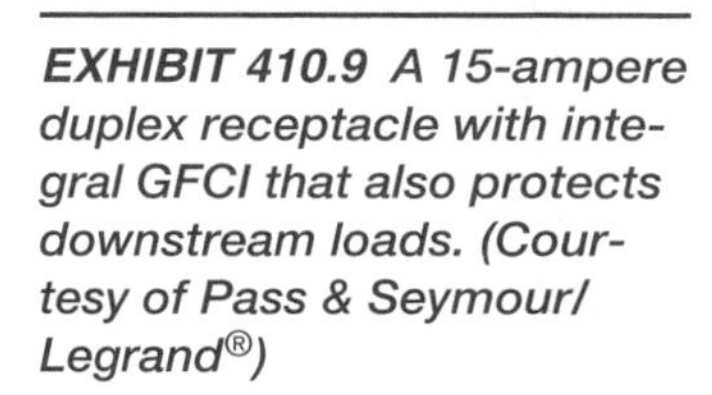

***EXHIBIT 410.9*** *A 15-ampere duplex receptacle with integral GFCI that also protects downstream loads. (Courtesy of Pass & Seymour/Legrand®)*

ing of not less than 660 watts if of the admedium type and not less than 750 watts if of any other type.

**(B) Receptacles.**

**(1) Single Receptacle on an Individual Branch Circuit.** A single receptacle installed on an individual branch circuit shall have an ampere rating of not less than that of the branch circuit.

**(2) Total Cord-and-Plug-Connected Load.** Where connected to a branch circuit supplying two or more receptacles or outlets, a receptacle shall not supply a total cord-and-plug-connected load in excess of the maximum specified in Table 410.5(B)(2).

*TABLE 410.5(B)(2)* *Maximum Cord-and-Plug-Connected Load to Receptacle*

| Circuit Rating (Amperes) | Receptacle Rating (Amperes) | Maximum Load (Amperes) |
|---|---|---|
| 15 or 20 | 15 | 12 |
| 20 | 20 | 16 |
| 30 | 30 | 24 |

**(3) Receptacle Ratings.** Where connected to a branch circuit supplying two or more receptacles or outlets, receptacle ratings shall conform to the values listed in Table 410.5(B)(3), or, where larger than 50 amperes, the receptacle rating shall not be less than the branch-circuit rating.

*Exception: Receptacles for one or more cord-and-plug-connected arc welders shall be permitted to have ampere ratings not less than the minimum branch-circuit conductor ampacity.*

*TABLE 410.5(B)(3)* *Receptacle Ratings for Various Size Circuits*

| Circuit Rating (Amperes) | Receptacle Rating (Amperes) |
|---|---|
| 15 | Not over 15 |
| 20 | 15 or 20 |
| 30 | 30 |
| 40 | 40 or 50 |
| 50 | 50 |

### 410.6 Cord Connections

A receptacle outlet shall be installed wherever flexible cords with attachment plugs are used. Where flexible cords are permitted to be permanently connected, receptacles shall be permitted to be omitted for such cords.

### 410.7 Outside Branch Circuit, Feeder, and Service Conductors, 600 Volts, Nominal, or Less

Sections 410.7(A), 410.7(B), 410.7(C), and 410.7(D) shall apply to branch circuit, feeder, and service conductors run outdoors as open conductors.

**(A) Conductors on Poles.** Conductors on poles shall have a separation of not less than 300 mm (1 ft) where not placed on racks or brackets. Conductors supported on poles shall provide a horizontal climbing space not less than the following:

(1) Power conductors below communications conductors: 750 mm (30 in.)
(2) Power conductors alone or above communications conductors: 300 volts or less—600 mm (24 in.); over 300 volts—750 mm (30 in.)
(3) Communications conductors below power conductors: same as power conductors
(4) Communications conductors alone: no requirement

**(B) Clearance from Ground.** Overhead spans of open conductors, open multiconductor cables, and service-drop conductors not exceeding 600 volts, nominal, shall conform to the following clearances:

(1) 3.0 m (10 ft)—above finished grade, sidewalks, or from any platform or projection from which they might be reached where the voltage does not exceed 150 volts to ground and accessible to pedestrians only
(2) 3.7 m (12 ft)—over residential property and driveways, and those commercial areas not subject to truck traffic where the voltage does not exceed 300 volts to ground
(3) 4.5 m (15 ft)—for those areas listed in item (2) where the voltage exceeds 300 volts to ground
(4) 5.5 m (18 ft)—over public streets, alleys, roads, parking areas subject to truck traffic, driveways on other than residential property, and other land traversed by vehicles, such as cultivated, grazing, forest, and orchard

**(C) Clearance from Building Openings.** Open conductors, open multiconductor cables, service-drop conductors, and final spans shall comply with 410.7(C)(1), 410.7(C)(2), and 410.7(C)(3).

**(1) Clearance from Windows.** Service conductors installed as open conductors or multiconductor cable without an overall outer jacket shall have a clearance of not less than 900 mm (3–ft) from windows designed to be opened, doors, porches, balconies, ladders, stairs, fire escapes, or similar locations.

*Exception: Conductors run above the top level of a window shall be permitted to be less than the 900-mm (3-ft) requirement.*

**(2) Vertical Clearance.** The vertical clearance of final spans above, or within 900 mm (3 ft) measured horizontally of, platforms, projections, or surfaces from which they might be reached shall be maintained in accordance with 410.7(B).

**(3) Building Openings.** Overhead service conductors shall not be installed beneath openings through which materials may be moved, such as openings in farm and commercial buildings, and shall not be installed where they obstruct entrance to these building openings.

**(D) Clearances from Buildings for Conductors of Not Over 600 Volts, Nominal, Above Roofs.**

**(1) General.** Overhead spans of open conductors and open multiconductor cables shall have a vertical clearance of not less than 2.5 m (8 ft) above the roof surface.

**(2) Vertical Clearance.** The vertical clearance above the roof level shall be maintained for a distance not less than 900 mm (3 ft) in all directions from the edge of the roof.

*Exception No. 1: The area above a roof surface subject to pedestrian or vehicular traffic shall have a vertical clearance from the roof surface in accordance with the clearance requirements of 410.7(B).*

*Exception No. 2: Where the voltage between conductors does not exceed 300, and the roof has a slope of 100 mm (4 in.) in 300 mm (12 in.) or greater, a reduction in clearance to 900 mm (3 ft) shall be permitted.*

*Exception No. 3: Where the voltage between conductors does not exceed 300, a reduction in clearance above only the overhanging portion of the roof to not less than 450 mm (18 in.) shall be permitted if (1) not more than 1.8 m (6 ft) of the conductors, 1.2 m (4 ft) horizontally, pass above the roof overhang and (2) they are terminated at a through-the-roof raceway or approved support.*

*Exception No. 4: The requirement for maintaining the vertical clearance 900 mm (3 ft) from the edge of the roof shall not apply to the final conductor span where the conductors are attached to the side of a building.*

**(E) Location of Outdoor Lamps.** Locations of lamps for outdoor lighting shall be below all energized conductors, transformers, or other electric utilization equipment, unless either of the following applies:

(1) Clearances or other safeguards are provided for relamping operations.
(2) Equipment is controlled by a disconnecting means that can be locked in the open position.

### 410.8 Services

**(A) Service Equipment—Disconnecting Means.** Means shall be provided to disconnect all conductors in a building or other structure from the service-entrance conductors.

No maximum distance is specified from the service entrance to a readily accessible location for the service disconnecting means. The authority enforcing NFPA 70E has the responsibility for, and is charged with, making the decision on how far inside the building the service-entrance conductors are allowed to travel to the main disconnecting means. The length of service-entrance conductors should be kept to a minimum inside buildings, because power utilities provide limited overcurrent protection. In event of a fault, the service conductors could ignite nearby combustible materials.

Some local ordinances allow service-entrance conductors to run within the building up to a specified length before terminating at the service disconnecting means. The authority having jurisdiction may permit service conductors to bypass fuel storage tanks or gas meters and similar equipment, permitting the service disconnecting means to be located in a readily accessible location. However, if the authority judges the distance as being excessive, the disconnecting means may be required to be located on the outside of the building or near the building at a readily accessible location that is not necessarily nearest the point of entrance of the conductors. See also 230.6 of the *NEC* and Exhibit 410.10 in this standard for conductors considered to be outside a building.

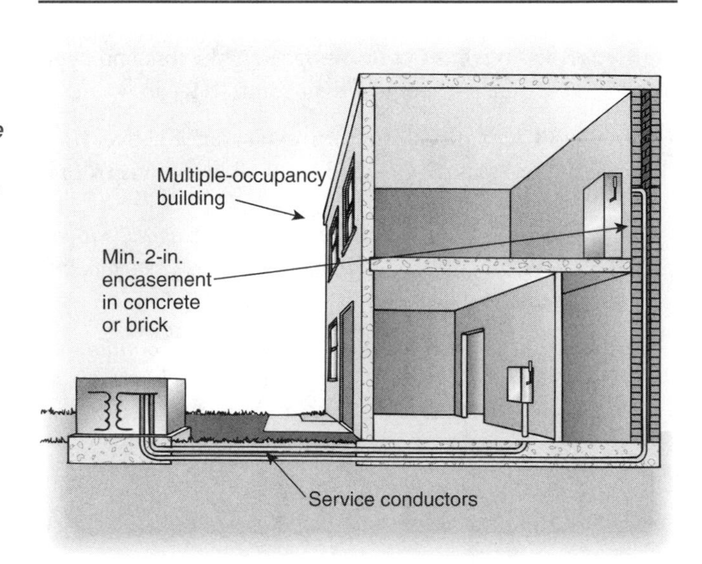


***EXHIBIT 410.10** Service conductors considered outside a building where installed under not less than 2 in. of concrete beneath the building or in a raceway encased by 2 in. of concrete or brick within the building.*

See 404.8(A) of the *NEC* for mounting-height restrictions for switches and for circuit breakers used as switches.

**(1) Location.** The service disconnecting means shall be installed in accordance with (a), (b), and (c) of this section.

(a) Readily Accessible Location. The service disconnecting means shall be installed at a readily accessible location either outside of a building or structure or inside nearest the point of entrance of the service conductors.

(b) Bathrooms. Service disconnecting means shall not be installed in bathrooms.

(c) Remote Control. Where a remote control device(s) is used to actuate the service disconnecting means, the service disconnecting means shall be located in accordance with 410.8(A)(1)(a)

**(2) Marking.** Each service disconnect shall be permanently marked to identify it as a service disconnect.

**(3) Suitable for Use.** Each service disconnecting means shall be suitable for the prevailing conditions.

**(B) Services Exceeding 600 Volts, Nominal.**

**(1) Locked Rooms or Enclosures.** The entrances to all buildings, rooms, or enclosures containing exposed live parts or exposed conductors operating at over 600 volts, nominal, shall be kept locked unless such entrances are under the observation of a qualified person at all times.

**(2) Warning Signs.** Where the voltage exceeds 600 volts, nominal, permanent and conspicuous warning signs shall be provided, reading as follows:

DANGER—HIGH VOLTAGE—KEEP OUT

FPN: For further information on hazard signs and labels, see ANSI Z535-4, *Product Signs and Safety Labels.*

**410.9 Overcurrent Protection**

**(A) 600 Volts, Nominal, or Less.**

**(1) Protection of Conductors and Equipment.** Conductors and equipment shall be protected from overcurrent in accordance with their ability to safely conduct current.

**(2) Grounded Conductors.** No overcurrent device shall be connected in series with any conductor that is intentionally grounded, unless one of the following two conditions is met:

(1) The overcurrent device opens all conductors of the circuit, including the grounded conductor, and is designed so that no pole can operate independently.
(2) Where required for motor overload protection.

**(3) Disconnecting Means for Fuses.** A disconnecting means shall be provided on the supply side of all fuses in circuits over 150 volts to ground and of cartridge fuses in circuits of any voltage where accessible to other than qualified persons so that each individual circuit containing fuses can be independently disconnected from the source of power. A current-limiting device without a disconnecting means shall be permitted on the supply side of the service disconnecting means. A single disconnecting means shall be permitted on the supply side of more than one set of fuses as permitted by the exception to 420.10(E)(2)(e), for group operation of motors and for fixed electric space-heating equipment.

**(4) Arcing or Suddenly Moving Parts.** Arcing or suddenly moving parts shall comply with (a) and (b) of this section.

(a) Location. Fuses and circuit breakers shall be located or shielded so that persons will not be burned or otherwise injured by their operation.

(b) Suddenly Moving Parts. Handles or levers of circuit breakers, and similar parts that may move suddenly in such a way that persons in the vicinity are likely to be injured by being struck by them, shall be guarded or isolated.

**(5) Circuit Breakers.** Circuit breakers shall clearly indicate whether they are in the open off or closed on position. Where circuit breaker handles are operated vertically rather than rotationally or horizontally, the up position of the handle shall be the on position.

(a) Used as Switches. Circuit breakers used as switches in 120 volt and 277 volt fluorescent lighting circuits shall be listed and shall be marked SWD or HID. Circuit breakers used as switches in high-intensity discharge lighting circuits shall be listed and shall be marked as HID.

(b) Applications. A circuit breaker with a straight voltage rating, such as 240 V or 480 V, shall be permitted to be applied in a circuit in which the nominal voltage between any two conductors does not exceed the circuit breaker's voltage rating. A two-pole circuit breaker shall not be used for protecting a 3-phase, corner-grounded delta circuit unless the circuit breaker is marked 1f–3f to indicate such suitability. A circuit breaker with a slash rating, such as 120/240 V or 480Y/277 V, shall be permitted to be applied in a solidly grounded circuit where the nominal voltage of any conductor to ground does not exceed the lower of the two values of the circuit breaker's voltage rating and the nominal voltage between any two conductors does not exceed the higher value of the circuit breaker's voltage rating.

FPN: Proper application of molded case circuit breakers on 3-phase systems, other than solidly grounded wye, particularly on corner-grounded delta systems, considers the circuit breaker's individual pole-interrupting capability.

**(B) Overcurrent Protection, Over 600 Volts, Nominal Feeders and Branch Circuits.**

**(1) Location and Type of Protection.** Feeder and branch-circuit conductors shall have overcurrent protection in each ungrounded conductor located at the point where the conductor receives its supply or at an alternative location in the circuit when designed under engineering supervision that includes but is not limited to consideration of the appropriate fault studies and time-current coordination analysis of the protective devices and the conductor damage curves. The overcurrent protection shall be permitted to be provided by either 410.9(B)(1)(a) or 410.9(B)(1)(b).

(a) Overcurrent Relays and Current Transformers. Circuit breakers used for overcurrent protection of 3-phase circuits shall have a minimum of three overcurrent relay elements operated from three current transformers. The separate overcurrent relay elements (or protective functions) shall be permitted to be part of a single electronic protective relay unit. On 3-phase, 3-wire circuits, an overcurrent relay in the residual circuit of the current transformers shall be permitted to replace one of the phase relays. An overcurrent relay element, operated from a current transformer that links all phases of a 3-phase, 3-wire circuit, shall be permitted to replace the residual relay element and one of the phase-conductor current transformers. Where the neutral is not regrounded on the load side of the circuit, the current transformer shall be permitted to link all 3-phase conductors and the grounded circuit conductor (neutral).

(b) Fuses. A fuse shall be connected in series with each ungrounded conductor.

**(2) Protective Devices.** The protective device(s) shall be capable of detecting and interrupting all values of current that can occur at their location in excess of their trip setting or melting point.

**(3) Conductor Protection.** The operating time of the protective device, the available short-circuit current, and the conductor used shall be coordinated to prevent damaging or dangerous temperatures in conductors or conductor insulation under short-circuit conditions.

**(C) Additional Requirements for Feeders.**

**(1) Rating or Setting of Overcurrent Protective Devices.** The continuous ampere rating of a fuse shall not exceed three times the ampacity of the conductors. The long-time trip element setting of a breaker or the minimum trip setting of an electronically actuated fuse shall not exceed six times the ampacity of the conductor. For fire pumps, conductors shall be permitted to be protected for overcurrent.

**(2) Feeder Taps.** Conductors tapped to a feeder or connection to a transformer secondary shall be permitted to be protected by the feeder overcurrent device where that overcurrent device also protects the tap conductor.

### 410.10 Grounding

Sections 410.10(A) through 410.10(G) cover grounding requirements for systems, circuits, and equipment.

**(A) Grounding Path.** The path to ground from circuits, equipment, and enclosures shall be permanent, continuous, and effective.

**(B) General Bonding.** Bonding shall be provided where necessary to ensure electrical continuity and the capacity to conduct safely any fault current likely to be imposed.

**(C) Systems to Be Grounded.** The following systems, which supply premises wiring, shall be grounded:

**(1) Three-Wire, Direct-Current Systems.** The neutral conductor of all 3-wire, dc systems supplying premises wiring shall be grounded.

**(2) Two-Wire, Direct-Current Systems.** A 2-wire, dc system supplying premises wiring and operating at greater than 50 volts but not greater than 300 volts shall be grounded.

*Exception No. 1: A system equipped with a ground detector and supplying only industrial equipment in limited areas shall not be required to be grounded.*

*Exception No. 2: A rectifier-derived dc system supplied from an ac system complying with 410.10(C) shall not be required.*

*Exception No. 3: Direct-current fire alarm circuits having a maximum current of 0.030 ampere shall not be required to be grounded.*

**(3) Alternating-Current Circuits of Less Than 50 Volts.** Alternating-current circuits of less than 50 volts shall be grounded under any of the following conditions:

(1) Where supplied by transformers, if the transformer supply systems exceeds 150 volts to ground
(2) Where supplied by transformers, if the transformer supply system is ungrounded
(3) Where installed as overhead conductors outside of buildings

**(4) Alternating-Current Systems of 50 Volts to 1000 Volts.** AC systems of 50 volts to 1000 volts that supply premises wiring systems shall be grounded under any of the following conditions:

(1) Where the system can be grounded so that the maximum voltage to ground on the ungrounded conductors does not exceed 150 volts

Exhibit 410.11 illustrates grounding requirements defined by 250.20(B)(1) of the *NEC* for a 120-volt, single-phase, 2-wire system and for a 120/240-volt, single-phase, 3-wire system. The selection of which conductor to be grounded is covered by 250.26 of the *NEC*.

***EXHIBIT 410.11*** *Typical systems required to be grounded by 250.20(B)(1) of the NEC. The conductor to be grounded is in accordance with 250.26 of the NEC.*

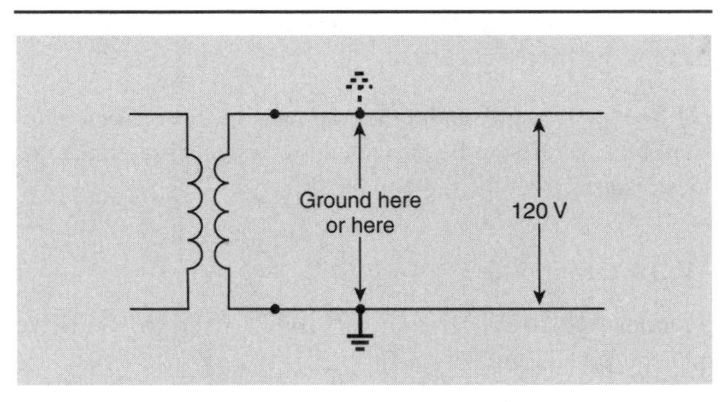


**120-V, single-phase, 2-wire system**

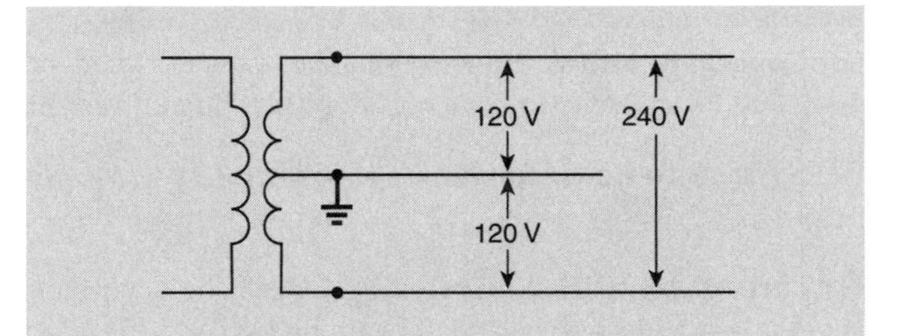


**120/240-V, single-phase, 3-wire system**

(2) Where the system is rated 3-phase, 4-wire, wye connected in which the neutral is used as a circuit conductor

(3) Where the system is rated 3-phase, 4-wire, delta connected in which the midpoint of one phase winding is used as a circuit conductor

Exhibit 410.12 illustrates which conductor is required to be grounded for all wye-connected systems that use the neutral conductor of a circuit conductor. Where the midpoint of one phase of a 3-phase, 4-wire delta system is used as a circuit conductor, the midpoint of the phase must be grounded, and the high-leg conductor must be identified. See 250.20(B)(2) and (B)(3) and 250.26 of the *NEC*.

**(5) Alternating-Current Systems of 50 Volts to 1000 Volts Not Required to Be Grounded.** The following ac systems of 50 volts to 1000 volts shall be permitted to be grounded but shall not be required to be grounded:

(1) Electrical systems used exclusively to supply industrial electric furnaces for melting, refining, tempering, and the like

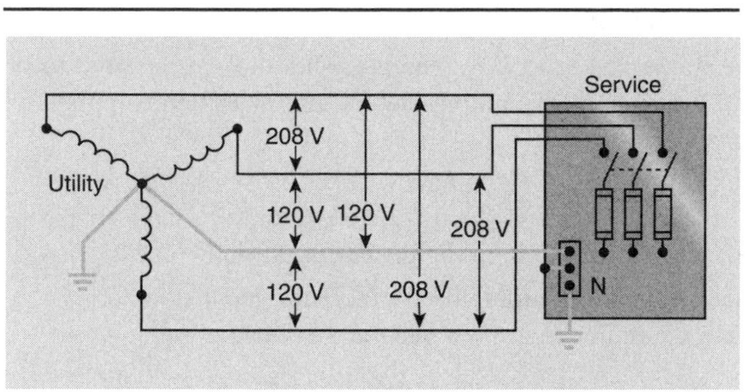


208Y/120-V, 3-phase, 4-wire wye system

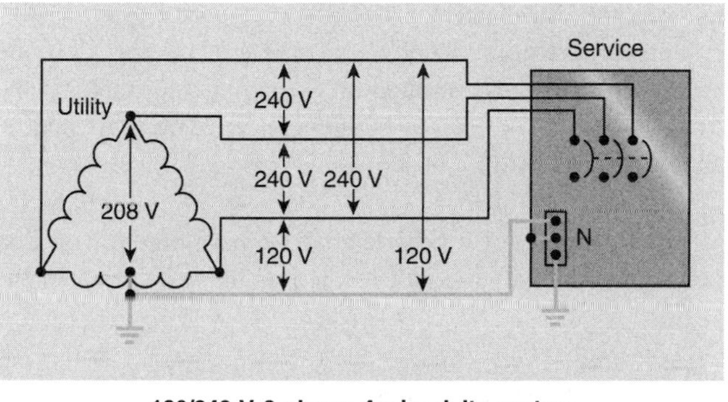


120/240-V, 3-phase, 4-wire delta system

***EXHIBIT 410.12*** *Typical systems required to be grounded by 250.20(B)(2) and (B)(3) of the NEC. The conductor to be grounded must be in accordance with NEC 250.26.*

(2) Separately derived systems used exclusively for rectifiers that supply only adjustable speed industrial drives

(3) Separately derived systems supplied by transformers that have a primary voltage rating less than 1000 volts, provided that all of the following conditions are met:

a. The system is used exclusively for control circuits.

b. The conditions of maintenance and supervision ensure that only qualified persons service the installation.

c. Continuity of control power is required.

d. Ground detectors are installed on the control system.

(4) Where high-impedance grounded neutral systems in which a grounding impedance, usually a resistor, limits the ground-fault current to a low value shall be permitted for 3-phase ac systems of 480 volts to 1000 volts where all of the following conditions are met:

a. The conditions of maintenance and supervision ensure that only qualified persons service the installation.

b. Continuity of power is required.

c. Ground detectors are installed on the system.

d. Line-to-neutral loads are not served.

(5) Other systems that are not required to be grounded in accordance with the requirements of 410.10(C).

**(6) Alternating Current Systems of 1 kV and Over.** Alternating-current systems supplying mobile or portable equipment shall be grounded. Where supplying other than mobile or portable equipment, such systems shall be permitted to be grounded.

**(7) Portable and Vehicle-Mounted Generators.**

(a) Portable Generators. The frame of a portable generator shall not be required to be grounded and shall be permitted to serve as the grounding electrode for a system supplied by the generator under the following conditions:

(1) The generator supplies only equipment mounted on the generator, cord-and-plug-connected equipment through receptacles mounted on the generator, or both.
(2) The non–current-carrying metal parts of equipment and the equipment grounding conductor terminals of the receptacles are bonded to the generator frame.

The word *portable* means that the equipment is easily carried by personnel from one location to another. The word *mobile* describes equipment, such as vehicle-mounted generators, that is capable of being moved, on wheels or rollers, for example.

The frame of a portable generator is not required to be connected to earth (by a ground rod, water pipe, etc.) if the generator has receptacles mounted on the generator panel and the receptacles have equipment-grounding terminals bonded to the generator frame. In this case, the generator frame serves the same purpose as earth.

(b) Vehicle-Mounted Generators. The frame of a vehicle shall be permitted to serve as the grounding electrode for a system supplied by a generator located on the vehicle under the following conditions:

(1) The frame of the generator is bonded to the vehicle frame.
(2) The generator supplies only equipment located on the vehicle or cord-and-plug-connected equipment through receptacles mounted on the vehicle, or both equipment located on the vehicle and cord-and-plug-connected equipment through receptacles mounted on the vehicle or on the generator.
(3) The non–current-carrying metal parts of equipment and the equipment grounding conductor terminals of the receptacles are bonded to the generator frame.
(4) The system complies with all other provisions of 410.10.

The neutral point of vehicle-mounted generators that have a neutral conductor is required to be bonded both to the generator frame and the vehicle frame. These generators serve as separately derived systems. All non–current-carrying parts of the equipment must be bonded to the generator frame.

(c) Grounded Conductor Bonding. A system conductor that is required to be grounded shall be bonded to the generator frame where the generator is a component of a separately derived system.

Portable and vehicle-mounted generators that are installed as separately derived systems and that provide a neutral conductor (such as 3-phase, 4-wire, wye-connected; single-phase 240/120 volt; or 3-phase, 4-wire, delta-connected) are required to have the neutral point bonded to the generator frame.

**(D) Grounding Connections.**

**(1) For Grounded Systems.** The connection shall be made by bonding the equipment grounding conductor to the grounded service conductor and the grounding electrode conductor.

**(2) For Ungrounded Systems.** The connection shall be made by bonding the equipment grounding conductor to the grounding electrode conductor.

**(3) Nongrounding Receptacle Replacement or Branch-Circuit Extension.** The equipment grounding conductor of a grounding-type receptacle or a branch-circuit extension shall be permitted to be connected to any of the following:

(1) Any accessible point on the grounding electrode system
(2) Any accessible point on the grounding electrode conductor
(3) The equipment grounding terminal bar within the enclosure where the branch circuit for the receptacle or branch circuit originates
(4) For grounded systems, the grounded service conductor within the service equipment enclosure
(5) For ungrounded systems, the grounding terminal bar within the service equipment enclosure

**(E) Enclosure, Raceway, and Service Cable Grounding.**

**(1) Service Raceways and Enclosures.** Metal enclosures and raceways for service conductors and equipment shall be grounded.

*Exception: A metal elbow that is installed in an underground installation of rigid nonmetallic conduit and is isolated from possible contact by a minimum cover of 450 mm (18 in.) to any part of the elbow shall not be required to be grounded.*

**(2) Service Equipment Enclosures.** Metal enclosures for service equipment shall be grounded.

**(3) Frames of Ranges and Clothes Dryers.** Frames of electric ranges, wall-mounted ovens, counter-mounted cooking units, clothes dryers, and outlet or junction boxes that are part of the circuit for these appliances shall be grounded.

**(4) Fixed Equipment.** Exposed non current-carrying metal parts of fixed equipment likely to become energized shall be grounded under any of the following conditions:

(1) Where within 2.5 m (8 ft) vertically or 1.5 m (5 ft) horizontally of ground or grounded metal objects and subject to contact by persons
(2) Where located in a wet or damp location and not isolated
(3) Where in electrical contact with metal
(4) Where in a hazardous (classified) location
(5) Where supplied by a metal-clad, metal-sheathed, metal raceway, or other wiring method that provides an equipment ground, except for short sections of metal enclosures
(6) Where equipment operates with any terminal at over 150 volts to ground

*Exception No. 1: Metal frames of electrically heated appliances, exempted by special permission, in which case the frames shall be permanently and effectively insulated from ground.*

*Exception No. 2: Distribution apparatus, such as transformer and capacitor cases, mounted on wooden poles, at a height exceeding 2.5 m (8 ft) above ground or grade level.*

*Exception No. 3: Listed equipment protected by a system of double insulation, or its equivalent, shall not be required to be grounded. Where such a system is employed, the equipment shall be distinctively marked.*

**(5) Equipment Connected by Cord and Plug.** Under any of the conditions described in (1) through (3), exposed non–current-carrying metal parts of cord-and-plug-connected equipment likely to become energized shall be grounded.

*Exception: Listed tools, listed appliances, and listed equipment covered in (2) and (3) shall not be required to be grounded where protected by a system of double insulation or its equivalent. Double insulated equipment shall be distinctively marked.*

(1) In hazardous (classified) locations *(see Article 440)*
(2) Where operated at over 150 volts to ground

*Exception No. 1: Motors, where guarded, shall not required to be grounded.*

*Exception No. 2: Metal frames of electrically heated appliances, exempted by special permission, shall not be required to be grounded, in which case the frames shall be permanently and effectively insulated from ground.*

(3) For the following in other than residential occupancies:
   a. Refrigerators, freezers, and air conditioners
   b. Clothes-washing, clothes-drying, dishwashing machines, information technology equipment, sump pumps, and electric aquarium equipment
   c. Hand-held motor-operated tools, stationary and fixed motor-operated tools, and light industrial motor-operated tools
   d. Motor-operated appliances of the following types: hedge clippers, lawn mowers, snow blowers, and wet scrubbers
   e. Cord-and-plug-connected appliances used in damp or wet locations or by persons standing on the ground or on metal floors or working inside of metal tanks or boilers
   f. Tools likely to be used in wet and conductive locations

*Exception to (f): Tools and portable hand lamps likely to be used in wet or conductive locations shall not be required to be grounded where supplied through an isolating transformer with an ungrounded secondary of not over 50 volts.*

**(6) Nonelectric Equipment.** The metal parts of nonelectrical equipment described in this section shall be grounded:

(1) Frames and tracks of electrically operated cranes and hoists
(2) Frames of nonelectrically driven elevator cars to which electric conductors are attached
(3) Hand-operated metal shifting ropes or cables of electric elevators

FPN: Where extensive metal in or on buildings could become energized and is subject to personal contact, adequate bonding and grounding will provide additional safety.

**(F) Equipment Considered Effectively Grounded.** Under the conditions in 410.10(F)(1) and 410.10(F)(2), the non–current-carrying metal parts of the equipment shall be considered effectively grounded.

Exhibit 410.13 illustrates electrical equipment secured to and in electrical contact with a metal rack that is effectively grounded in accordance with 250.136(A) of the *NEC.*

**(1) Equipment Secured to Grounded Metal Supports.** Electric equipment secured to and in electrical contact with a metal rack or structure provided for its support shall be considered to be effectively grounded. The structural metal frame of a building shall not be used as the required equipment grounding conductor for ac equipment.

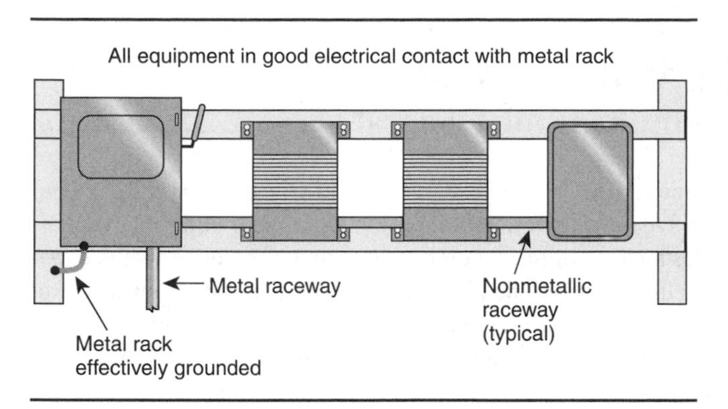


***EXHIBIT 410.13*** *An example of equipment grounding on a metal rack.*

**(2) Metal Car Frames.** Metal car frames supported by metal hoisting cables attached to or running over metal sheaves or drums of elevator machines shall also be considered to be effectively grounded.

**(G) Grounding of Systems and Circuits of 1 kV and Over (High Voltage)**

**(1) General.** Where high-voltage systems are grounded, they shall comply with all applicable provisions of 410.10 and with 410.10(F)(2) and 410.10(F)(3), which supplement and modify the preceding sections.

**(2) Grounding of Systems Supplying Portable or Mobile Equipment.** Systems supplying portable or mobile high-voltage equipment, other than substations installed on a temporary basis, shall comply with 410.10(G)(2)(a) through 410.10(G)(2)(e).

(a) Portable or Mobile Equipment. Portable or mobile high-voltage equipment shall be supplied from a system having its neutral grounded through an impedance. Where a delta-connected high-voltage system is used to supply portable or mobile equipment, a system neutral shall be derived.

(b) Exposed Non–Current-Carrying Metal Parts. Exposed non–current-carrying metal parts of portable or mobile equipment shall be connected by an equipment grounding conductor to the point at which the system neutral impedance is grounded.

(c) Ground-Fault Current. The voltage developed between the portable or mobile equipment frame and ground by the flow of maximum ground-fault current shall not exceed 100 volts.

(d) Ground-Fault Detection and Relaying. Ground-fault detection and relaying shall be provided to automatically deenergize any high-voltage system component that has developed a ground fault. The continuity of the equipment grounding conductor shall be continuously monitored so as to deenergize automatically the high-voltage circuit to the portable or mobile equipment upon loss of continuity of the equipment grounding conductor.

(e) Isolation. The grounding electrode to which the portable or mobile equipment system neutral impedance is connected shall be isolated from and separated in the ground by at least 6.0 m (20 ft) from any other system or equipment grounding electrode, and there shall be no direct connection between the grounding electrodes, such as buried pipe and fence.

**(3) Grounding of Equipment.** All non–current-carrying metal parts of fixed, portable, and mobile equipment and associated fences, housings, enclosures, and supporting structures shall be grounded.

*Exception: Where isolated from ground and located so as to prevent any person who can make contact with ground from contacting such metal parts when the equipment is energized.*

# ARTICLE 420 Wiring Methods, Components, and Equipment for General Use

Common wiring method requirements in Chapter 3 of NFPA 70 (the *National Electrical Code*®, *NEC*®, or *Code*) are aligned using a common numbering system. Circular raceway articles and many of the cable articles have a common numbering format. The objective is to assist users in locating the wiring methods within the chapter. Users will find it easier to learn, compare, and understand the many common requirements of various wiring methods. This format allows similar articles to be grouped and provides space for new wiring method articles that might be developed in the future.

Annex F of the *NEC* contains three cross-reference tables to guide the user through Chapter 3 since it was reorganized.

### 420.1 Wiring Methods

The provisions of this chapter are not intended to apply to the conductors that form an integral part of equipment, such as motors, controllers, motor control centers, factory assembled control equipment, or listed utilization equipment.

Generally, wiring internal to electrical equipment is covered by product standards. For example, internal wiring of motors is covered by NEMA *MG-1, Motors and Generators;* internal wiring of industrial control equipment is covered by UL 508, *Standard for Safety of Industrial Control Equipment;* and internal wiring of industrial machinery is covered by NFPA 79, *Electrical Standard for Industrial Machinery.*

**(A) Bonding Other Enclosures.**

**(1) General.** Metal raceways, cable trays, cable armor, cable sheath, enclosures, frames, fittings, and other metal non–current-carrying parts that are to serve as grounding conductors, with or without the use of supplementary equipment grounding conductors, shall be effectively bonded where necessary to ensure electrical continuity and the capacity to conduct safely any fault current likely to be imposed on them. Any nonconductive paint, enamel, or similar coating shall be removed at threads, contact points, and contact surfaces or be connected by means of fittings designed so as to make such removal unnecessary.

**(2) Isolated Grounding Circuits.** Where required for the reduction of electrical noise (electromagnetic interference) of the grounding circuit, an equipment enclosure supplied by a branch circuit shall be permitted to be isolated from a raceway containing circuits that supply only that equipment by one or more listed nonmetallic raceway fittings located at the point of attachment of the raceway to the equipment enclosure. The metal raceway shall comply with provisions of this standard and shall be supplemented by an internal insulated equipment grounding conductor installed to ground the equipment enclosure.

FPN: Use of an isolated equipment grounding conductor does not relieve the requirement for grounding the raceway system.

**(3) Ducts for Dust, Loose Stock, or Vapor Removal.** No wiring systems of any type shall be installed in ducts used to transport dust, loose stock, or flammable vapors. No wiring system of any type shall be installed in any duct, or any shaft containing only such ducts, used for vapor removal or for ventilation of commercial-type cooking equipment.

**(B) Temporary Wiring.** Temporary electrical power and lighting wiring methods may be of a class less than would be required for a permanent installation. Except as specifically modified in 420.1(B)(1)(a) through 420.1(B)(1)(d), all other requirements of this standard for permanent wiring shall apply to temporary wiring installations.

**(1) Time Constraints.**

(a) During the Period of Construction. Temporary electrical power and lighting installations shall be permitted during the period of construction, remodeling, maintenance, repair, or demolition of buildings, structures, equipment, or similar activities.

(b) 90 Days. Temporary electrical power and lighting installations shall be permitted for a period not to exceed 90 days for holiday decorative lighting and similar purposes.

The 90-day time limit required by 420.1(B) applies only to temporary electrical installations associated with holiday displays. Installations for construction, emergency, and test wiring are not bound by this time limit.

(c) Emergencies and Tests. Temporary electrical power and lighting installations shall be permitted during emergencies and for tests, experiments, and developmental work.

(d) Removal. Temporary wiring shall be removed immediately upon completion of construction or purpose for which the wiring was installed.

Due to the modifications permitted by Article 527 of the *NEC,* temporary wiring installations might not meet all of the requirements for a permanent installation. Therefore, all temporary wiring not only must be disconnected, but it also must be removed from the building, structure, or other location of installation when the need for the temporary wiring has expired.

**(2) General Requirements for Temporary Wiring.**

(a) Feeders. Feeders shall be protected as provided in 410.9. They shall originate in an approved distribution center. Conductors shall be permitted within cable assemblies or within multiconductor cords or cables of a type identified for hard usage or extra-hard usage. For the purpose of this section, Type NM and Type NMC cables shall be permitted to be used in any dwelling, building, or structure without any height limitation.

Temporary feeders are permitted to be (1) cable assemblies, (2) multiconductor cords, or (3) single-conductor cords. Cords used as feeders must be identified for hard or extra-hard usage according to Table 400.4 of the *NEC.* Individual conductors, as described in Table 310.13 of the *NEC,* are not permitted as open conductors. They must be part of a cable assembly or used in a raceway system. Open or individual conductor feeders are permitted only during emergencies or testing operations.

The authority having jurisdiction must approve temporary wiring methods. [See 527.2(B) of the *NEC.*]

*Exception: Single insulated conductors shall be permitted where installed for the purpose(s) specified in 420.1(B)(3)(c) where accessible only to qualified persons.*

(b) Branch Circuits. All branch circuits shall originate in an approved power outlet or panelboard. Conductors shall be permitted within cable assemblies or within multiconductor cord or cable of a type identified for hard usage or extra-hard usage. All conductors shall be protected as provided in 410.9. For the purpose of this section, Type NM and NMC cables shall be permitted to be used in any dwelling, building, or structure without any height limitation.

The basic requirement for safety in 420.1(B)(2)(b) is that temporary wiring must be installed such that it will not be physically damaged. Section 527.2(A) of the *NEC* requires that temporary wiring comply with the requirements of the appropriate Chapter 3 article for the wiring method employed unless modified in Article 527.

Note that hard-usage or extra-hard-usage extension cords are permitted to be laid on the floor.

*Exception: Branch circuits installed for the purposes specified in 420.1(B)(3)(b) or 420.1(B)(3)(c) shall be permitted to be run as single insulated conductors. Where the wiring is installed in accordance with 420.1(B)(1)(c), the voltage to ground shall not exceed 150 volts, the wiring shall not be subject to physical damage, and the conductors shall be supported on insulators at intervals of not more than 3.0 m (10 ft); or, for festoon lighting, the conductors shall be arranged so that excessive strain is not transmitted to the lampholders.*

(c) Receptacles. All receptacles shall be of the grounding type. Unless installed in a continuous grounded metal raceway or metal-covered cable, all branch circuits shall contain a separate equipment grounding conductor, and all receptacles shall be electrically connected to the equipment grounding conductors. Receptacles on construction sites shall not be installed on branch circuits that supply temporary lighting. Receptacles shall not be connected to the same ungrounded conductor of multiwire circuits that supply temporary lighting.

Section 420.1(B)(2)(c) requires individual hot conductors for lighting and receptacle loads. The objective is that a blown fuse, tripped circuit breaker, or ground-fault circuit interrupter will not deenergize the lighting circuit in the event of a fault or equipment overload and leave the area without lights.

(d) Disconnecting Means. Suitable disconnecting switches or plug connectors shall be installed to permit the disconnection of all ungrounded conductors of each temporary circuit. Multiwire branch circuits shall be provided with a means to disconnect simultaneously all ungrounded conductors at the power outlet or panelboard where the branch circuit originated. Approved handle ties shall be permitted.

(e) Lamp Protection. All lamps for general illumination shall be protected from accidental contact or breakage by a suitable fixture or lampholder with a guard. Brass shell, paper-lined sockets, or other metal-cased sockets shall not be used unless the shell is grounded.

(f) Splices. On construction sites, a box shall not be required for splices or junction connections where the circuit conductors are multiconductor cord or cable assemblies, provided that the equipment grounding continuity is maintained with or without the box. A box, conduit body, or terminal fitting having a separately bushed hole for each conductor shall be used wherever a change is made to a conduit or tubing system or a metal-sheathed cable system.

(g) Protection from Accidental Damage. Flexible cords and cables shall be protected from accidental damage. Sharp corners and projections shall be avoided. Where passing through doorways or other pinch points, protection shall be provided to avoid damage.

Flexible cords and cables are permitted to pass through doorways, provided the doors are blocked in such a way that prevents damage by the door to the cord or cable.

(h) Termination(s) at Devices. Flexible cords and cables entering enclosures containing devices requiring termination shall be secured to the box with fittings designed for the purpose.

(i) Support. Cable assemblies and flexible cords and cables shall be supported in place at intervals that ensure that they will be protected from physical damage. Support shall be in the form of staples, cable ties, straps, or similar type fittings installed so as not to cause damage. Vegetation shall not be used for support of overhead spans of branch circuits or feeders.

Temporary wiring is not required to be supported as frequently as necessary for a permanent installation as would be required by Chapter 3 of the *NEC* for the particular wiring method. The expectation is that temporary wiring will be removed as soon as the construction is complete. Support for temporary wiring is needed only to minimize the chance of damage during the temporary period. However, vegetation must not be used to support branch-circuit and feeder conductors.

**(C) Cable Trays.**

**(1) Uses Permitted.** Cable tray shall be permitted to be used as a support system for services, feeders, branch circuits, communications circuits, control circuits, and signaling circuits. Cable tray installations shall not be limited to industrial establishments. Where exposed to direct rays of the sun, insulated conductors and jacketed cables shall be identified as being sunlight resistant. Cable trays and their associated fittings shall be identified for the intended use.

Typically, cable trays are an industrial-type wiring method. However, cable tray installations are not restricted to industrial installations. Cable tray support methods are applicable to commercial installations, especially as a wire-and-cable management system for telecommunications/data installations.

**(2) Wiring Methods.** The following wiring methods shall be permitted to be installed in cable tray systems: armored cable; communication raceways; electrical metallic tubing; electrical nonmetallic tubing; fire alarm cables; flexible metal conduit; flexible metallic tubing; instrumentation tray cable; intermediate metal conduit; liquidtight flexible metal conduit and liquidtight flexible nonmetallic conduit; metal-clad cable; mineral-insulated, metal-sheathed cable; multiconductor service-entrance cable; multiconductor underground feeder and branch-circuit cable; multipurpose and communications cables; nonmetallic-sheathed cable; power and control tray cable; power-limited tray cable; optical fiber cables; optical fiber raceways; other factory-assembled, multiconductor control, signal, or power cables that are specifically approved for installation in cable trays; rigid metal conduit; and rigid nonmetallic conduit.

Section 420.1(C)(2) identifies raceways and many cable types that are permitted to be supported in commercial and industrial cable tray installations. In the United States, cable tray is rarely used as a major raceway support system. For raceway support, strut systems are normally more versatile than cable tray.

This section does not identify all the specific cable types that are permitted to be installed in cable tray systems. Other factory assembled, multiconductor control, signal, and power cables that are specifically approved for installation in cable trays are permitted by 420.1(C)(1). Commentary Table 420.1 is a useful guide for locating requirements for various wiring methods in the *NEC*.

**(3) In Industrial Establishments.** The wiring methods in 420.1(C)(2) shall be permitted to be used in any industrial establishment under the conditions described in their respective articles. In industrial establishments only, where conditions of maintenance and supervision ensure that only qualified persons service the installed cable tray system, any of the cables in 420.1(C)(2)(a) and 420.1(C)(2)(b) shall be permitted to be installed in ladder, ventilated trough, solid bottom, or ventilated channel cable trays.

Section 420.1(C)(3) permits single-conductor cables (rated 0 to 2000 volts) and Type MV cables to be installed in ladder, ventilated-trough, or ventilated-channel cable trays, provided the installation is located in a qualifying industrial facility. Single-conductor and Type MV cable installations are not permitted in solid-bottom cable trays.

***COMMENTARY TABLE 420.1*** *Wiring Methods Covered by the NEC*

| *Wiring Method* | *Article* | *Section* |
|---|---|---|
| Armored cable | 320 | |
| Communication raceways | 800 | |
| Electrical metallic tubing | 358 | |
| Electrical nonmetallic tubing | 362 | |
| Fire alarm cables | 760 | |
| Flexible metal conduit | 348 | |
| Flexible metallic tubing | 360 | |
| Instrumentation tray cable | 727 | |
| Intermediate metal conduit | 342 | |
| Liquidtight flexible metal conduit | 350 | |
| Liquidtight flexible nonmetallic conduit | 356 | |
| Metal-clad cable | 330 | |
| Mineral-insulated, metal-sheathed cable | 332 | |
| Multiconductor service-entrance cable | 338 | |
| Multiconductor underground feeder and branch-circuit cable | 340 | |
| Multipurpose and communications cables | 800 | |
| Nonmetallic-sheathed cable | 334 | |
| Power and control tray cable | 336 | |
| Power-limited tray cable | | 725.61(C) and 725.71(F) |
| Optical fiber cables | 770 | |
| Optical fiber raceways | 770 | |
| Other factory-assembled, multiconductor control, signal, or power cables that are specifically approved for installation in cable trays | 336 | |
| Rigid metal conduit | 344 | |
| Rigid nonmetallic conduit | 352 | |

(a) Single Conductors. Single conductor cables shall be permitted to be installed in accordance with the following:

(1) Single conductor cable shall be 1/0 AWG or larger and shall be of a type listed and marked on the surface for use in cable trays. Where 1/0 AWG through 4/0 AWG single conductor cables are installed in ladder cable tray, the maximum allowable run spacing for the ladder cable tray shall be 230 mm (9 in.).
(2) Welding cables shall be installed in dedicated cable trays, as permitted.

Cable trays used to support welding cables are required to be dedicated for welding cable installation. (See *NEC* 630.42.)

(3) Single conductors used as equipment grounding conductors shall be insulated, covered, or bare, and they shall be 4 AWG or larger.

(b) Medium Voltage. Single- and multiconductor medium voltage cables shall be Type MV cable. Single conductors shall be installed in accordance with 420.1(C)(1).

(c) Equipment Grounding Conductors. Metallic cable trays shall be permitted to be used as equipment grounding conductors where continuous maintenance and supervision ensure that qualified persons service the installed cable tray system.

Section 392.3(C) of the *NEC* permits cable tray to serve as an equipment grounding conductor. This practice historically has been permitted in qualifying industrial installations. The limitation to qualifying industrial installations has been removed, and the practice now applies to all metallic cable tray systems. To qualify as an equipment-grounding conductor, the cable tray system must meet all four requirements of 392.7(B) of the *NEC.*

(d) Hazardous (Classified) Locations. Hazardous (classified) locations as permitted.

(e) Nonmetallic Cable Tray. Nonmetallic cable tray shall be permitted in corrosive areas and in areas requiring voltage isolation.

**(4) Uses Not Permitted.** Cable tray systems shall not be used in hoistways or where subject to severe physical damage. Cable tray systems shall not be used in environmental airspaces, except as permitted in 420.1(A)(2) to support wiring methods recognized for use in such spaces.

Section 300.22(C) of the *NEC* limits the wiring methods permitted to be used within "other spaces used for environmental air." One example of such spaces is underfloor wiring for computer rooms. Metallic cable trays may be used within these spaces only to support wiring methods permitted in these spaces. Permitted cable tray types include ladder, ventilated trough, ventilated channel, or solid bottom. The limiting factor is the wiring method, rather than the type of cable trays.

**(D) Open Wiring on Insulators.**

**(1) Uses Permitted.** Open wiring on insulators shall be permitted on systems of 600 volts, nominal, or less, as follows:

(1) Indoors or outdoors
(2) In wet or dry locations
(3) Where subject to corrosive vapors
(4) For services

See Tables 310.17 and 310.19 of the *NEC* for ampacities of conductors.

**(2) Securing and Supporting Conductor Sizes Smaller Than 8 AWG.** Conductors smaller than 8 AWG shall be rigidly supported on noncombustible, nonabsorbent insulating materials and shall not contact any other objects. Supports shall be installed as follows:

(1) Within 150 mm (6 in.) from a tap or splice
(2) Within 300 mm (12 in.) of a dead-end connection to a lampholder or receptacle
(3) At intervals not exceeding 1.4 m (4½ ft) and at closer intervals sufficient to provide adequate support where likely to be disturbed

Mill construction is considered to be a building in which floors and ceilings are supported by wood timbers or beams or wood cross members spaced approximately 15 ft apart. This type of construction is sometimes referred to as plank-on-timber construction. Section 420.1(D)(2) permits 8 AWG and larger conductors to span this distance where the ceilings are high and free of obstructions and where the conductors are unlikely to come into contact with other objects.

**(3) Exposed Work.**

(a) Dry Locations. In dry locations, where not exposed to severe physical damage, conductors shall be permitted to be separately enclosed in flexible nonmetallic tubing. The tubing shall be in continuous lengths not exceeding 4.5 m (15 ft) and secured to the surface by straps at intervals not exceeding 1.4 m (4½ ft).

(b) Entering Spaces Subject to Dampness, Wetness, or Corrosive Vapors. Conductors entering or leaving locations subject to dampness, wetness, or corrosive vapors shall have drip loops formed on them and shall then pass upward and inward from the outside of the buildings, or from the damp, wet, or corrosive location, through non-combustible, nonabsorbent insulating tubes.

(c) Exposed to Physical Damage. Conductors within 2.1 m (7 ft) from the floor shall be considered exposed to physical damage. Where open conductors cross ceiling joists and wall studs and are exposed to physical damage, they shall be protected by one of the following methods:

(1) Guard strips not less than 25 mm (1 in.) nominal in thickness and at least as high as the insulating supports, placed on each side of and close to the wiring.
(2) A substantial running board at least 13 mm (½ in.) thick in back of the conductors with side protections. Running boards shall extend at least 25 mm (1 in.) outside the conductors, but not more than 50 mm (2 in.), and the protecting sides shall be at least 50 mm (2 in.) high and at least 25 mm (1 in.) nominal in thickness.
(3) Boxing made in accordance with items (1) and (2) and furnished with a cover kept at least 25 mm (1 in.) away from the conductors within. Where protecting vertical conductors on side walls, the boxing shall be closed at the top and the holes through which the conductors pass shall be bushed.
(4) Rigid metal conduit, intermediate metal conduit, rigid nonmetallic conduit, or electrical metallic tubing, or by metal piping, in which case the conductors shall be encased in continuous lengths of approved flexible tubing.

**(4) Through or Parallel to Framing Members.** Open conductors shall be separated from contact with walls, floors, wood cross members, or partitions through which they pass by tubes or bushings of noncombustible, nonabsorbent insulating material. Where the bushing is shorter than the hole, a waterproof sleeve of noninductive material shall be inserted in the hole and an insulating bushing slipped into the sleeve at each end in such a manner as to keep the conductors absolutely out of contact with the sleeve. Each conductor shall be carried through a separate tube or sleeve.

### 420.2 Cabinets, Cutout Boxes, and Meter Socket Enclosures

**(A) Cabinets, Cutout Boxes, and Meter Socket Enclosures.** Conductors entering enclosures within the scope of this standard shall be protected from abrasion and shall comply with 420.2(A)(1) through 420.2(A)(3).

**(1) Openings to Be Closed.** Opening through which conductors enter shall be adequately closed.

**(2) Metal Cabinets, Cutout Boxes, and Meter Socket Enclosures.** Where metal enclosures within the scope of this standard are installed with open wiring or concealed knob-and-tube wiring, conductors shall enter through insulating bushings or, in dry locations, through flexible tubing extending from the last insulating support and firmly secured to the enclosure.

**(3) Cables.** Where cable is used, each cable shall be secured to the cabinet, cutout box, or meter socket enclosure.

The intent of 420.2(A)(3) is to prohibit the installation of several cables bunched together and run through a knockout or chase nipple. Individual cable clamps or connectors are required to be used with only one cable per clamp or connector, unless the clamp or connector is identified for more than a single cable.

*Exception: Cables with entirely nonmetallic sheaths shall be permitted to enter the top of a surface-mounted enclosure through one or more nonflexible raceways not less than 450 mm (18 in.) or more than 3.0 m (10 ft) in length, provided all the following conditions are met:*

*(a) Each cable is fastened within 300 mm (12 in.), measured along the sheath, of the outer end of the raceway.*

*(b) The raceway extends directly above the enclosure and does not penetrate a structural ceiling.*

*(c) A fitting is provided on each end of the raceway to protect the cable(s) from abrasion and the fittings remain accessible after installation.*

*(d) The raceway is sealed or plugged at the outer end using approved means so as to prevent access to the enclosure through the raceway.*

*(e) The cable sheath is continuous through the raceway and extends into the enclosure beyond the fitting not less than 6 mm (¼ in.).*

*(f) The raceway is fastened at its outer end and at other points in accordance with applicable section.*

*(g) Where installed as conduit or tubing, the allowable cable fill does not exceed that permitted for complete conduit or tubing systems.*

The exception to 420.2(A), which was added for the 1999 *NEC,* spells out requirements that allow multiple nonmetallic cables, such as Type NM, NMC, NMS, UF, SE, and USE, to enter the top of a surface-mounted enclosure through a nonflexible raceway sleeve or nipple. These sleeves or nipples are permitted to be 18 in. to 10 ft in length. However, if the nipple length exceeds 24 in., ampacity of the conductors must be adjusted by the factors specified in 310.15(B)(2) of the *NEC.*

**(B) Covers and Canopies.** In completed installations, each box shall have a cover, faceplate, or fixture canopy.

**(1) Nonmetallic or Metal Covers and Plates.** Nonmetallic or metal covers and plates shall be permitted. Where metal covers and plates are used, they shall comply with grounding requirements.

**(2) Exposed Combustible Wall or Ceiling Finish.** Where a luminaire (fixture) canopy or pan is used, any combustible wall or ceiling finish exposed between the edge of the canopy or pan and the outlet box shall be covered with noncombustible material.

Heat resulting from a short circuit, a ground fault, or over-lamping could create a fire hazard within a fixture canopy or pan. Therefore, according to 420.2(B)(2), any exposed combustible wall or ceiling space between the edge of the outlet box and the perimeter of the luminaire must be covered with noncombustible material. The noncombustible material is not required to be metal. Glass fiber pads commonly provided as thermal barriers within the ceiling pan of luminaires may be used to fulfill this requirement. Where the wall or ceiling finish is concrete, tile, gypsum, plaster, or similar noncombustible material, the requirements of this section do not apply. Section 314.20 of the *NEC* identifies requirements for setback from finish surfaces for flush mounted boxes and boxes.

**(3) Flexible Cord Pendants.** Covers of outlet boxes and conduit bodies having holes through which flexible cord pendants pass shall be provided with bushings designed for the purpose or shall have smooth, well-rounded surfaces on which the cords may bear. So-called hard rubber or composition bushings shall not be used.

**(C) Pull and Junction Boxes for Use on Systems Over 600 Volts, Nominal.** In addition to other requirements in this standard for pull and junction boxes, (1) and (2) shall apply:

(1) Boxes shall provide a complete enclosure for the contained conductors or cables.
(2) Boxes shall be closed by suitable covers securely fastened in place. Underground box covers that weigh over 45 kg (100 lb) shall be considered as meeting this requirement. Covers for boxes shall be permanently marked "DANGER—HIGH VOLTAGE—KEEP OUT." The marking shall be on the outside of the box cover and shall be readily visible. Letters shall be block type and at least 13 mm (½ in.) in height.

FPN: For further information on hazard signs and labels, see ANSI Z535–4, *Product Signs and Safety Labels.*

**420.3 Position and Connection of Switches**

**(A) Single-Throw Knife Switches.** Single-throw knife switches shall be placed so that gravity will not tend to close them. Single-throw knife switches approved for use in the inverted position shall be provided with a locking device that ensures that the blades remain in the open position when so set.

**(B) Double-Throw Knife Switches.** Double-throw knife switches shall be permitted to be mounted so that the throw is either vertical or horizontal. Where the throw is vertical, a locking device shall be provided to hold the blades in the open position when so set.

**(C) Connection of Switches.** Single-throw knife switches and switches with butt contacts shall be connected so that their blades are de-energized when the switch is in the open position. Bolted pressure contact switches shall have barriers that prevent inadvertent contact with energized blades. Single-throw knife switches, bolted pressure contact switches, molded-case switches, switches with butt contacts, and circuit breakers used as switches shall be connected so that the terminals supplying the load are de-energized when the switch is in the open position.

Bolted pressure switches with blades that are energized when open, such as bottom-feed designs, must be provided with barriers or a means to guard against inadvertent contact with the energized blades. Section 420.3(C) is intended to provide protection against accidental contact with live parts in those cases where personnel are working on energized equipment. If the blades are energized when the switch is open, a label is necessary to warn of the unusual condition.

*Exception: The blades and terminals supplying the load of a switch shall be permitted to be energized when the switch is in the open position where the switch is connected to circuits or equipment inherently capable of providing a backfeed source of power. For such installations, a permanent sign shall be installed on the switch enclosure or immediately adjacent to open switches with the following words or equivalent:*

WARNING
LOAD SIDE TERMINALS MAY BE ENERGIZED BY BACKFEED

Batteries, generators, and double-ended switchboard ties are typical backfeed sources. These sources can cause the load side of the switch or circuit breaker to be energized when it is in the open position, a condition inherent to the circuit construction.

**(D) Provisions for Snap Switch Faceplates.**

**(1) Faceplates.** Faceplates provided for snap switches mounted in boxes and other enclosures shall be installed so as to completely cover the opening and, where the switch is flush mounted, seat against the finished surface.

**(2) Grounding.** Snap switches, including dimmer and similar control switches, shall be effectively grounded and shall provide a means to ground metal faceplates, whether or not a metal faceplate is installed. Snap switches shall be considered effectively grounded if either of the following conditions is met:

(1) The switch is mounted with metal screws to a metal box or to a nonmetallic box with integral means for grounding devices.

(2) An equipment grounding conductor or equipment bonding jumper is connected to an equipment grounding termination of the snap switch.

*Exception to (2): Where no grounding means exists within the snap-switch enclosure or where the wiring method does not include or provide an equipment ground, a snap switch without a grounding connection shall be permitted for replacement purposes only. A snap switch wired under the provisions of this exception and located within reach of earth, grade conducting floors, or other conducting surfaces shall be provided with a faceplate of nonconducting, noncombustible material.*

This section requires that switching devices, including snap switches, dimmers, and similar control devices, be grounded. Although the non–current-carrying metal parts of these devices typically are not subject to contact by personnel, there is concern about the use of metal faceplates, which do pose a shock hazard (touch potential) if they become energized. Therefore, the grounded switch must provide a means for grounding the metal faceplate.

The requirements in (1) or (2) of 420.3(D) describe the provisions to satisfy the main requirement. Switch plates in existing installations attached to switches in boxes without an equipment grounding conductor must be made of insulating material. Exhibit 420.1 illustrates how the effective grounding of a metal faceplate can be accomplished by connecting the equipment grounding conductor to a grounding terminal on a metal yoke or strap.

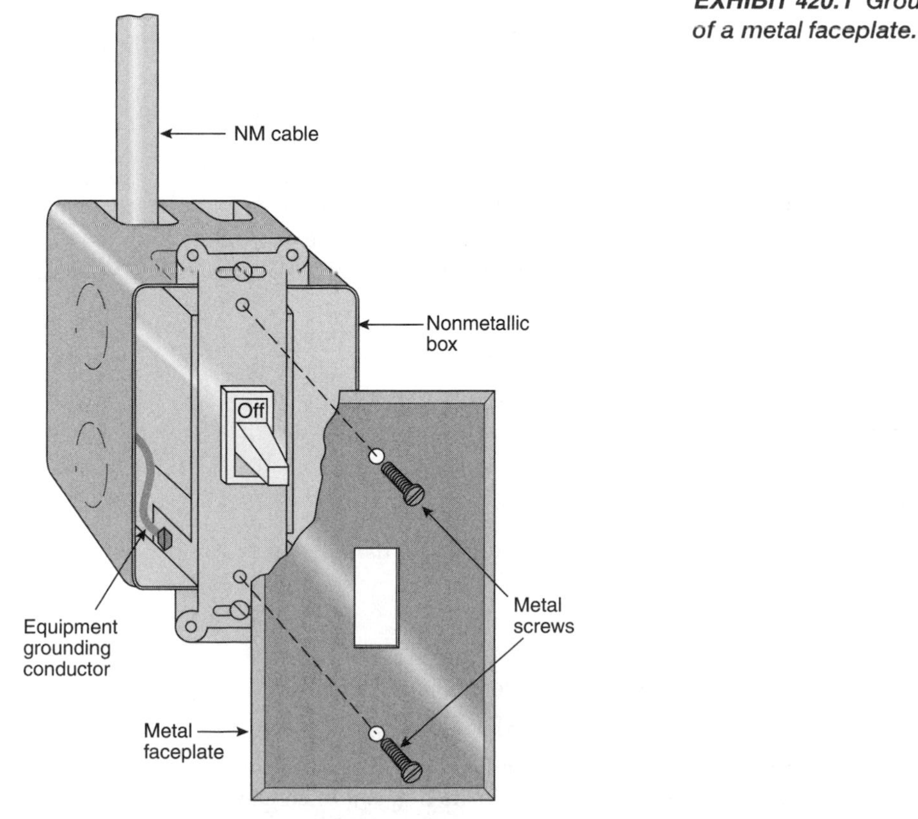


***EXHIBIT 420.1*** *Grounding of a metal faceplate.*

### 420.4 Switchboards and Panelboards

Switchboards that have any exposed live parts operating at 50 volts or more shall be located in permanently dry locations and then only where under competent supervision and accessible only to qualified persons. Switchboards shall be located so that the probability of damage from equipment or processes is reduced to a minimum. Panelboards shall be mounted in cabinets, cutout boxes, or enclosures designed for the purpose and shall be dead front.

*Exception: Panelboards other than of the dead-front, externally operable type shall be permitted where accessible only to qualified persons.*

### 420.5 Enclosures for Damp or Wet Locations

**(A) Damp or Wet Locations.** In damp or wet locations, surface-type enclosures within the scope of this standard shall be placed or equipped so as to prevent moisture or water from entering and accumulating within the cabinet or cutout box, and shall be mounted so there is at least a 6-mm (¼-in.) airspace between the enclosure and the wall or other supporting surface. Enclosures installed in wet locations shall be weatherproof.

*Exception: Nonmetallic enclosures shall be permitted to be installed without the airspace on a concrete, masonry, tile, or similar surface.*

**(B) Switchboards and Panelboards in Damp or Wet Locations.** Switchboards or panelboards in a wet location or outside of a building shall be enclosed in a weatherproof enclosure or cabinet that shall comply with 420.5(A).

### 420.6 Conductor Identification

**(A) Grounded Conductors.** Insulated or covered grounded conductors shall be identified in accordance with this standard.

**(B) Equipment Grounding Conductors.** Equipment grounding conductors shall be identified in accordance with this standard.

**(C) Ungrounded Conductors.** Conductors that are intended for use as ungrounded conductors, whether used as a single conductor or in multiconductor cables, shall be finished to be clearly distinguishable from grounded and grounding conductors. Distinguishing markings shall not conflict in any manner with the surface markings.

### 420.7 Flexible Cords and Cables

**(A) Suitability.** Flexible cords and cables and their associated fittings shall be suitable for the conditions of use and location.

**(B) Uses Permitted.**

**(1) Uses.** Flexible cords and cables shall be used only for the following:

(1) Pendants
(2) Wiring on luminaires (fixtures)
(3) Connection of portable lamps, portable and mobile signs, or appliances
(4) Elevator cables
(5) Wiring on cranes and hoists
(6) Connection of utilization equipment to facilitate frequent interchange
(7) Prevention of the transmission of noise or vibration

(8) Appliances where the fastening means and mechanical connections are specifically designed to permit ready removal for maintenance and repair, and the appliance is intended or identified for flexible cord connection
(9) Data processing cables
(10) Connection of moving parts
(11) Temporary wiring

**(2) Attachment Plugs.** Where used as permitted in 420.7(B)(1)(3), (9), and (11), each flexible cord shall be equipped with an attachment plug and shall be energized from a receptacle outlet.

Flexible cords are permitted to be hard-wired into a junction box if the cord is used for the following:

- Luminaires and fixtures mentioned in 400.7(A) of NFPA 70E
- Pendant pushbutton stations for cranes
- Portable lamp (droplight) connections

**(C) Uses Not Permitted.** Unless specifically permitted in 420.7(B), flexible cords and cables shall not be used for the following:

(1) As a substitute for the fixed wiring of a structure
(2) Where run through holes in walls, structural ceilings, suspended ceilings, dropped ceilings, or floors
(3) Where run through doorways, windows, or similar openings
(4) Where attached to building surfaces

*Exception to (4): Flexible cord and cable shall be permitted to be attached to building surfaces in accordance with the provisions of NEC Article 368.8.*

(5) Where concealed by walls, floors, or ceilings or where located above suspended or dropped ceilings
(6) Where installed in raceways, except as otherwise permitted in this standard

Flexible cords and cables may not be used as a substitute for fixed wiring or where they are concealed behind building walls, floors, or ceilings (including structural, suspended, or dropped-type ceilings). See 240.5 and 527.4(B) and (C) of the *NEC* for use of multiconductor flexible cords for feeder and branch-circuit installations and for overcurrent protection requirements for flexible cord. See 410.30 of the *NEC* for cord-connected luminaires.

**(D) In Show Windows and Show Cases.** Flexible cords used in show windows and show cases shall be Type S, SE, SEO, SEOO, SJ, SJE, SJEO, SJEOO, SJO, SJOO, SJT, SJTO, SJTOO, SO, SOO, ST, STO, STOO, SEW, SEOW, SEOOW, SJEW, SJEOW, SJEOOW, SJOW, SJOOW, SJTW, SJTOW, SJTOOW, SOW, SOOW, STW, STOW, or STOOW.

*Exception No. 1: For the wiring of chain-supported luminaires (lighting fixtures).*

*Exception No. 2: As supply cords for portable lamps and other merchandise being displayed or exhibited.*

Flexible cords listed for hard or extra-hard usage must be used in show windows and show cases because such cords might come in contact with combustible materials such as fabrics or paper products usually present at these locations. These cords are exposed to wear and tear from continual housekeeping and display changes. Flexible cords used in show windows and show cases must be maintained in good condition.

**(E) Markings, Splices, and Pull at Joints and Terminals.**

**(1) Standard Markings.** Flexible cords and cables shall be marked by means of a printed tag attached to the coil reel or carton. The tag shall contain the required information. Types S, SC, SCE, SCT, SE, SEO, SEOO, SJ, SJE, SJEO, SJEOO, SJO, SJT, SJTO, SJTOO, SO, SOO, ST, STO, STOO, SEW, SEOW, SEOOW, SJEW, SJEOW, SJEOOW, SJOW, SJTW, SJTOW, SJTOOW, SOW, SOOW, STW, STWO, and STOOW flexible cords and G, G-GC, PPE, and W flexible cables shall be durably marked on the surface at intervals not exceeding 610 mm (24 in.) with the type designation, size, and number of conductors.

**(2) Splices.** Flexible cord shall be used only in continuous lengths without splice or tap where initially installed in applications permitted by this section. The repair of hard-service cord and junior hard-service cord 14 AWG and larger shall be permitted if conductors are spliced in accordance with this standard and the completed splice retains the insulation, outer sheath properties, and usage characteristics of the cord being spliced.

Section 420.7(E)(2) is intended to ensure that when flexible cords and cables are first installed, as permitted by 420.7(B)(1), they are in their original or near-original condition. The provisions of this section permit repair of a cord in such a manner that the cord will retain the integrity of its original operating condition. However, if the repaired cord is reused or reinstalled at a new location, the in-line repair is no longer permitted. Flexible cords can be used only in lengths without a splice.

**(3) Pull at Joints and Terminals.** Flexible cords and cables shall be connected to devices and to fittings so that tension is not transmitted to joints or terminals.

*Exception: Listed portable single-pole devices that are intended to accommodate such tension at their terminals shall be permitted to be used with single conductor flexible cable.*

### 420.8 Portable Cables Over 600 Volts, Nominal

**(A) Construction.**

**(1) Conductors.** The conductors shall be 8 AWG copper or larger and shall employ flexible stranding.

*Exception: The size of the insulated ground-check conductor of Type G-GC cables shall be not smaller than 10 AWG.*

**(2) Shields.** Cables operated at over 2000 volts shall be shielded. Shielding shall be for the purpose of confining the voltage stresses to the insulation.

**(3) Equipment Grounding Conductor(s).** An equipment grounding conductor(s) shall be provided. The total area shall not be less than that of the size of the equipment grounding conductor required in this standard.

**(B) Shielding.** All shields shall be grounded.

**(C) Grounding.** Grounding conductors shall be connected in accordance with this standard.

**(D) Minimum Bending Radii.** The minimum bending radii for portable cables during installation and handling in service shall be adequate to prevent damage to the cable.

**(E) Fittings.** Connectors used to connect lengths of cable in a run shall be of a type that lock firmly together. Provisions shall be made to prevent opening or closing these connectors while energized. Suitable means shall be used to eliminate tension at connectors and terminations.

**(F) Splices and Terminations.** Portable cables shall not contain splices unless the splices are of the permanent molded, vulcanized types in accordance with this standard. Terminations on portable cables rated over 600 volts, nominal, shall be accessible only to authorized and qualified personnel.

### 420.9 Fixture Wires

**(A) General.** Fixture wires shall be a type approved for the voltage, temperature, and location of use. One conductor of fixture wires that is intended to be used as a grounded conductor shall be identified by means of stripes or by a colored braid, tracer in braid, colored insulation, colored separator, or tinned conductors.

**(B) Uses Permitted.** Fixture wires shall be permitted (1) for installation in luminaires (lighting fixtures) and in similar equipment where enclosed or protected and not subject to bending or twisting in use, or (2) for connecting luminaires (lighting fixtures) to the branch-circuit conductors supplying the luminaires (fixtures).

**(C) Uses Not Permitted.** Fixture wires shall not be used as branch-circuit conductors.

### 420.10 Equipment for General Use

**(A) Live Parts.** Luminaires (fixtures), lampholders, and lamps shall have no live parts operating at 50 volts or more normally exposed to contact. Exposed accessible terminals in lampholders and switches shall not be installed in metal luminaire (fixture) canopies or in open bases or portable table or floor lamps.

*Exception: Cleat-type lampholders located at least 2.5 m (8 ft) above the floor shall be permitted to have exposed terminals.*

**(1) Portable Handlamps.** Portable handlamps shall comply with the following:

(1) Metal shell, paper-lined lampholders shall not be used.
(2) Handlamps shall be equipped with a handle of molded composition or other insulating material.
(3) Handlamps shall be equipped with a substantial guard attached to the lampholder or handle.
(4) Metallic guards shall be grounded by means of an equipment grounding conductor run with circuit conductors within the power-supply cord.
(5) Portable handlamps shall not be required to be grounded where supplied through an isolating transformer with an ungrounded secondary not over 50 volts.

**(B) Installation of Lampholders.**

**(1) Screw-Shell Type.** Lampholders of the screw-shell type shall be installed for use as lampholders only. Where supplied by a circuit having a grounded conductor, the grounded conductor shall be connected to the screw shell. Lampholders installed in wet or damp locations shall be of the weatherproof type.

The practice of installing screw-shell lampholders with screw-shell adapters in baseboards is prohibited. Screw-shell lampholders without lamps expose workers to exposed live parts. See 410.5 in NFPA 70E for permitted uses of receptacles.

**(2) Double-Pole Switched Lampholders.** Where supplied by the ungrounded conductors of a circuit, the switching device of lampholders of the switched type shall simultaneously disconnect both conductors of the circuit.

Single-pole switching may be used to interrupt the ungrounded conductor of a 2-wire circuit that has one conductor grounded. The grounded conductor must be connected to the screw shell of the lampholder. Where a 2-wire circuit is derived from the two ungrounded conductors of a multiwire circuit (3- or 4-wire system) or from the two ungrounded conductors of a 2-wire circuit (3-wire system) and used with switched lampholders, the switching device must be double-pole and must disconnect both ungrounded conductors of the circuit simultaneously. See 410.52 of the *NEC* for information on the construction of switched lampholders.

**(3) Lampholders in Wet and Damp Locations.** Lampholders installed in wet or damp locations shall be of the weatherproof type.

**(C) Receptacles, Cord Connectors, and Attachment Plugs (Caps).**

**(1) Attachment Plugs.** All attachment plugs and cord connectors shall be listed for the purpose and marked with the manufacturer's name or identification and voltage and ampere ratings.

(a) Construction. Attachment plugs and cord connectors shall be constructed so that there are no exposed current-carrying parts except the prongs, blades, or pins. The cover for wire terminations shall be a part that is essential for the operation of an attachment plug or connector (dead-front construction).

(b) Installation. Attachment plugs shall be installed so that their prongs, blades, or pins are not energized unless inserted into an energized receptacle. No receptacle shall be installed so as to require an energized attachment plug as its source of supply.

The design requirements found in 406.6(B) of the *NEC* (referred to as dead-front construction) minimize electrical faults between metal plates and attachment plugs with terminal screws exposed on the face of the plug. Information formerly found only in product information is now clearly stated in the *NEC*. [See *NEC* 406.6(B).]

An improperly installed attachment plug cap presents an extreme risk of electrocution. The blades of an attachment plug cap must never be energized while exposed.

(c) Attachment Plug Ejector Mechanisms. Attachment plug ejector mechanisms shall not adversely affect engagement of the blades of the attachment plug with the contacts of the receptacle.

Section 420.10(C)(1)(c) permits the use of a device that minimizes the chance of damage when the cord is pulled to remove the plug. These devices are designed for people with physical mobility limitations or impaired vision.

**(2) Noninterchangeability.** Receptacles, cord connectors, and attachment plugs shall be constructed so that receptacle or cord connectors do not accept an attachment plug with a different voltage or current rating from that for which the device is intended. However, a 20-ampere T-slot receptacle or cord connector shall be permitted to accept a 15-ampere attachment plug of the same voltage rating. Non–grounding-type receptacles and connectors shall not accept grounding-type attachment plugs.

For information on receptacle and attachment cap configurations, see Exhibits 420.2 and 420.3 of standard configuration charts defined in ANSI C73, *Dimensions of Attachment Plugs and Receptacles*.

| DESCRIPTION | | NEMA NUMBER | 15 AMPERE | | 20 AMPERE | | 30 AMPERE | | 50 AMPERE | | 60 AMPERE | |
|---|---|---|---|---|---|---|---|---|---|---|---|---|
| | | | RECEPTACLE | PLUG | RECEPTACLE | PLUG | RECEPTACLE | PLUG | RECEPTACLE | PLUG | RECEPTACLE | PLUG |
| 2-POLE 2-WIRE | 125V | 1 | 1-15R | 1-15P POLARIZED / NON POLAR. | | 2-20P MATES WITH 5-20R | | 1-30P MATES WITH 5-30R | | | | |
| | 250V | 2 | | 2-15P MATES WITH 6-15R | 2-20R | 2-20P | 2-30R | 2-30P | | | | |
| | 277V AC | 3 | | | | | | | | | | |
| | 600V | 4 | | | | | | | | | | |
| 2-POLE 3-WIRE GROUNDING | 125V | 5 | 5-15R | 5-15P | 5-20R | 5-20P | 5-30R | 5-30P | 5-50R | 5-50P | | |
| | 125V | 5ALT | | | 5ALT-20R | | | | | | | |
| | 250V | 6 | 6-15R | 6-15P | 6-20R | 6-20P | 6-30R | 6-30P | 6-50R | 6-50P | | |
| | 250V | 6ALT | | | 6ALT-20R | | | | | | | |
| | 277V AC | 7 | 7-15R | 7-15P | 7-20R | 7-20P | 7-30R | 7-30P | 7-50R | 7-50P | | |
| | 347V AC | 24 | 24-15R | 24-15P | 24-20R | 24-20P | 24-30R | 24-30P | 24-50R | 24-50P | | |
| | 480V AC | 8 | | | | | | | | | | |
| | 600V AC | 9 | | | | | | | | | | |
| 3-POLE 3-WIRE | 125 / 250V | 10 | | | 10-20R | 10-20P | 10-30R | 10-30P | 10-50R | 10-50P | | |
| | 3 Ø 250V | 11 | 11-15R | 11-15P | 11-20R | 11-20P | 11-30R | 11-30P | 11-50R | 11-50P | | |
| | 3 Ø 480V | 12 | | | | | | | | | | |
| | 3 Ø 600V | 13 | | | | | | | | | | |
| 3-POLE 4-WIRE GROUNDING | 125 / 250V | 14 | 14-15R | 14-15P | 14-20R | 14-20P | 14-30R | 14-30P | 14-50R | 14-50P | 14-60R | 14-60P |
| | 3 Ø 250V | 15 | 15-15R | 15-15P | 15-20R | 15-20P | 15-30R | 15-30P | 15-50R | 15-50P | 15-60R | 15-60P |
| | 3 Ø 480V | 16 | | | | | | | | | | |
| | 3 Ø 600V | 17 | | | | | | | | | | |
| 4-POLE 4-WIRE | 3 Ø Y 120 / 280V | 18 | 18-15R | 18-15P | 18-20R | 18-20P | 18-30R | 18-30P | 18-50R | 18-50P | 18-60R | 18-60P |
| | 3 Ø Y 277 / 480V | 19 | | | | | | | | | | |
| | 3 Ø Y 347 / 600V | 20 | | | | | | | | | | |
| 4-POLE 5-WIRE GROUNDING | 3 Ø Y 120 / 208V | 21 | | | | | | | | | | |
| | 3 Ø Y 277 / 480V | 22 | | | | | | | | | | |
| | 3 Ø Y 347 / 600V | 23 | | | | | | | | | | |

*Note: Blank spaces reserved for future configurations.*

***EXHIBIT 420.2*** *Configuration chart for general-purpose locking plugs and receptacles. (Reproduced from Wiring, Devices—Dimensional Requirements, NEMA WD 6–1997)*

| DESCRIPTION | | NEMA NUMBER | 15 AMPERE | | 20 AMPERE | | 30 AMPERE | | 50 AMPERE | | 60 AMPERE | |
|---|---|---|---|---|---|---|---|---|---|---|---|---|
| | | | RECEPTACLE | PLUG | RECEPTACLE | PLUG | RECEPTACLE | PLUG | RECEPTACLE | PLUG | RECEPTACLE | PLUG |
| 2-POLE 2-WIRE | 125V | 1 | L 1-15R | L 1-15P | | | | | | | | |
| | 250V | 2 | | | L 2-20R | L 2-20P | | | | | | |
| | 277V AC | 3 | | | | | | | | | | |
| | 600V | 4 | | | | | | | | | | |
| 2-POLE 3-WIRE GROUNDING | 125V | 5 | L 5-15R | L 5-15P | L 5-20R | L 5-20P | L 5-30R | L 5-30P | L 5-50R | L 5-50P | L 5-60R | L 5-60P |
| | 250V | 6 | L 6-15R | L 6-15P | L 6-20R | L 6-20P | L 6-30R | L 6-30P | L 6-50R | L 6-50P | L 6-60R | L 6-60P |
| | 277V AC | 7 | L 7-15R | L 7-15P | L 7-20R | L 7-20P | L 7-30R | L 7-30P | L 7-50R | L 7-50P | L 7-60R | L 7-60P |
| | 347V AC | 24 | | | L 24-20R | L 24-20P | | | | | | |
| | 480V AC | 8 | | | L 8-20R | L 8-20P | L 8-30R | L 8-30P | L 8-50R | L 8-50P | L 8-60R | L 8-60P |
| | 600V AC | 9 | | | L 9-20R | L 9-20P | L 9-30R | L 9-30P | L 9-50R | L 9-50P | L 9-60R | L 9-60P |
| 3-POLE 3-WIRE | 125 / 250V | 10 | | | L 10-20R | L 10-20P | L 10-30R | L 10-30P | | | | |
| | 3 Ø 250V | 11 | 11-15R | 11-15P | L 11-20R | L 11-20P | L 11-30R | L 11-30P | | | | |
| | 3 Ø 480V | 12 | | | L 12-20R | L 12-20P | L 12-30R | L 12-30P | | | | |
| | 3 Ø 600V | 13 | | | | | L 13-30R | L 13-30P | | | | |
| 3-POLE 4-WIRE GROUNDING | 125 / 250V | 14 | | | L 14-20R | L 14-20P | L 14-30R | L 14-30P | L 14-50R | L 14-50P | L 14-60R | L 14-60P |
| | 3 Ø 250V | 15 | | | L 15-20R | L 15-20P | L 15-30R | L 15-30P | L 15-50R | L 15-50P | L 15-60R | L 15-60P |
| | 3 Ø 480V | 16 | | | L 16-20R | L 16-20P | L 16-30R | L 16-30P | L 16-50R | L 16-50P | L 16-60R | L 16-60P |
| | 3 Ø 600V | 17 | | | | | L17-30R | L17-30P | L17-50R | L17-50P | L17-60R | L17-60P |
| 4-POLE 4-WIRE | 3 Ø Y 120 / 208V | 18 | | | L 18-20R | L 18-20P | L 18-30R | L 18-30P | | | | |
| | 3 Ø Y 277 / 480V | 19 | | | L 19-20R | L 19-20P | L 19-30R | L 19-30P | | | | |
| | 3 Ø Y 347 / 600V | 20 | | | L 20-20R | L 20-20P | L 20-30R | L 20-30P | | | | |
| 4-POLE 5-WIRE GROUNDING | 3 Ø Y 120 / 208V | 21 | | | L 21-20R | L 21-20P | L 21-30R | L 21-30P | L 21-50R | L 21-50P | L 21-60R | L 21-60P |
| | 3 Ø Y 277 / 480V | 22 | | | L 22-20R | L 22-20P | L 22-30R | L 22-30P | L 22-50R | L 22-50P | L 22-60R | L 22-60P |
| | 3 Ø Y 347 / 600V | 23 | | | L 23-20R | L 23-20P | L 23-30R | L 23-30P | L 23-50R | L 23-50P | L 23-60R | L 23-60P |

*Note: Blank spaces reserved for future configurations.*

***EXHIBIT 420.3*** *Configuration chart for specific-purpose locking plugs and receptacles. (Reproduced from Wiring Devices—Dimensional Requirements, NEMA WD 6–1997)*

**(3) Receptacles in Damp or Wet Locations**

(a) Receptacles installed outdoors in a location protected from the weather or in other damp locations shall have an enclosure for the receptacle that is weatherproof when the receptacle is covered (attachment plug cap not inserted and receptacle covers closed).

(b) Installations suitable for wet locations shall also be considered suitable for damp locations.

(c) Receptacles shall be considered to be in a location protected from the weather where located under roofed open porches, canopies, marquees, and the like, and will not be subjected to a beating rain or water runoff.

**(4) Receptacles in Wet Locations.**

(a) 15- and 20-Ampere Outdoor Receptacles. All 15- and 20-ampere and 125- and 250-volt receptacles installed outdoors in a wet location shall have an enclosure that is weatherproof whether or not the attachment plug cap is inserted.

To ensure the integrity of each weatherproof cord-and-plug connection to a receptacle located in an outdoor wet location, 420.10(C)(4)(a) requires receptacle covers that ensure water tightness at all times. This requirement is not contingent on the anticipated use of the receptacle. The requirement applies to all 15- and 20-ampere, 125- and 250-volt receptacles that are installed in outdoor wet locations, including those receptacle outlets at dwelling units specified by 210.52(E) of the *NEC*. Exhibits 420.4 and 420.5 are examples of the type of receptacle enclosure required by 406.8(B)(1) of the *NEC*.

***EXHIBIT 420.4** A single-gang weatherproof cover suitable for use in wet locations. (Courtesy of L. E. Mason Co.)*

(b) Other Receptacles. Receptacles installed in a wet location shall comply with either of the following:

(1) A receptacle installed in a wet location where the product intended to be plugged into it is not attended while in use (e.g., sprinkler system controller, landscape lighting, holiday

***EXHIBIT 420.5*** *A two-gang weatherproof cover suitable for use in outdoor or indoor wet locations. (Courtesy of Carlon®/Lamson and Sessions)*

lights, and so forth) shall have an enclosure that is weatherproof with the attachment plug cap inserted or removed.

Section 420.10(C)(4)(b) applies to receptacles other than those rated 15 and 20 amperes, 125- and 250-volt, that supply cord-and-plug-connected equipment likely to be used outdoors or in a wet location for long periods of time. A portable pump motor is an example of such equipment. Receptacles for this application must remain weatherproof while they are in use.

(2) A receptacle installed in a wet location where the product intended to be plugged into it will be attended while in use (e.g., portable tools, and so forth) shall have an enclosure that is weatherproof when the attachment plug is removed.

Section 420.10(C)(4)(b)(2) applies to receptacles other than those rated 15 and 20 amperes, 125 and 250 volt, that supply cord-and-plug-connected portable tools or other portable equipment that is likely to be used outdoors for a specific purpose and then removed.

**(D) Appliances.**

**(1) Live Parts.** Appliances shall have no live parts operating at 50 volts or more normally exposed to contact other than those parts functioning as open-resistance heating elements, such as the heating element of a toaster, which are necessarily exposed.

**(2) Disconnecting Means.** A means shall be provided to disconnect each appliance from all ungrounded conductors. If an appliance is supplied by more than one source, the disconnecting means shall be grouped and identified.

**(3) Nameplate.**

(a) Nameplate Marking. Each electric appliance shall be provided with a nameplate giving the identifying name and the rating in volts and amperes or in volts and watts. If the appliance is to be used on a specific frequency or frequencies, it shall be so marked. Where motor overload protection external to the appliance is required, the appliance shall be so marked.

(b) To Be Visible. Marking shall be located so as to be visible or easily accessible after installation.

**(E) Motors.**

**(1) In Sight From (Within Sight From, Within Sight).** Where one piece of equipment shall be "in sight from," "within sight from," or "within sight," and so forth, of another equipment, the specified equipment is to be visible and not more than 15 m (50 ft) distant from the other.

**(2) Disconnecting Means Location.**

(a) Controller. An individual disconnecting means shall be provided for each controller and shall disconnect the controller. The disconnecting means shall be located in sight from the controller location.

*Exception No. 1: For motor circuits over 600 volts, nominal, a controller disconnecting means capable of being locked in the open position shall be permitted to be out of sight of the controller, provided the controller is marked with a warning label giving the location of the disconnecting means.*

*Exception No. 2: A single disconnecting means shall be permitted for a group of coordinated controllers that drive several parts of a single machine or piece of apparatus. The disconnecting means shall be located in sight from the controllers, and both the disconnecting means and the controllers shall be located in sight from the machine or apparatus.*

(b) Motor. A separate disconnecting means shall be located in sight from the motor location and the driven machinery location. The disconnecting means required in accordance with 420.10(E)(1) shall be permitted to serve as the disconnecting means for the motor if it is located in sight from the motor location and the driven machinery location.

*Exception: The disconnecting means shall not be required to be in sight from the motor and the driven machinery location under either condition (a) or (b), provided the disconnecting means required in accordance with 420.10(E)(2) is individually capable of being locked in the open position. The provision for locking or adding a lock to the disconnecting means shall be permanently installed on or at the switch or circuit breaker used as the disconnecting means.*

*(a) Where such a location of the disconnecting means is impracticable or introduces additional or increased hazard to persons or property*

*(b) In industrial installations, with written safety procedures, where conditions of maintenance and supervision ensure that only qualified persons service the equipment*

FPN: Some examples of increased or additional hazards include, but are not limited to, motors rated in excess of 100 hp, multimotor equipment, submersible motors, motors associated with variable frequency drives, and motors located in hazardous (classified) locations.

The primary intent of 420.10(E)(2)(b) is to require a disconnecting means within sight of the controller, the motor location, and the driven machinery. For motors over 600 volts, the controller disconnecting means is not required to be within sight of the controller, provided the controller has a warning label indicating the location and identifying labels on the disconnecting means (see Exhibit 420.6). In this instance, the disconnecting means must be capable of being locked in the open position.

***EXHIBIT 420.6*** *A motor installation over 600 volts, where the motor controller is not located within sight of its disconnecting means.*

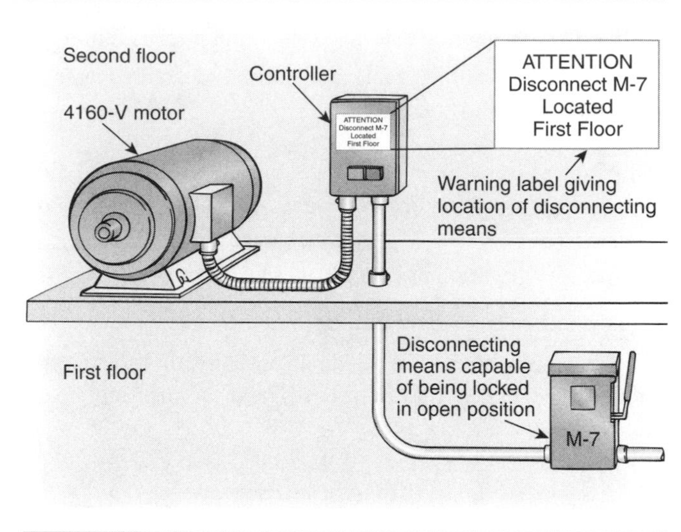


A single disconnecting means may be located adjacent to a group of coordinated controllers, as illustrated in Exhibit 420.7, where the controllers are mounted on a multimotor continuous process machine.

***EXHIBIT 420.7*** *A single disconnecting means located adjacent to a group of coordinated controllers mounted on a multimotor continuous process machine.*

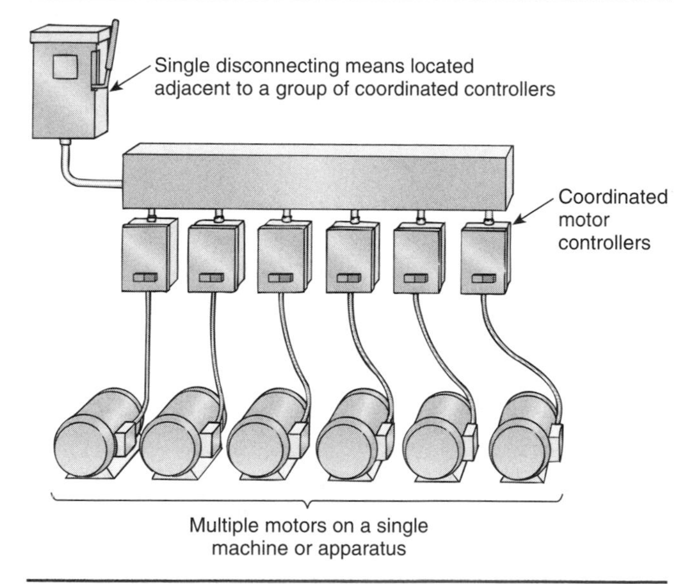


Section 430.102 of the *NEC* requires individual disconnecting means (disconnect switch or circuit breaker) to be capable of being locked in the open position. Disconnect switches or circuit breakers that are located only behind the locked door of a panelboard or within locked rooms do not meet this requirement (see Exhibit 420.8).

(c) To Be Indicating. The disconnecting means shall plainly indicate whether it is in the open (off) or closed (on) position.

(d) Readily Accessible. At least one of the disconnecting means shall be readily accessible.

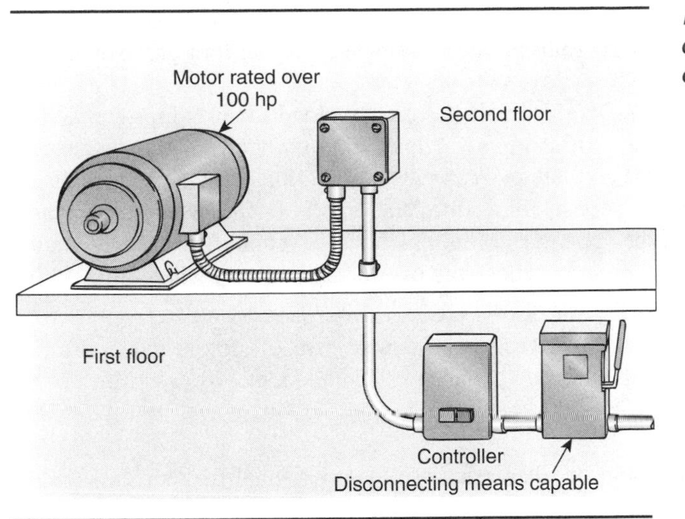


***Exhibit 420.8*** *A controller disconnecting means that is out of sight of the motor.*

(e) Motors Served by Single Disconnecting Means. Each motor shall be provided with an individual disconnecting means.

*Exception: A single disconnecting means shall be permitted to serve a group of motors under any one of the following conditions:*

*(a) Where a number of motors drive several parts of a single machine or piece of apparatus, such as metal and woodworking machines, cranes, and hoists*

*(b) Where a group of motors is under the protection of one set of branch-circuit protective devices*

*(c) Where a group of motors is in a single room within sight from the location of the disconnecting means*

The exception following 420.10(E)(2)(e) permits a single disconnecting means to serve as the disconnecting means for a group of motors. The disconnecting means must have a rating equal to the sum of the horsepower or current of each motor in the group. If the sum is over 2 hp, a motor circuit switch (horsepower-rated) must be used; thus, for five 2-hp motors, the disconnecting means should be a motor circuit switch rated at not less than 10 hp.

Part (a) of the exception to 420.10(E)(2)(e) indicates that a single disconnecting means may be used where a number of motors drive several parts of a single machine, such as cranes (see 610.31, 610.32, and 610.33 of the *NEC*), metal or woodworking machines, or steel rolling mill machinery. The single disconnecting means for multimotor machinery provides a positive means of simultaneously deenergizing all motor branch circuits, including remote control circuits, interlocking circuits, limit-switch circuits, and operator control stations.

Part (b) of the exception to 420.10(E)(2)(e) refers to 430.53(A) of the *NEC*. This *NEC* section permits a group of motors to be protected by the same branch-circuit device, provided the device is rated not more than 20 amperes at 125 volts or 15 amperes at more than 125 volts but not more than 600 volts. The motors must be rated 1 hp or less, and the full-load current for each motor is not permitted to exceed 6 amperes. A single disconnecting means is both practical and economical for a group of such small motors.

Part (c) of the exception to 420.10(E)(2)(e) covers a common situation in which a group of motors is located in one room, such as a pump room, a compressor room, or a mixer room. In this situation, a layout can be designed with a single disconnecting means that has an unobstructed view (not more than 50 ft) from each motor.

These conditions for an individual disconnecting means are similar to those specified in 430.87, Exception, of the *NEC,* which permits the use of a single controller for a group of motors.

(f) Motor and Branch-Circuit Overload Protection. Overload devices are intended to protect motors, motor-control apparatus, and motor branch-circuit conductors against excessive heating due to motor overloads and failure to start. Overload in electrical apparatus is an operating overcurrent that, when it persists for a sufficient length of time, would cause damage or dangerous overheating of the apparatus. It does include short circuits or ground faults. These provisions shall not be interpreted as requiring overload protection where it might introduce additional or increased hazards, as in the case of fire pumps.

(g) Protection of Live Parts—All Voltages. Exposed live parts of motors and controllers operating at 50 volts or more shall be guarded against accidental contact by enclosures or by location as follows:

(1) By installation in a room or enclosure that is accessible only to qualified persons
(2) By installation on a suitable balcony, gallery, or platform elevated and arranged so as to exclude unqualified persons
(3) By elevation 2.5 m (8 ft) or more above the floor

*Exception to (g): Live parts of motors operating at more than 50 volts between terminals shall not require additional guarding for stationary motors that have commutators, collectors, and brush rigging located inside of motor-end brackets and not conductively connected to supply circuits operating at more than 150 volts to ground.*

(h) Guards for Attendants. Where live parts of motors or controllers operating at over 150 volts to ground are guarded against accidental contact only by location, and where adjustment or other attendance might be necessary during the operation of the apparatus, suitable insulating mats or platforms shall be provided so that the attendant cannot readily touch live parts unless standing on the mats or platforms.

**(F) Transformers.**

**(1) General.** Sections 420.10(F)(2) through 420.10(F)(8) cover the installation of all transformers. The following transformers are not covered by 420.10(F):

(1) Current transformers
(2) Dry-type transformers that constitute a component part of other apparatus and comply with requirements for such apparatus
(3) Transformers that are an integral part of an X-ray, high-frequency, or electrostatic-coating apparatus
(4) Transformers used with Class 2 and Class 3 circuits
(5) Transformers for sign and outline lighting
(6) Transformers for electric-discharge lighting
(7) Transformers used for power-limited fire alarm circuits
(8) Transformers used for research, development, or testing, where effective arrangements are provided to safeguard persons from contacting energized parts

**(2) Voltage Warning.** The operating voltage of exposed live parts operating at 50 volts or more of transformer installations shall be indicated by signs or visible markings on the equipment or structures.

**(3) Dry-Type Transformers Installed Indoors.** Dry-type transformers installed indoors and rated over 35 kV shall be installed in a vault.

Dry-type transformers depend on the surrounding air for adequate ventilation. Where the transformers are rated 112½ kilovolt-amperes or less, they are not required to be installed in a fire-resistant transformer room; however, the installation must comply with the requirements of 450.9 of the *NEC*.

Dry-type transformers or gas-filled or less-flammable liquid-insulated transformers (see *NEC* 450.23) installed indoors with a primary voltage of not more than 35,000 are commonly used because a transformer vault is not required. For the same reason, askarel-filled transformers have been extensively used indoors in the past. Askarel, which contains a polychlorinated biphenyl (PCB), is no longer being manufactured. Acceptable substitutes that comply with *NEC* 450.23 are readily available.

Exhibit 420.9 shows a dry-type transformer with the outside casing in place and with the latest core and coil design for a typical dry-type power transformer rated at 1000 kilovolt-amperes, 13,800 volts to 480 volts, 3-phase, and 60 Hz. This transformer has a high-voltage and low-voltage flange for connection to switchgear and a high-voltage, 2-position (double-throw), 3-pole-load air-break switch that may be attached to the case and arranged as a selector switch for the connection of the transformer primary to either of two feeder sources.

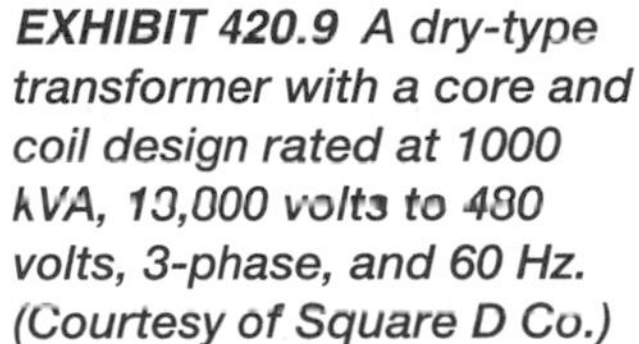

***EXHIBIT 420.9*** *A dry-type transformer with a core and coil design rated at 1000 kVA, 13,000 volts to 480 volts, 3-phase, and 60 Hz. (Courtesy of Square D Co.)*

Dry-type transformers rated 112½ kilovolt-amperes or less require 12 in. of separation from combustible material or separation by fire-resistant barriers. Transformers rated less than 600 volts and completely enclosed, except for ventilating openings, are exempt from this requirement unless the manufacturer's installation instructions specify clearance distances. Noncombustible insulation used in transformers, such as mica, porcelain, and glass, can withstand high temperatures and permits larger dry-type transformers. Combustible materials, however, such as varnishes, might have been used with those insulations, and under short-circuit conditions, flames can escape from the transformer enclosure. Transformers rated over 112½ kilovolt-amperes must be located in fire-resistant transformer rooms or vaults unless either of the exceptions to 450.21(B) of the *NEC* apply.

**(4) Oil-Insulated Transformers Installed Indoors.** Oil-insulated transformers installed indoors shall be installed in a vault.

**(5) Oil-Insulated Transformers Installed Outdoors.** Combustible material, combustible buildings, and parts of buildings, fire escapes, and door and window openings shall be safeguarded from fires originating in oil-insulated transformers installed on roofs, attached to or adjacent to a building or combustible material.

**(6) Doorways.** Vault doorways shall be protected in accordance with 420.10(F)(6)(a), 420.10(F)(6)(b), and 420.10(F)(6)(c):

(a) Type of Door. Each doorway leading into a vault from the building interior shall be provided with a tight-fitting door that has a minimum fire rating of 3 hours. The authority having jurisdiction shall be permitted to require such a door for an exterior wall opening where conditions warrant.

*Exception: Where transformers are protected with automatic sprinkler, water spray, carbon dioxide, or halon, construction of 1-hour rating shall be permitted.*

(b) Sills. A door sill or curb that is of sufficient height to confine the oil from the largest transformer within the vault shall be provided, and in no case shall the height be less than 100 mm (4 in.).

(c) Locks. Doors shall be equipped with locks, and doors shall be kept locked, access being allowed only to qualified persons. Personnel doors shall swing out and be equipped with panic bars, pressure plates, or other devices that are normally latched but open under simple pressure.

Section 420.10(F)(6)(c) prohibits conventional rotation-type doorknobs on transformer vault doors. An injured worker attempting to escape from a transformer vault might not be able to operate a rotating-type doorknob, but he or she would be able to operate panic-type door hardware.

**(7) Water Pipes and Accessories.** Any pipe or duct system foreign to the electrical installation shall not enter or pass through a transformer vault. Piping or other facilities provided for vault fire protection, or for transformer cooling, shall not be considered foreign to the electrical installation.

Section 420.10(F)(7) permits automatic sprinkler protection for transformer vaults. Piping or ductwork for cooling the transformer is also permitted in a transformer vault. No other piping or ductwork is permitted to enter or pass through a transformer vault.

**(8) Storage in Vaults.** Materials shall not be stored in transformer vaults.

**(G) Capacitors.**

**(1) Switching: Load Current.** Group-operated switches shall be used for capacitor switching and shall be capable of the following:

(1) Carrying continuously not less than 135 percent of the rated current of the capacitor installation
(2) Interrupting the maximum continuous load current of each capacitor, capacitor bank, or capacitor installation that will be switched as a unit

(3) Withstanding the maximum inrush current, including contributions from adjacent capacitor installations
(4) Carrying currents due to faults on capacitor side of switch

**(2) Isolation.**

(a) General. A means shall be installed to isolate from all sources of voltage each capacitor, capacitor bank, or capacitor installation that will be removed from service as a unit. The isolating means shall provide a visible gap in the electrical circuit adequate for the operating voltage.

(b) Isolating or Disconnecting Switches with No Interrupting Rating. Isolating or disconnecting switches (with no interrupting rating) shall be interlocked with the load-interrupting device or shall be provided with prominently displayed caution signs to prevent switching load current.

**(3) Additional Requirements for Series Capacitors.** The proper switching sequence shall be ensured by use of one of the following:

(1) Mechanically sequenced isolating and bypass switches
(2) Interlocks
(3) Switching procedure prominently displayed at the switching location

**(H) Storage Batteries.** Provisions shall be made for sufficient diffusion and ventilation of the gases from the battery to prevent the accumulation of an explosive mixture.

# ARTICLE 430
# Specific Purpose Equipment and Installations

NFPA 70E is about safe work practices for workers. This chapter, however, provides information for owners of specific-purpose equipment. Installation of equipment also is important to the safety of workers. To facilitate compliance, the requirements of Article 430 modify basic installation requirements in other articles in Chapter 4. In some cases, however, this article adds additional requirements.

### 430.1 Electric Signs and Outline Lighting

**(A) Disconnects.** Each sign and outline lighting system, or feeder circuit or branch circuit supplying a sign or outline lighting system, shall be controlled by an externally operable switch or circuit breaker that opens all ungrounded conductors. The disconnecting means for signs and outline lighting systems shall be accessible and within sight from its equipment.

*Exception No. 1: A disconnecting means shall not be required for an exit directional sign located within a building.*

*Exception No. 2: A disconnecting means shall not be required for cord-connected signs with an attachment plug.*

**(B) Location.**

**(1) Within Sight of the Sign.** The disconnecting means shall be within sight of the sign or outline lighting system that it controls. Where the disconnecting means is out of the line of sight from any section that may be energized, the disconnecting means shall be capable of being locked in the open position.

Section 430.1(B)(1) covers sign installations where the branch circuit or feeder is run directly to the sign. Each branch circuit or feeder supplying a sign must have an externally operable switch or circuit breaker to open the ungrounded conductors. Two options are permitted for locating the sign disconnecting means. The disconnecting means is required either to be located within sight of the sign or to be equipped with the provision to lock it in the open position.

Exhibit 430.1 depicts a sign with two supply circuits. These circuits could be feeders or branch circuits. Each circuit is provided with an externally operable switch that is located within sight of the sign.

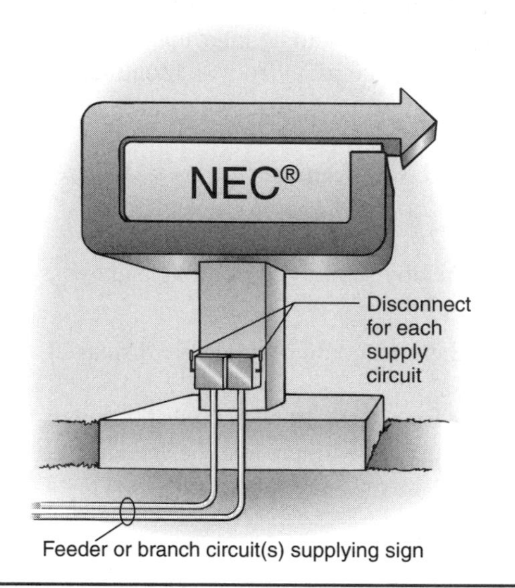


***EXHIBIT 430.1*** *Supply circuit-disconnecting means located at or on an electric sign.*

Exhibit 430.2 illustrates three compliance alternatives. The supply circuit disconnecting means shown in Example 1 is externally operable and located at and within sight of the sign. The disconnecting means in Example 2 is externally operable, and its location, although not at or on the sign, is acceptable because it meets the definition of "within sight." Where the disconnecting means is not located within sight of the sign, as shown in Example 3, it is required to be located within sight of the controller and must be capable of being locked in the open position.

**(2) Within Sight of the Controller.** The following shall apply for signs or outline lighting systems operated by electronic or electromechanical controllers located external to the sign or outline lighting system:

(1) The disconnecting means shall be permitted to be located within sight of the controller or in the same enclosure with the controller.
(2) The disconnecting means shall disconnect the sign or outline lighting system and the controller from all ungrounded supply conductors.
(3) The disconnecting means shall be designed so that no pole can be operated independently and shall be capable of being locked in the open position.

For signs or outline lighting systems operated by mechanical or electromechanical controllers located externally to the sign, the disconnecting means is required to be located within sight of or in the same enclosure as the controller and must be capable of being locked in the open position. This requirement enhances safe working conditions for persons servicing the controller or the sign.

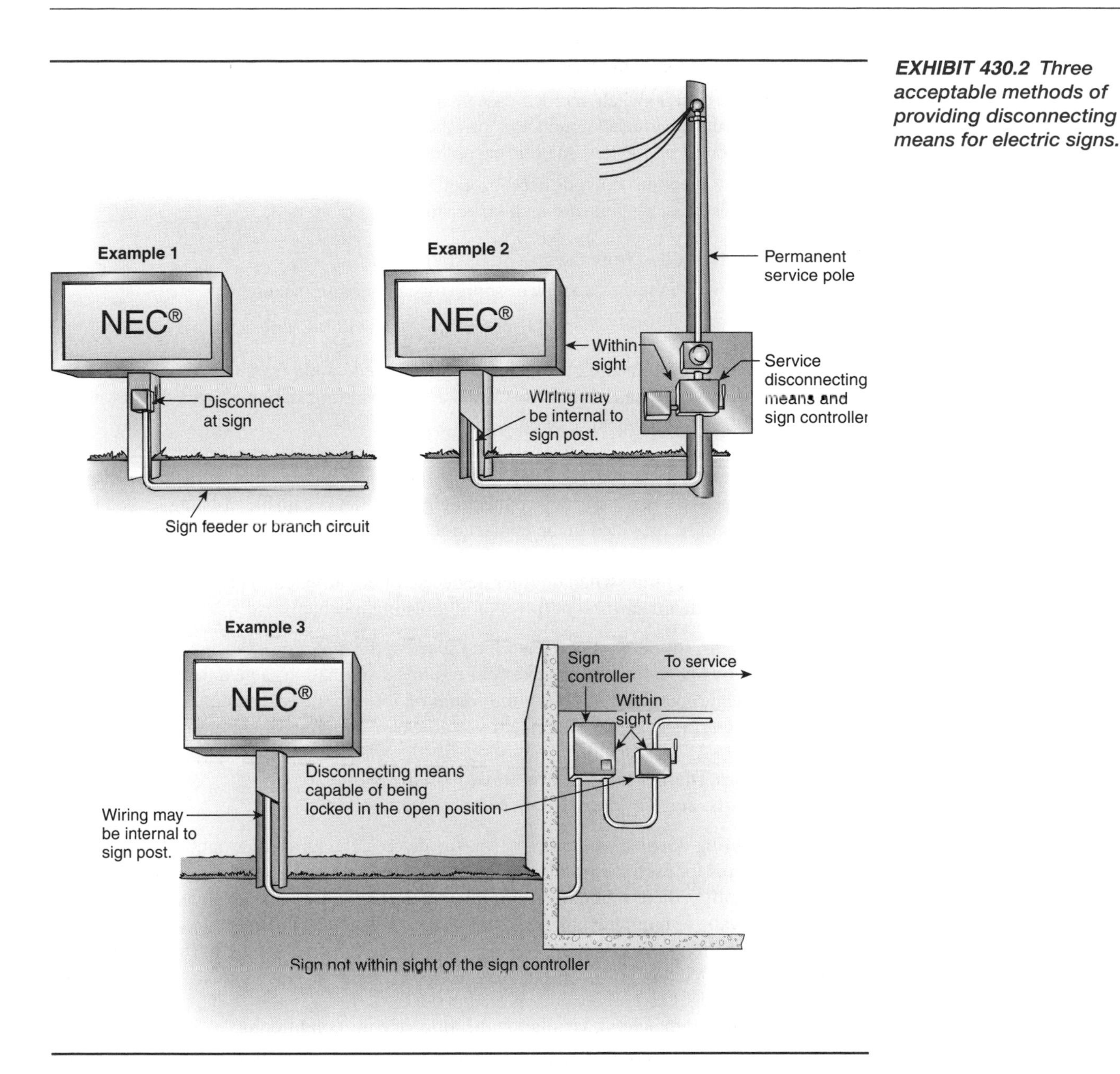


***EXHIBIT 430.2*** *Three acceptable methods of providing disconnecting means for electric signs.*

## 430.2 Cranes and Hoists

### (A) Disconnecting Means.

**(1) Runway Conductor Disconnecting Means.** A disconnecting means that has a continuous ampere rating shall be provided between the runway contact conductors and the power supply. Such disconnecting means shall consist of a motor-circuit switch, circuit breaker, or molded case switch. The disconnecting means shall be as follows:

(1) Readily accessible and operable from the ground or floor level
(2) Capable of being locked in the open position
(3) Open all ungrounded conductors simultaneously
(4) Placed within view of the runway contact conductors

**(2) Disconnecting Means for Cranes and Monorail Hoists.**

(a) A motor circuit switch, molded-case switch, or circuit breaker shall be provided in the leads from the runway contact conductors or other power supply on all cranes and monorail hoists. The disconnecting means shall be capable of being locked in the open position.

(b) Where a monorail hoist or hand-propelled crane bridge installation meets all of the following, the disconnecting means shall be permitted to be omitted:

(1) The unit is controlled from the ground or floor level.
(2) The unit is within view of the power supply disconnecting means.
(3) No fixed work platform has been provided for servicing the unit.

(c) Where the disconnecting means is not readily accessible from the crane or monorail hoist operating station, means shall be provided at the operating station to open the power circuit to all motors of the crane or monorail hoist.

Many crane installations are arranged so that the unit is not within sight of the power supply disconnecting means. Therefore, a disconnecting means that is capable of being locked in the open position must be provided in the contact conductors.

**(B) Limit Switch.** A limit switch or other device shall be provided to prevent the load block from passing the safe upper limit of travel of all hoisting mechanisms.

**(C) Clearance.** The dimension of the working space in the direction of access to live parts operating at 50 volts or more that are likely to require examination, adjustment, servicing, or maintenance while energized shall be a minimum of 750 mm (2½ ft). Where controls are enclosed in cabinets, the door(s) shall either open at least 90 degrees or be removable.

### 430.3 Elevators, Dumbwaiters, Escalators, Moving Walks, Wheelchair Lifts, and Stairway Chair Lifts

**(A) Disconnecting Means.** A single means for disconnecting all ungrounded main power supply conductors for each unit shall be provided and be designed so that no pole can be operated independently. Where multiple driving machines are connected to a single elevator, escalator, moving walk, or pumping unit, there shall be one disconnecting means to disconnect the motor(s) and control valve operating magnets. The disconnecting means for the main power supply conductors shall not disconnect a branch circuit supplying such items as the following:

(1) Car lighting, receptacle(s), ventilation, heating, and air conditioning.
(2) Machine room or control room/machinery space for control spacing lighting and receptacle(s).
(3) Hoistway pit lighting and receptacle(s).

The branch circuits that supply elevator car lighting, receptacles, ventilation, air conditioning, and heating must be independent of the elevator control circuit(s). The main elevator power disconnect switch must not interrupt the branch circuits supplying lighting and receptacles in the elevator pit, lighting and receptacles in the machine room, or lighting and receptacles in the control room lights.

Section 430.3(A)(3) provides for passenger safety and comfort during an inadvertent or emergency shutdown of the main power circuit. It also enables maintenance personnel to service the elevator while it is not operating.

**(1) Type.** The disconnecting means shall be an enclosed externally operable fused motor circuit switch or circuit breaker capable of being locked in the open position. The disconnecting means shall be a listed device.

**(2) Operation.** No provision shall be made to open or close this disconnecting means from any other part of the premises. If sprinklers are installed in hoistways, machine rooms, control rooms, machinery spaces, or control spaces, the disconnecting means shall be permitted to automatically open the power supply to the affected elevator(s) prior to the application of water. No provision shall be made to automatically close this disconnecting means. Power shall only be restored by manual means.

ANSI/ASME A17.1-1998, *Safety Code for Elevators and Escalators,* Rule 102.2(c), requires a means to automatically disconnect the main line power supply upon or prior to the application of water where sprinklers are installed in hoistways, machine rooms, or machinery spaces. Water can result in hazards such as uncontrolled car movement (wet machine brakes), movement of an elevator with open doors (water on safety circuits bypassing car and/or hoistway door interlocks), and shock hazards.

Where sprinklers are not provided for hoistways and machine rooms, ANSI/ASME A17.1 does not require automatic operation of the disconnecting means. NFPA 13, *Standard for the Installation of Sprinkler Systems,* defines requirements for the installation of sprinklers in machine rooms, hoistways, and pits.

Heat detectors located near sprinkler heads generally initiate elevator shutdown. The heat detectors actuate sprinkler heads to discharge water and simultaneously generate an alarm signal. An output control relay powered by the fire alarm system provides a signal to activate the shunt trip of the main disconnecting means. This method ensures that components have secondary power and are monitored for integrity, as required by *NFPA 72®, National Fire Alarm Code®*. Stand-alone heat detectors connected directly to the elevator disconnecting means control circuit are not monitored for integrity, have no secondary power supply, and are not permitted by *NFPA 72.*

Elevator shutdown can occur, even if the car is not at a landing. To avoid trapping people in the car(s), the elevator should be recalled to the designated landing before disconnecting the main line power. Most fires produce detectable quantities of smoke before sufficient heat is generated to activate a sprinkler head. Therefore, ANSI/ASME A17.1, Rule 102.2(c), requires smoke detectors to be installed in hoistways that have sprinklers installed. The smoke detector is intended to recall the elevator car(s) to the designated floor before the main line power is disconnected. See 3.9.4 of *NFPA 72* for additional fire alarm system and elevator shutdown requirements.

Exhibit 430.3 illustrates a typical method of supervising control power using a fire alarm system. Loss of control power would produce a supervisory signal at the fire alarm control unit that then would be investigated.

**(3) Location.** The disconnecting means shall be located where it is readily accessible to qualified persons.

(a) On Elevators Without Generator Field Control. On elevators without generator field control, the disconnecting means shall be located within sight of the motor controller. Driving machines or motion and operation controllers not within sight of the disconnecting means shall be provided with a manually operated switch installed in the control circuit to prevent starting. The manually operated switch(es) shall be installed adjacent to this equipment. Where the driving machine of an electric elevator or the hydraulic machine of a hydraulic elevator is located in a remote machinery room or machinery space, a single means for disconnecting all ungrounded main power supply conductors shall be provided and be capable of being locked in the open position.

See Exhibit 430.4 for disconnecting means for driving machines or motion and operation controllers not within sight of the main line disconnecting means.

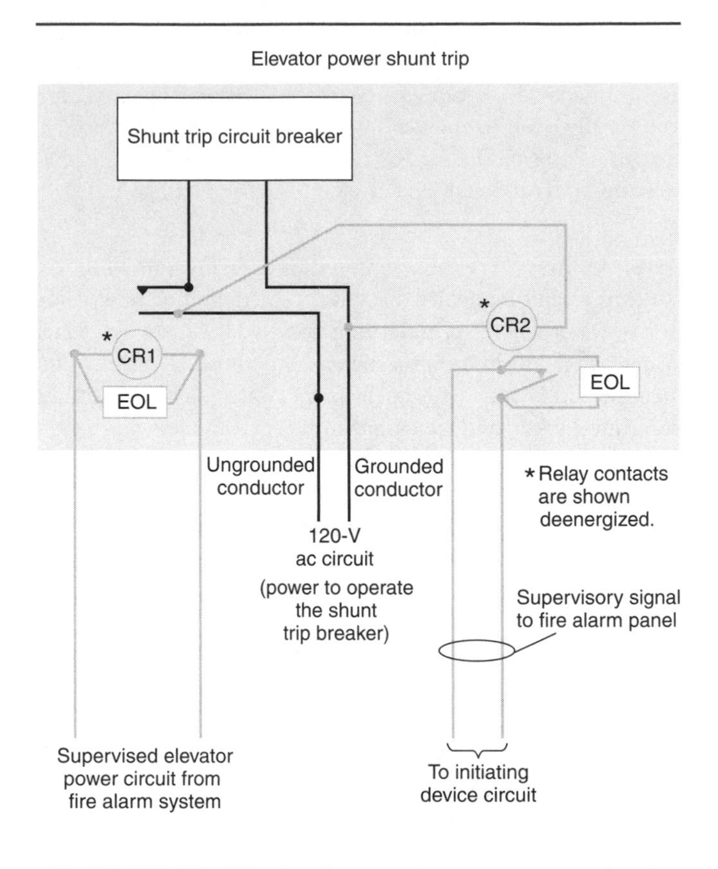


***EXHIBIT 430.3*** *Typical method of control power supervision using a fire alarm control unit.*

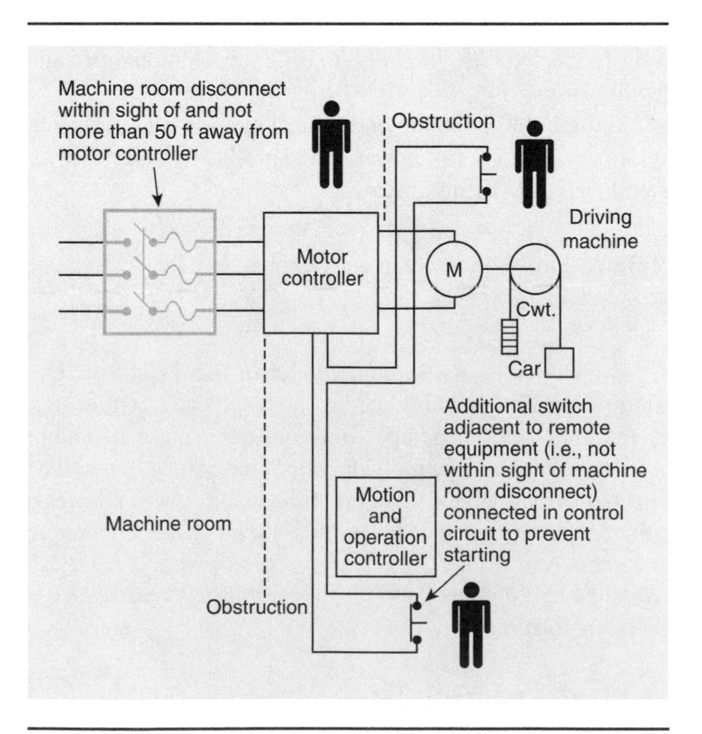


***EXHIBIT 430.4*** *Disconnecting means for driving machines or motion and operation controllers not within sight of the main line disconnecting means. (Redrawn from ASME)*

(b) On Elevators with Generator Field Control. On elevators with generator field control, the disconnecting means shall be located within sight of the motor controller for the driving motor of the motor-generator set. Driving machines, motor-generator sets, or motion and operation controllers not within sight of the disconnecting means shall be provided with a manually operated switch installed in the control circuit to prevent starting. The manually operated switch(es) shall be installed adjacent to this equipment. Where the driving machine or the motor-generator set is located in a remote machine room or machinery space, a single means for disconnecting all ungrounded main power supply conductors shall be provided and be capable of being locked in the open position.

See Exhibits 430.5 and 430.6 for examples of disconnecting means for a motor-generator set and for driving machines in remote locations.

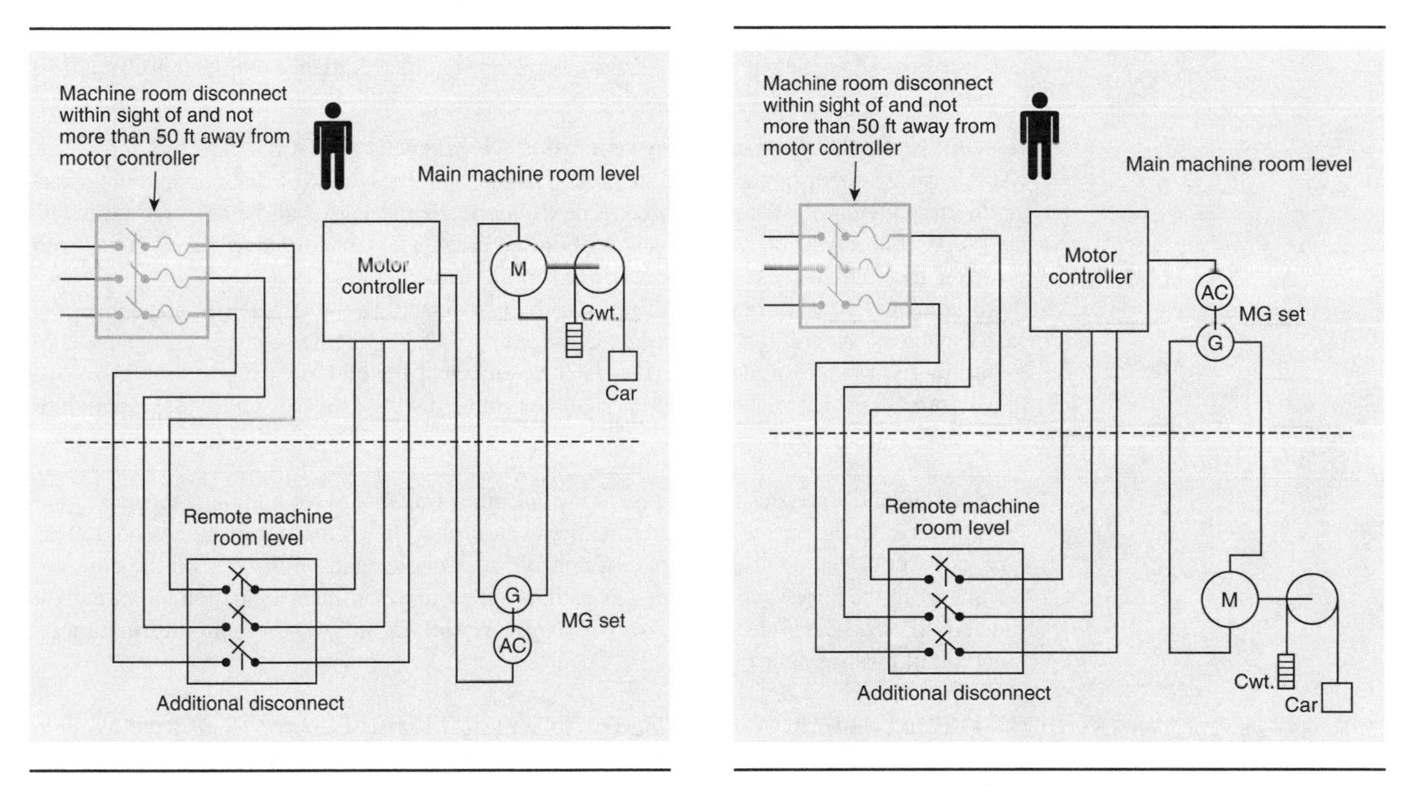


***EXHIBIT 430.5** Disconnecting means for a motor-generator (MG) set in a remote location. (Redrawn from ASME)*

***EXHIBIT 430.6** Disconnecting means for driving machines in a remote location. (Redrawn from ASME)*

(c) On Escalators and Moving Walks. On escalators and moving walks, the disconnecting means shall be installed in the space where the controller is located.

(d) On Wheelchair Lifts and Stairway Chair Lifts. On wheelchair lifts and stairway chair lifts, the disconnecting means shall be located within sight of the motor controller.

**(4) Identification and Signs.** Where there is more than one driving machine in a machine room, the disconnecting means shall be numbered to correspond to the identifying number of the driving machine that they control. The disconnecting means shall be provided with a sign to identify the location of the supply side overcurrent protective device.

Sign requirements for the location of supply-side overcurrent devices assist the elevator mechanic in troubleshooting during a power loss.

**(B) Power from More Than One Source.** On single-car and multicar installations, equipment receiving electrical power from more than one source shall be provided with a disconnecting means for each source of electrical power. The disconnecting means shall be within sight of the equipment served.

**(C) Warning Sign for Multiple Disconnecting Means.** Where multiple disconnecting means are used and parts of the controllers remain energized from a source other than the one disconnected, a warning sign shall be mounted on or next to the disconnecting means. The sign shall be clearly legible and shall read as follows:

WARNING
PARTS OF THE CONTROLLER ARE NOT DEENERGIZED BY THIS SWITCH

**(D) Interconnection Multicar Controllers.** Where interconnections between controllers are necessary for the operation of the system on multicar installations that remain energized from a source other than the one disconnected, a warning sign in accordance with 430.3(C) shall be mounted on or next to the disconnecting means.

**(E) Car Light, Receptacle(s), and Ventilation Disconnecting Means.** Elevators shall have a single means for disconnecting all ungrounded car light, receptacle(s), and ventilation power-supply conductors for the elevator car. The disconnecting means shall be an enclosed externally operable fuses motor circuit switch or circuit breaker capable of being locked in the open position and shall be located in the machine room or control room for that elevator car. Where there is no machine room or control room, the disconnecting means shall be located in the same space as the disconnecting means required by 430.3(A)(2). Disconnecting means shall be numbered to correspond to the identifying number of the elevator car whose light source they control. The disconnecting means shall be provided with a sign to identify the location of the supply side overcurrent protective device.

Section 430.3(E) specifies the location of the disconnecting means for lighting, receptacle, and ventilation branch circuits associated with elevators that do not have a machine room. This type of installation includes designs using drive systems located on the car, on the counterweight, or in the hoistway. Such designs include screw-drive or linear-induction motor drives. (See ANSI/ASME A17.1, *Safety Code for Elevators and Escalators,* for more information on this type of arrangement.)

**(F) Heating and Air-Conditioning Disconnecting Means.** Elevators shall have a single means for disconnecting all ungrounded car heating and air-conditioning power-supply conductors for that elevator car. The disconnecting means shall be an enclosed externally operable fused motor circuit switch or circuit breaker capable of being locked in the open position and shall be located in the machine room or control room for that elevator car. Where there is no machine room or control room, the disconnecting means shall be located in the same space as the disconnecting means required by 430.3(A)(2). Where there is equipment for more than one elevator car in the machine room, the disconnecting means shall be numbered to correspond to the identifying number of the elevator car whose heating and air-conditioning source they control. The disconnecting means shall be provided with a sign to identify the location of the supply side overcurrent protective device.

**(G) Utilization Equipment Disconnecting Means.** Each branch circuit for other utilization equipment shall have a single means for disconnecting all ungrounded conductors. The disconnecting means shall be capable of being locked in the open position and shall be located in the machine room or control room/machine space or control space. Where there is more than one branch circuit for other utilization equipment, the disconnecting means shall be numbered to correspond to the identifying number of the equipment served. The disconnecting means shall be provided with a sign to identify the location of the supply side overcurrent protective device.

**(H) Motor Controllers.** Motor controllers shall be permitted outside the spaces herein specified, provided they are in enclosures with doors or removable panels that are capable of being locked in the closed position and the disconnecting means is located adjacent to or is an integral part of the motor controller. Motor controller enclosures for escalator or moving walks shall be permitted in the balustrade on the side located away from the moving steps or moving treadway. If the disconnecting means is an integral part of the motor controller, it shall be operable without opening the enclosure.

### 430.4 Electric Welders—Disconnecting Means

**(A) Arc Welders.** A disconnecting means shall be provided in the supply circuit for each arc welder that is not equipped with a disconnect mounted as an integral part of the welder. The disconnecting means shall be a switch or circuit breaker, and its rating shall not be less than that necessary to accommodate overcurrent protection.

**(B) Resistance Welder.** A switch or circuit breaker shall be provided by which each resistance welder and its control equipment can be disconnected from the supply circuit. The ampere rating of this disconnecting means shall not be less than the supply conductor ampacity. The supply circuit switch shall be permitted as the welder disconnecting means where the circuit supplies only one welder.

### 430.5 Information Technology Equipment—Disconnecting Means

A means shall be provided to disconnect power to all electronic equipment in the information technology equipment room. There shall be a similar means to disconnect the power to all dedicated HVAC systems serving the room and cause all required fire/smoke dampers to close. The control for these disconnecting means shall be grouped and identified and shall be readily accessible at the principal exit doors. A single means to control both the electronic equipment and HVAC system shall be permitted. Where a pushbutton is used as a means to disconnect power, pushing the button in shall disconnect the power.

Two separate disconnecting means are required by 430.5. However, a single control device, such as a pushbutton, is permitted to operate both disconnecting means. The disconnecting means must disconnect each circuit conductor from its supply source and close all required fire/smoke dampers. (See the definition of *disconnecting means* in Article 100.) The disconnecting means is permitted to be a remote-controlled switching device, such as a relay, with pushbutton stations at the principal exit doors. The *National Electrical Code*® (NFPA 70, *NEC*®, or *Code*) specifies that emergency pushbutton(s) be activated by pushing the button in, rather than pulling it out. This requirement recognizes that in an emergency situation, the intuitive reaction of a person to operate the control is to push, not pull, the button.

The requirements of 430.5 and those of 645.7 of the *NEC* for sealing penetrations are intended to minimize the passage of smoke or fire to other parts of the building.

### 430.6 X-Ray Equipment

**(A) Disconnecting Means.** A disconnecting means of adequate capacity for at least 50 percent of the input required for the momentary rating or 100 percent of the input required for the long-time rating of the X-ray equipment, whichever is greater, shall be provided in the supply circuit. The disconnecting means shall be operable from a location readily accessible from the X-ray control. For equipment connected to a 120-volt, nominal, branch circuit of 30 amperes or less, a grounding-type attachment plug cap and receptacle of proper rating shall be permitted to serve as a disconnecting means.

**(B) Independent Control.** Where more than one piece of equipment is operated from the same high-voltage circuit, each piece or each group of equipment as a unit shall be provided with a high-voltage switch or equivalent disconnecting means. This disconnecting means shall be constructed, enclosed, or located so as to avoid contact by persons with its live parts operating at 50 volts or more.

If only one piece of equipment is supplied by a high-voltage circuit, only one disconnecting means is necessary. However, if the same high-voltage circuit supplies two or more pieces of equipment, each piece of equipment must have its own discrete disconnecting means.

**(C) Control—Industrial and Commercial Laboratory Equipment.**

**(1) Radiographic and Fluoroscopic Types.** All radiographic- and fluoroscopic-type equipment shall be effectively enclosed or shall have interlocks that de-energize the equipment automatically to prevent ready access to live parts operating at 50 volts or more.

**(2) Diffraction and Irradiation Types.** Diffraction- and irradiation-type equipment or installations not effectively enclosed or provided with interlocks to prevent access to live current-carrying parts during operation shall be provided with a positive means to indicate when they are energized. The indicator shall be a pilot light, readable meter deflection, or equivalent means.

### 430.7 Induction and Dielectric Heating Equipment

**(A) General.** Sections 430.7(B) and 430.7(C) cover the construction and installation of dielectric heating, induction heating, induction melting, and induction welding equipment and accessories for industrial and scientific applications. Medical and dental applications, appliances, or line frequency pipeline and vessel heating are not covered in this section.

**(B) Guarding, Grounding, and Labeling.**

**(1) Enclosures.** The converting device (excluding the component interconnections) shall be completely contained within an enclosure(s) of noncombustible material.

**(2) Panel Controls.** All panel controls shall be of dead-front construction.

**(3) Access to Internal Equipment.** Access doors or detachable access panels shall be employed for internal access to heating equipment. Access doors to internal compartments containing equipment employing voltages from 150 to 1000 volts ac or dc shall be capable of being locked closed or shall be interlocked to prevent the supply circuit from being energized while the door(s) is open. Access doors to internal compartments containing equipment employing voltages exceeding 1000 volts ac or dc shall be provided with a disconnecting means equipped with mechanical lockouts to prevent access while the heating equipment is energized, or the access doors shall be capable of being locked closed and interlocked to prevent the supply circuit from being energized while the door(s) is open. Detachable panels not normally used for access to such parts shall be fastened in a manner that will make them inconvenient to remove.

**(4) Warning Labels or Signs.** Warning labels or signs that read "DANGER—HIGH VOLTAGE—KEEP OUT" shall be attached to the equipment and shall be plainly visible where persons might come in contact with energized parts when doors are open or closed or when panels are removed from compartments containing over 150 volts ac or dc.

**(5) Dielectric Heating Applicator Shielding.** Protective cages or adequate shielding shall be used to guard dielectric heating applicators. Interlock switches shall be used on all hinged access doors, sliding panels, or other easy means of access to the applicator. All interlock switches shall be connected in such a manner as to remove all power from the applicator when any one of the access doors or panels is open.

**(6) Disconnecting Means.** A readily accessible disconnecting means shall be provided to disconnect each heating equipment from its supply circuit. The disconnecting means shall be located within sight from the controller or be capable of being locked in the open position. The rating of the disconnecting means shall be not less than the nameplate rating of the heating equipment. The supply circuit disconnecting means shall be permitted to serve as the heating equipment disconnecting means where only one heating equipment is supplied.

**(C) Remote Control.**

**(1) Multiple Control Point.** Where multiple control points are used for applicator energization, a means shall be provided and interlocked so that the applicator can be energized from only one control point at a time. A means for deenergizing the applicator shall be provided at each control point.

**(2) Foot Switches.** Switches operated by foot pressure shall be provided with a shield over the contact button to avoid accidental closing of a foot switch.

### 430.8 Electrolytic Cells

**(A) Scope.** The provisions for this section shall apply to the installation of the electrical components and accessory equipment of electrolytic cells, electrolytic cell lines, and process power supply for the production of aluminum, cadmium, chlorine, copper, fluorine, hydrogen peroxide, magnesium, sodium, sodium chlorate, and zinc. Not covered by this section are cells used as a source of electric energy and for electroplating processes and cells used for the production of hydrogen.

**(B) Definitions.** For the purposes of 430.8, the following definitions shall apply.

**Cell Line.** An assembly of electrically interconnected electrolytic cells supplied by a source of direct current power.

**Cell Line Attachments and Auxiliary Equipment.** As applied to this section, a term that includes, but is not limited to, auxiliary tanks; process piping; ductwork; structural supports; exposed cell line conductors; conduits and other raceways; pumps, positioning equipment, and cell cutout or bypass electrical devices. Auxiliary equipment includes tools, welding machines, crucibles, and other portable equipment used for operation and maintenance within the electrolytic cell line working zone. In the cell line working zone, auxiliary equipment includes the exposed conductive surfaces of ungrounded cranes and crane-mounted cell-servicing equipment.

**Electrolytic Cell.** A tank or vat in which electrochemical reactions are caused by applying electrical energy for the purpose of refining or producing usable materials.

**Electrolytic Cell Line Working Zone.** The space envelope wherein operation or maintenance is normally performed on or in the vicinity of exposed energized surfaces of electrolytic cell lines or their attachments.

**(C) Electrolytic Cell Lines.** Electrolytic cell lines shall comply with the provisions of Articles 400, 410, 420, and 430, except as exempted by 430.8(C)(1) through 430.8(C)(4).

**(1) Conductors.** The electrolytic cell line conductors shall not be required to comply with the provisions of Article 400 and 410.2 and 410.3.

**(2) Overcurrent Protection.** Overcurrent protection of electrolytic cell dc process power circuits shall not be required to comply with the requirements of 410.5.

**(3) Grounding.** Equipment located or used within the electrolytic cell line working zone or associated with the cell line dc power circuits shall not be required to comply with the provisions of Section 410.6.

**(4) Working Zone.** The electrolytic cells, cell line attachments, and the wiring of auxiliary equipments and devices within the cell line working zone shall not be required to comply with the provisions of Article 400 and 410.2 and 410.3.

**(D) Disconnecting Means.**

**(1) More Than One Process Power Supply.** Where more than one dc cell line process power supply serves the same cell line, a disconnecting means shall be provided on the cell line circuit side of each power supply to disconnect it from the cell line circuit.

**(2) Removable Links or Conductors.** Removable links or removable conductors shall be permitted to be used as the disconnecting means.

**(E) Portable Electric Equipment.**

**(1) Portable Electrical Equipment Not to Be Grounded.** The frames and enclosures of portable electric equipment used within the cell line working zone shall not be grounded.

*Exception No. 1: Where the cell line voltage does not exceed 200 volts dc, these frames and enclosures shall be permitted to be grounded.*

*Exception No. 2: These frames and enclosures shall be permitted to be grounded where guarded.*

**(2) Marking.** Ungrounded portable electric equipment shall be distinctively marked and shall employ plugs and receptacles of a configuration that prevents connection of this equipment to grounding receptacles and that prevents inadvertent interchange of ungrounded and grounded portable electrical equipment.

**(F) Power Supply Circuits and Receptacles for Portable Electric Equipment.**

**(1) Isolated Circuits.** Circuits supplying power to ungrounded receptacles for hand-held, cord-connected equipments shall be electrically isolated from any distribution system supplying areas other than the cell line working zone and shall be ungrounded. Power for these circuits shall be supplied through isolating transformers. Primaries of such transformers shall operate at not more than 600 volts between conductors and shall be provided with proper overcurrent protection. The secondary voltage of such transformers shall not exceed 300 volts between conductors, and all circuits supplied from such secondaries shall be ungrounded and shall have an approved overcurrent device of proper rating in each conductor.

**(2) Noninterchangeability.** Receptacles and their mating plugs for ungrounded equipment shall not have provision for a grounding conductor and shall be of a configuration that prevents their use for equipment required to be grounded.

**(3) Marking.** Receptacles on circuits supplied by an isolating transformer with an ungrounded secondary shall be a distinctive configuration, distinctively marked, and shall not be used in any other location in the plant.

**(G) Fixed and Portable Electrical Equipment.**

**(1) Electric Equipment Not Required to Be Grounded.** Alternating-current systems supplying fixed and portable electric equipments within the cell line working zone shall not be required to be grounded.

**(2) Exposed Conductive Surfaces Not Required to Be Grounded.** Exposed conductive surfaces, such as electrical equipment housings, cabinets, boxes, motors, raceways, and the like, that are within the cell line working zone shall not be required to be grounded.

**(3) Wiring Methods.** Auxiliary electrical equipment such as motors, transducers, sensors, control devices, and alarms, mounted on an electrolytic cell or other energized surface, shall be connected to premises wiring systems by any of the following means:

(1) Multiconductor hard usage cord.
(2) Wire or cable in suitable raceways or metal or nonmetallic cable trays. If metal conduit, cable tray, armored cable, or similar metallic systems are used, they shall be installed with insulating breaks such that they do not cause a potentially hazardous electrical condition.

**(4) Circuit Protection.** Circuit protection shall not be required for control and instrumentation that are totally within the cell line working zone.

**(5) Bonding.** Bonding of fixed electric equipment to the energized conductive surfaces of the cell line, its attachments, or auxiliaries shall be permitted. Where fixed electric equipment is mounted on an energized conductive surface, it shall be bonded to that surface.

**(H) Auxiliary Nonelectric Connections.** Auxiliary nonelectric connections, such as air hoses, water hoses, and the like, to an electrolytic cell, its attachments, or auxiliary equipments shall not have continuous conductive reinforcing wire, armor, braids, and the like. Hoses shall be of a nonconductive material.

**(I) Cranes and Hoists.**

**(1) Conductive Surfaces to Be Insulated from Ground.** The conductive surfaces of cranes and hoists that enter the cell line working zone shall not be required to be grounded. The portion of an overhead crane or hoist that contacts an energized electrolytic cell or energized attachments shall be insulated from ground.

**(2) Hazardous Electrical Conditions.** Remote crane or hoist controls that may introduce hazardous electrical conditions into the cell line working zone shall employ one or more of the following systems:

(1) Isolated and ungrounded control circuit in accordance with 430.8(F)(1)
(2) Nonconductive rope operator
(3) Pendant pushbutton with nonconductive supporting means and having nonconductive surfaces or ungrounded exposed conductive surfaces
(4) Radio

### 430.9 Electrically Driven or Controlled Irrigation Machines

**(A) Lightning Protection.** If an irrigation machine has a stationary point, a grounding electrode system shall be connected to the machine at the stationary point for lightning protection.

Where the electrical power supply to irrigation machine equipment is a service, the requirements of Article 250 of the *NEC* for grounding the system and equipment are applicable. Due to the physical location of irrigation equipment, the most likely grounding electrode of the types covered in 250.52 of the *NEC* is a driven ground rod or ground plate.

Consideration should be given to the requirements of 250.60 of the *NEC* and NFPA 780, *Standard for the Installation of Lightning Protection Systems,* in areas where lightning protection is crucial. A common electrode system is not permitted to be used for the dual function of grounding the electric service and grounding the lightning protection system. The separate electrode systems are required to be bonded together.

**(B) Main Disconnecting Means.** The main disconnecting means for the machine shall provide overcurrent protection, and shall be at the point of connection of electrical power to the machine or shall be visible and not more than 15 m (50 ft) from the machine, and shall be readily accessible and capable of being locked in the open position. The disconnecting means shall have a horsepower and current rating not less than required for the main controller.

In accordance with 430.9(B), the main disconnecting means is permitted to be up to 50 ft from the machine but must be readily accessible and capable of being locked in the open position. This requirement eliminates one set of overcurrent protective devices and one disconnecting means where the circuit originates at the motor control panel for the irrigation pump and the panel is located within 50 ft of the center pivot machine. It also alleviates some potential problems with machines designed to be towed to a second site.

**430.10 Swimming Pools, Fountains, and Similar Installations**

**(A) Scope.** Sections 430.10(B) through 430.10(F) shall apply to the construction and installation of electric wiring for and equipment in or adjacent to all swimming, wading, therapeutic, and decorative pools, fountains, hot tubs, spas, and hydro-massage bathtubs, whether permanently installed or storable, and to metallic auxiliary equipment, such as pumps, filters, and similar equipment.

Article 680 of the *NEC* applies to decorative pools and fountains; swimming, wading, and wave pools; therapeutic tubs and tanks; hot tubs; spas; hydro-massage bathtubs; and similar installations. The installations covered by this article can be indoors or outdoors, permanent or storable, and may or may not be supplied directly by electrical circuits of any nature.

Studies conducted by Underwriters Laboratories®, various manufacturers, and others indicate that a person in a swimming pool can receive a severe electric shock by reaching over and touching the energized casing of a faulty appliance, such as a radio or a hair dryer, as the person's body establishes a conductive path through the water to earth. Also, a person not in contact with a faulty appliance or any grounded object can receive an electric shock and be rendered immobile by a potential gradient in the water itself. Accordingly, the requirements of Article 680 of the *NEC* that cover such items as effective bonding and grounding, installation of receptacles and luminaires (lighting fixtures), use of ground-fault circuit interrupters, and modified wiring methods apply not only to the installation of the pool but also to installations and equipment adjacent to or associated with the pool.

**(B) Receptacles.**

**(1) Circulation and Sanitation System, Location.** Receptacles that provide power for water-pump motors for or other loads directly related to the circulation and sanitation system shall be located at least 3.0 m (10 ft) from the inside walls of the pool, or not less than 1.5 m (5 ft) from the inside walls of the pool if they meet all of the following conditions:

(1) Consist of single receptacles
(2) Employ a locking configuration
(3) Are of the grounding type
(4) Have GFCI protection

**(2) Other Receptacles, Location.** Other receptacles shall be not less than 3.0 m (10 ft) from the inside walls of a pool.

**(3) GFCI Protection.** All 125-volt receptacles located within 6.0 m (20 ft) of the inside walls of a pool or fountain shall be protected by a ground-fault circuit interrupter. Receptacles that supply pool pump motors and that are rated 15 or 20 amperes, 120 volts through 240 volts, single phase, shall be provided with GFCI protection.

All single-phase, 15- and 20-ampere, 120 through 240 receptacles that supply swimming pool pump motors are required to have ground-fault circuit interrupter (GFCI) protection. While this requirement applied only to installations at other than dwellings in the 1999 *Code,* the 2002 *Code* has been revised to require GFCI protection of these receptacles for all swimming pool installations. It should be noted that this requirement applies to these receptacles regardless of their proximity to the swimming pool, and it applies only to cord-and-plug-connected pump motors.

**(4) Measurements.** In determining the dimensions in 430.10 addressing receptacle spacings, the distance to be measured shall be the shortest path the supply cord of an appliance connected to the receptacle would follow without piercing a floor, wall, ceiling, doorway with hinged or sliding door, window opening, or other effective permanent barrier.

The requirements of 680.22(A) of the *NEC* apply to receptacles located near a permanently installed pool or fountain. They do not apply to direct-connected equipment. Permission is given in 680.22(A)(1) of the *NEC* to allow a single locking- and grounding-type receptacle to supply a recirculation pump motor where the receptacle is located not less than 5 ft from the inside walls of the pool or fountain and is protected by a GFCI.

**(C) Luminaires (Lighting Fixtures), Lighting Outlets, and Ceiling-Suspended (Paddle) Fans.**

**(1) New Outdoor Installation Clearances.** In outdoor pool areas, luminaires (lighting fixtures), lighting outlets, and ceiling-suspended (paddle) fans installed above the pool area extending 1.5 m (5 ft) horizontally from the inside walls of the pool shall be installed at a height not less than 3.7 m (12 ft) above the maximum water level of the pool.

**(2) Indoor Clearance.** For installations in indoor pool areas, the clearances shall be the same as for outdoor areas unless modified as provided in this paragraph. If the branch circuit supplying the equipment is protected by a ground fault circuit interrupter, in which case the following equipment shall be permitted at a height not less than 2.3 m (7 ft 6 in.) above the maximum pool water level:

(1) Totally enclosed luminaires (fixtures)
(2) Ceiling-suspended (paddle) fans identified for use beneath ceiling structures such as provided on porches or patios

**(3) Existing Installations.** Existing luminaires (lighting fixtures) and lighting outlets located less than 1.5 m (5 ft) measured horizontally from the inside walls of a pool shall be not less than 1.5 m (5 ft) above the surface of the maximum water level, shall be rigidly attached to the existing structure, and shall be protected by a ground-fault circuit interrupter.

**(4) GFCI Protection in Adjacent Areas.** Luminaires (lighting fixtures), lighting outlets, and ceiling-suspended (paddle) fans installed in the area extending between 1.5 m (5 ft) and 3.0 m (10 ft) horizontally from the inside walls of a pool shall be protected by a ground-fault circuit-interrupter unless installed not less than 1.5 m (5 ft) above the maximum water level and rigidly attached to the structure adjacent to or enclosing the pool.

**(5) Cord-and-Plug-Connected Luminaires (Lighting Fixtures).** Cord-and-plug-connected luminaires (lighting fixtures) shall comply with 430.10(C)(5)(a) through 430.10(C)(5)(c) where installed within 4.9 m (16 ft) of any point on the water surface, measured radially.

(a) Length. For other than storable pools, the flexible cord shall not exceed 900 mm (3 ft) in length.

(b) Equipment Grounding. The flexible cord shall have a copper equipment grounding conductor not smaller than 12 AWG. The cord shall terminate in a grounding-type attachment plug.

(c) Construction. The equipment grounding conductors shall be connected to a fixed metal part of the assembly. The removable part shall be mounted on or bonded to the fixed metal part.

See Exhibit 430.7 for diagrams that clarify the limitations applicable to certain zones surrounding outdoor and indoor pools.

**(D) Switching Devices.** Switching devices shall be located at least 1.5 m (5 ft) horizontally from the inside walls of a pool unless separated from the pool by a solid fence, wall, or other permanent barrier. Alternatively, a switch that is listed as being acceptable for use within 1.5 m (5 ft) shall be permitted.

***EXHIBIT 430.7*** *Limitations that apply to the placement of luminaires (lighting fixtures), lighting outlets, and ceiling-suspended fans in the area surrounding outdoor and indoor pools.*

**Outdoor Pools**

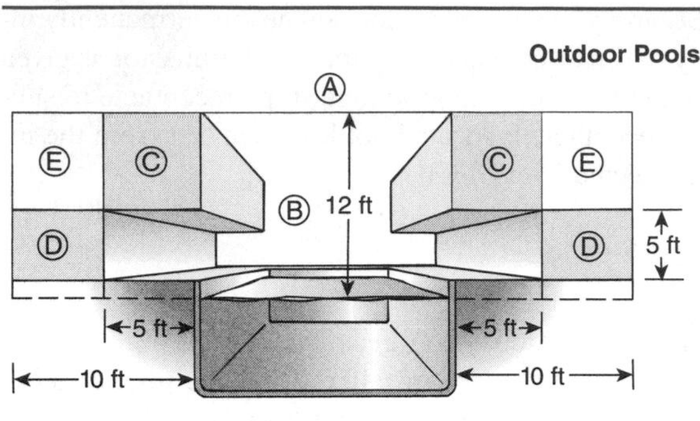


Ⓐ Luminaires, lighting outlets, and ceiling-suspended (paddle) fans permitted above 12 ft.
Ⓑ Luminaires, lighting outlets, and ceiling-suspended (paddle) fans not permitted below 12 ft.
Ⓒ Existing luminaires and lighting outlets permitted in this space if rigidly attached to existing structure (GFCI required).
Ⓓ Luminaires and lighting outlets permitted if protected by a GFCI.
Ⓔ Luminaires and lighting outlets permitted if rigidly attached.

**Indoor Pools**

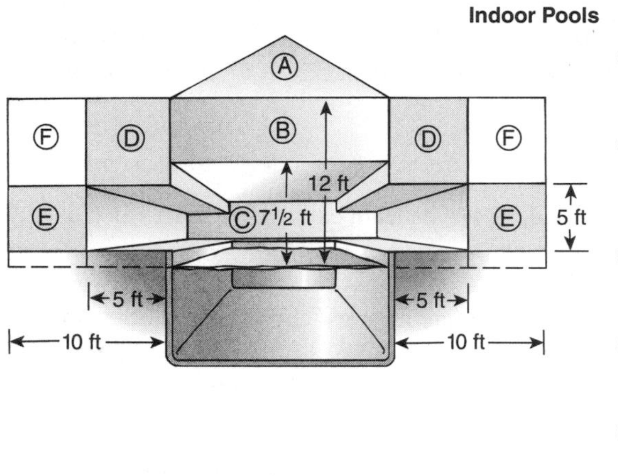


Ⓐ Luminaires, lighting outlets, and ceiling-suspended (paddle) fans permitted above 12 ft.
Ⓑ Totally enclosed luminaires protected by a GFCI and ceiling-suspended (paddle) fans protected by a GFCI permitted above $7^1/_2$ ft.
Ⓒ Luminaires, lighting outlets, and ceiling-suspended (paddle) fans not permitted below 5 ft.
Ⓓ Existing luminaires and lighting outlets permitted in this space if rigidly attached to existing structure (GFCI required).
Ⓔ Luminaires and lighting outlets permitted if protected by a GFCI.
Ⓕ Luminaires and lighting outlets permitted if rigidly attached.

Devices such as panelboards, time clocks, and pool light switches, where located not less than 5 ft horizontally from the inside walls of a pool without a solid fence, wall, or other permanent barrier, must be out of reach of persons who are in the pool, thereby preventing contact and possible shock hazards.

**(E) Underwater Equipment.**

**(1) Luminaire (Fixture) Design, Normal Operation.** The design of an underwater luminaire (lighting fixture) supplied from a branch circuit either directly or by way of a transformer meeting the requirements of this section shall be such that, where the luminaire (fixture) is properly installed without a ground-fault circuit interrupter, there is no shock hazard with any likely combination of fault conditions during normal use (not relamping).

Dry-niche, no-niche, or wet-niche underwater luminaires (lighting fixtures) operating at more than 15 volts require ground-fault circuit-interrupter protection. See the commentary following 680.5 of the *NEC Handbook.*

Branch-circuit conductors for dry-niche fixtures are required to be installed in approved rigid metal conduit, intermediate metal conduit, or rigid nonmetallic conduit from the fixture to a panelboard or the service equipment. Branch-circuit conductors for wet-niche fixtures leaving the pool junction box are required to be enclosed in rigid metal conduit, intermediate metal conduit, liquidtight flexible nonmetallic conduit, or rigid nonmetallic conduit. If the conductors are located in or on buildings, they are permitted to be installed in electrical metallic tubing or electrical nonmetallic tubing. Unlike wet-niche fixtures, a junction box is not required for dry-niche fixtures. If one is used, it is not required to be elevated or located as specified in 680.24(A)(2) of the *NEC* (see Exhibits 430.8 and 430.9).

Note that 314.23(D) of the *NEC* specifies the requirements for the support of threaded boxes that do not contain devices, and that 352.12 of the *NEC* does not permit luminaires (lighting fixtures) or most other electrical equipment to be supported by rigid nonmetallic conduit. Exhibit 430.8 shows a properly supported junction box for a wet-niche fixture.

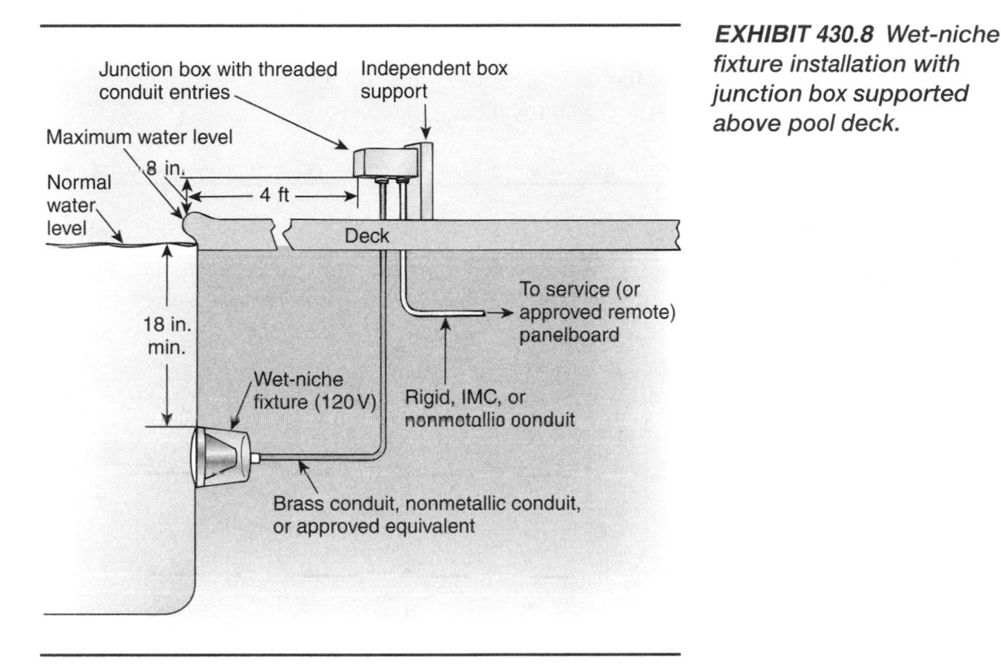


***EXHIBIT 430.8** Wet-niche fixture installation with junction box supported above pool deck.*

The requirements in 680.24(A) through (F) of the *NEC* cover the construction and installation of boxes and enclosures associated with underwater luminaires (lighting fixtures). Boxes and enclosures used for the supply wiring to wet-niche and no-niche underwater luminaires must be listed for the purpose by a recognized testing laboratory. The provisions of 680.24(D) of the *NEC* ensure the availability of integral grounding terminals necessary for the grounding and bonding of underwater luminaires (lighting fixtures). A box that is listed but not specifically for use with swimming pools does not provide the correct number of integral grounding and bonding terminals.

The number of grounding terminals in a box or enclosure is required to be one more than the number of conduit entries for which the box is designed. In an installation where nonmetallic conduit is the wiring method between the wet-niche forming shell and the deck (junction) box, two equipment grounding conductors in that conduit must be terminated in the junction box. The first equipment grounding conductor is covered by 680.23(B)(2)(b) of the *NEC.* The use of nonmetallic conduit requires the installation of an insulated, copper equipment grounding conductor in that section of conduit between the deck box and the wet-niche forming shell. This conductor can be solid or stranded and must not be smaller than 8 AWG.

The function of this conductor is twofold. It permanently bonds all non–current-carrying metal surfaces of the forming shell to any non–current-carrying parts of the deck box and to the equipment grounding conductor of the circuit that supplies the wet-niche luminaire (lighting fixture). In addition, this conductor serves as the path for ground-fault current in the event of a ground fault when the wet-niche luminaire is removed from the forming shell, as is typically done during relamping. Damage to the wet-niche luminaire supply cord could result in this ground-fault scenario.

The second equipment grounding conductor is the one contained in the flexible cord supplying the wet-niche luminaire. In accordance with 680.23(B)(3) of the *NEC,* this conductor is required to be insulated, copper, and sized no smaller than the circuit conductors within the cord, but not smaller than 16 AWG.

In addition to the two equipment grounding conductors contained in the section of nonmetallic conduit between the forming shell and the deck box, the wiring method from the deck box to the power source must contain a separate equipment grounding conductor. This equipment grounding conductor is required by 680.23(F)(2) of the *NEC* and must be insulated, copper, and no smaller than 12 AWG. The grounding terminals within the deck (junction) box are used to terminate and bond together all of these equipment grounding conductors.

Exhibit 430.9 illustrates an installation of a forming shell for a wet-niche luminaire and a flush junction (deck) box. (See Exhibit 430.8 for surface deck boxes.)

***EXHIBIT 430.9*** *A flush junction (deck) box and a forming shell for a wet-niche luminaire installed according to 680.24(A)(2) of the NEC.*

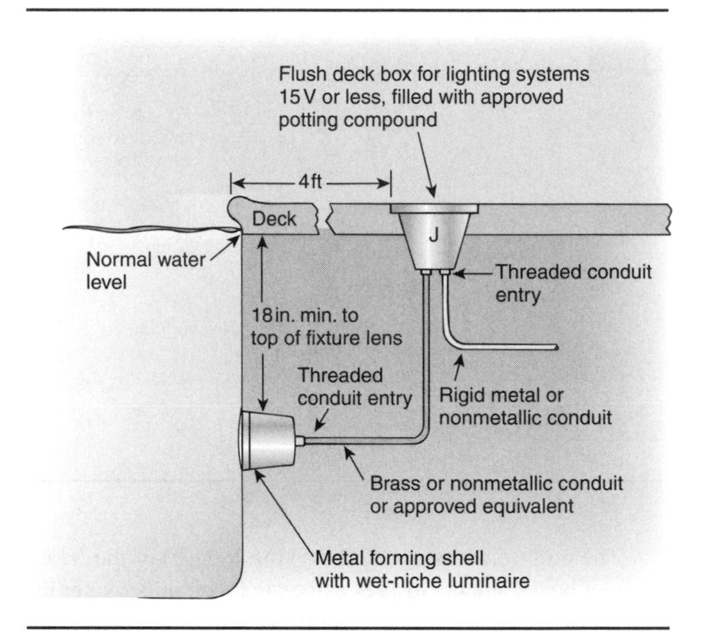


**(2) GFCI Protection, Relamping.** A ground-fault circuit interrupter shall be installed in the branch circuit supplying luminaires (fixtures) operating at more than 15 volts, so that there is no shock hazard during relamping. The installation of the ground-fault circuit interrupter shall be such that there is no shock hazard with any likely fault-condition combination that involves a person in a conductive path from any ungrounded part of the branch circuit or the luminaire (fixture) to ground.

**(3) Voltage Limitation.** No luminaire (lighting fixture) shall be installed for operation on supply circuits over 150 volts between conductors.

**(4) Location, Wall-Mounted Luminaires (Fixtures).** Luminaires (lighting fixtures) mounted in walls shall be installed with the top of the luminaire (fixture) lens not less than 450 mm (18 in.) below the normal water level of the pool, unless the luminaire (lighting fixture) is listed and identified for use at lesser depths. No luminaire (fixture) shall be installed less than 100 mm (4 in.) below the normal water level of the pool.

The reason for the 18-in. minimum submergence requirement is to reduce the likelihood that a person in the water and hanging onto the side of a pool directly over a fixture will have his or her chest in line with the fixture. This section covers fixtures that have been investigated and found acceptable for use where a person's chest might be directly in front of the fixture. The highest level of leakage current in the pool coming from a wet-niche fixture with a broken lens and bulb is found directly in front of the fixture.

**(5) Bottom-Mounted Luminaires (Fixtures).** Luminaires (lighting fixtures) facing upward shall have the lens adequately guarded to prevent contact by any person.

**(6) Dependence on Submersion.** Luminaires (fixtures) that depend on submersion for safe operation shall be inherently protected against the hazards of overheating when not submerged.

Fixtures that depend on submersion for safe operation are required to be inherently protected against the hazards of overheating when not submerged, for example, during a relamping process. Protection against overheating is required to be built into a fixture or to be a part of it. A remotely located low-water cutoff switch does not provide the intended protection.

**(7) Compliance.** Compliance with these requirements shall be obtained by the use of a listed underwater luminaire (lighting fixture) and by installation of a listed ground-fault circuit interrupter in the branch circuit or a listed transformer for luminaires (fixtures) operating at not more than 15 volts.

**(F) Fountains: Ground-Fault Circuit Interrupter.** Fountain equipment, unless listed for operation at 15 volts or less and supplied by a transformer shall be protected by a ground-fault circuit interrupter.

### 430.11 Carnivals, Circuses, Fairs, and Similar Events

**(A) Overhead Conductor Clearances.**

**(1) Vertical Clearances** Conductors shall have a vertical clearance to ground in accordance with 410.7(B). These clearances shall apply only to wiring installed outside of tents and concessions.

**(2) Clearance to Rides and Attractions.** Amusement rides and amusement attractions shall be maintained not less than 4.5 m (15 ft) in any direction from overhead conductors operating at 600 volts or less, except for the conductors supplying the amusement ride or attraction. Amusement rides or attractions shall not be located under or within 4.5 m (15 ft) horizontally of conductors operating in excess of 600 volts.

**(B) Protection of Electrical Equipment.** Electrical equipment and wiring methods in or on rides, concessions, or other units shall be provided with mechanical protection where such equipment or wiring methods are subject to physical damage.

**(C) Guarding—Services.** Service equipment shall not be installed in a location that is accessible to unqualified persons, unless the equipment is lockable.

Service equipment must be installed in accordance with Article 230 of the *NEC* and must be lockable where accessible by unqualified persons. At fairgrounds, carnivals, and similar events, significant pedestrian traffic throughout the site, including through those areas where electrical equipment is located, causes increased risk. Section 430.11(C) helps safeguard the general public from accidental contact with energized service equipment.

**(D) Wiring Methods.**

**(1) Type.** Where flexible cords or cables are used, they shall be listed for extra-hard usage. Where flexible cords or cables are used and are not subject to physical damage, they shall be permitted to be listed for hard usage. Where used outdoors, flexible cords and cables shall also be listed for wet locations and shall be sunlight resistant. Extra-hard-usage flexible cords or cables shall be permitted for use as permanent wiring on portable amusement rides and attractions where not subject to physical damage.

**(2) Single Conductor.** Single conductor cable shall be permitted only in size 2 AWG or larger.

**(3) Open Conductors.** Open conductors are prohibited except as part of a listed assembly or festoon lighting installed in accordance with Section 410.7.

**(4) Splices.** Flexible cords or cables shall be continuous without splice or tap between boxes or fittings. Cord connectors shall not be laid on ground unless listed for wet locations. Connectors and cable connections shall not be placed in audience traffic paths or within areas accessible to the public unless guarded.

**(5) Cord Connectors.** Cord connectors shall not be laid on the ground unless listed for wet locations. Connectors and cable connections shall not be placed in audience traffic paths or within areas accessible to the public unless guarded.

**(6) Support.** Wiring for an amusement ride, attraction, tent, or similar structure shall not be supported by any other ride or structure unless specifically designed for the purpose.

**(7) Protection.** Flexible cords or cables accessible to the public shall be arranged to minimize the tripping hazard and shall be permitted to be covered with nonconductive matting, provided that the matting does not constitute a greater tripping hazard than the uncovered cables. It shall be permitted to bury cables.

**(8) Boxes and Fittings.** A box or fitting shall be installed at each connection point, outlet, switchpoint, or junction point.

**(E) Rides, Tents, and Concessions.**

**(1) Disconnecting Means.** Each ride and concessions shall be provided with a fused disconnect switch or circuit breaker located within sight and within 1.8 m (6 ft) of the operator's station. The disconnecting means shall be readily accessible to the operator, including when the ride is in operation. Where accessible to unqualified persons, the enclosure for the switch or circuit breaker shall be of the lockable type. A shunt trip device that opens the fused disconnect or circuit breaker when a switch located in the ride operator's console is closed shall be a permissible method of opening the circuit.

**(2) Portable Wiring Inside Tents and Concessions.** Electrical wiring for lighting, where installed inside of tents and concessions, shall be securely installed and, where subject to physical damage, shall be provided with mechanical protection. All lamps for general illumination shall be protected from accidental breakage by a suitable fixture or lampholder with a guard.

**(F) Ground-Fault Circuit-Interrupter (GFCI) Protection for Personnel.**

**(1) General-Use 15- and 20-Ampere, 125-Volt Receptacles.** All 125-volt, single-phase, 15- and 20-ampere receptacle outlets that are in use by personnel shall have listed ground-fault circuit-interrupter protection for personnel. The ground-fault circuit interrupter shall be permitted to be an integral part of the attachment plug or located in the power-supply cord, within 300 mm (12 in.) of the attachment plug. For the purposes of this section, listed cord sets incorporating ground-fault circuit-interrupter protection for personnel shall be permitted. Egress lighting shall not be connected to the load side terminals of a ground-fault circuit-interrupter receptacle.

**(2) Appliance Receptacles.** Receptacles supplying items, such as cooking and refrigeration equipment, that are incompatible with ground-fault circuit-interrupter devices shall not be required to have ground-fault circuit-interrupter protection.

**(3) Other Receptacles.** Other receptacle outlets not covered in 430.10(F)(1) and 430.10(F)(2) shall be permitted to have ground-fault circuit-interrupter protection for personnel, or a written procedure shall be continuously enforced at the site by one or more designated persons to ensure the safety of equipment grounding conductors for all cord sets and receptacles.

**(G) Equipment Bonding.** The following equipment connected to the same source shall be bonded:

(1) Metal raceways and metal sheathed cable
(2) Metal enclosures of electrical equipment
(3) Metal frames and metal parts of rides, concessions, trailers, trucks, or other equipment that contain or support electrical equipment

**(H) Equipment Grounding.** All equipment requiring grounding shall be grounded by an equipment grounding conductor. The equipment grounding conductor shall be bonded to the system grounded conductor at the service disconnecting means, or in the case of a separately derived system such as a generator, at the generator or first disconnecting means supplied by the generator. The grounded circuit conductor shall not be connected to the equipment grounding conductor on the load side of the service disconnecting means or on the load side of a separately derived system disconnecting means.

**(I) Grounding Conductor Continuity Assurance.** The continuity of the grounding conductor system used to reduce electrical shock hazards shall be verified each time that portable electrical equipment is connected.

The transient nature of amusements, and, in some cases, the entire electrical distribution system associated with fairs, carnivals, and circuses, increases the possibility that continuity of the equipment grounding conductor system could be interrupted. Verification of the grounding system continuity helps ensure the safety of workers and the general public who might come in contact with exposed non–current-carrying surfaces of electrical equipment or equipment that is electrically powered.

The verification of the grounding system continuity is required each time that portable equipment is reconnected.

# ARTICLE 440 Hazardous (Classified) Locations, Class I, II, and III, Divisions 1 and 2 and Class I, Zones 0, 1, and 2

Work done in hazardous areas always requires special attention to safety, and working safely with electrical equipment and wiring within such areas requires unique practices to mitigate the potential explosion hazards. Article 440 defines requirements necessary to avoid igniting flammable materials that might be present in areas classified as electrically hazardous.

## 440.1 Scope

This article shall apply to the requirements for electric equipment and wiring in locations that are classified depending on the properties of the flammable vapors, liquids, or gases, or combustible dusts or fibers that might be present therein and the likelihood that a flammable or combustible concentration or quantity is present. Hazardous (classified) locations can be found in occupancies such as, but not limited to, aircraft hangars, gasoline dispensing and service stations, bulk storage plants for gasoline or other volatile flammable liquids, paint-finishing process plants, health care facilities, agricultural or other facilities where excessive combustible dusts might be present, marinas, boat yards, and petroleum and chemical processing plants. Each room, section, or area shall be considered individually in determining its classification.

### 440.2 Definition: Intrinsically Safe Equipment

Apparatus in which the circuits are not necessarily intrinsically safe themselves but that affect the energy in the intrinsically safe circuits and are relied on to maintain intrinsic safety. Associated apparatus may be either of the following:

(1) Electrical apparatus that has an alternative-type protection for use in the appropriate hazardous (classified) location
(2) Electrical apparatus not so protected that shall not be used within a hazardous (classified) location

### 440.3 General

The basic document that describes the limits of area classification is a plan drawing. Once the area classification has been established, a plan view drawing should be drawn that shows both the boundaries of each area and the classification. This drawing, then, should become the record document. Sometimes the drawing is a plot plan that illustrates a large area of a facility. If any other form of documentation is used, the documentation should be at least as effective as a clear, up-to-date drawing. Workers and authorities having jurisdiction must have sufficient documentation to determine whether the installation meets the requirements of consensus standards. If an area classification changes because of a change in the operating environment, the record drawing(s) must be revised to ensure that the records are accurate and up-to-date.

**(A) Documentation.** All areas designated as hazardous (classified) locations shall be properly documented. This documentation shall be available to those authorized to design, install, inspect, maintain, or operate electrical equipment at the location.

**(B) Approval for Class and Properties.**

**(1) Equipment Identification.** Equipment shall be identified not only for the class of location but also for the explosive, combustible, or ignitible properties of the specific gas, vapor, dust, fiber, or flyings that will be present. Class I equipment shall not have any exposed surface that operates at a temperature in excess of the ignition temperature of the specific gas or vapor.

FPN: Luminaires (lighting fixtures) and other heat-producing apparatus, switches, circuit breakers, and plugs and receptacles are potential sources of ignition and are investigated for suitability in classified locations. Such type of equipment, as well as cable terminations for entry into explosion proof enclosures, are available as listed for Class I, Division 2 locations. Fixed wiring, however, might utilize wiring methods that are not evaluated with respect to classified locations. Wiring products such as cable, raceways, boxes, and fittings, therefore, are not marked as being suitable for Class I, Division 2 locations.

Suitability of identified equipment shall be determined by any of the following:

(1) Equipment listing or labeling
(2) Evidence of equipment evaluation from a qualified testing laboratory or inspection agency concerned with product evaluation
(3) Evidence acceptable to the authority having jurisdiction such as manufacturer's self-evaluation or an owner's engineering judgment

**(2) Division Location.** Equipment has been identified for a Division 1 location shall be permitted in a Division 2 location of the same class and group.

**(3) General-Purpose Location.** Where specifically permitted, general-purpose equipment or equipment in general-purpose enclosures shall be permitted to be installed in Division 2 locations if the equipment does not constitute a source of ignition under normal operating conditions.

**(4) Equipment Requiring Sealing Means.** Equipment, regardless of the classification of the location in which it is installed, that depends on a single compression seal, diaphragm, or tube to prevent flammable or combustible fluids from entering the equipment shall be identified for a Class I, Division 2 location. Equipment installed in a Class I, Division 1 location shall be identified for the Class I, Division 1 location.

**(5) Normal Operating Conditions.** Unless otherwise specified, normal operating conditions for motors shall be assumed to be rated full-load steady conditions.

It is not intended that locked-rotor or other motor overload conditions, such as single phasing, be considered when evaluating motor operating temperatures (internal and external) in Class I, Division 2 locations. However, such abnormal load conditions must be considered when evaluating the external temperatures of explosion-proof motors for Class I, Division 1 locations and motors such as dust-ignition-proof motors for Class II, Division 1 locations.

**(6) Flammable Gases and Dusts.** Where flammable gases or combustible dusts are or may be present at the same time, the simultaneous presence of both shall be considered when determining the safe operation temperature of the electrical equipment.

FPN: The characteristics of various atmospheric mixtures of gases, vapors, and dusts depend on the specific material involved.

A coal-handling facility is one example of a location where both methane gas and coal dust might be present at the same time.

**(C) Conduits.** All threaded conduit or fittings shall be made wrenchtight to prevent sparking when fault current flows through the conduit system and to ensure the explosion proof and flameproof integrity of the conduit system where applicable. Equipment provided with threaded entries for field wiring connections shall be installed as applicable.

**(D) Marking.** All equipment shall be marked to show the class, group, and operating temperature or temperature class referenced to a 40°C (104°F) ambient.

The operating temperature or temperature range of equipment normally is referenced to a 40°C (104°F) ambient temperature. Unless the equipment is provided with thermally actuated sensors that limit the temperature to that marked on the equipment, operation in ambient temperatures higher than 40°C (104°F) will increase the operating temperature of the equipment. Many explosion-proof and dust-ignition-proof motors are equipped with thermal protectors. Similarly, operation in ambient temperatures lower than 40°C (104°F) usually will reduce the operating temperature.

*Exception No. 1: Equipment of the non–heat-producing type, such as junction boxes, conduit, and fittings, and equipment of the heat-producing type and having a maximum temperature not more than 100°C (212°F) shall not be required to have a marked operating temperature or temperature class.*

*Exception No. 2: Fixed luminaires (lighting fixtures) marked for use in Class I, Division 2 or Class II, Division 2 locations only shall not be required to be marked to indicate the group.*

*Exception No. 3: Fixed general-purpose equipment in Class I locations, other than fixed luminaires (lighting fixtures), that is acceptable for use in Class I, Division 2 locations shall not be required to be marked with the class, group, division, or operating temperature.*

*Exception No. 4: Fixed dusttight equipment other than fixed luminaires (lighting fixtures) that is acceptable for use in Class II, Division 2 and Class III locations shall not be required to be marked with class, group, division, or operating temperature.*

*Exception No. 5: Electric equipment suitable for ambient temperatures exceeding 40°C (104°F) shall be marked with both the maximum ambient temperature and the operating temperature or temperature class at that ambient temperature.*

A squirrel-cage induction motor without brushes, switching mechanisms, or similar arc-producing devices is an example of fixed general-purpose equipment. [See 501.8(B) of the *National Electrical Code®* (NFPA 70, *NEC®*, or *Code*) for more information on motors in Class I, Division 2 locations.]

### 440.4 Class I, Zone 0, 1, and 2 Locations

**(A) Scope** This article covers the requirements for the zone classification system as an alternative to the division classification system for electrical and electronic equipment and wiring for all voltage in Class I, Zone 0, Zone 1, and Zone 2 hazardous (classified) locations where fire or explosion hazards may exist due to flammable gases, vapors, or liquids.

FPN: Requirements for electric and electronic equipment and wiring for all voltages in Class I, Division 1 or Division 2; Class II, Division 1 or Division 2; and Class III, Division 1 or Division 2 hazardous (classified) locations where fire or explosion hazards may exist due to flammable gases or vapors, flammable liquids, or combustible dusts or fibers, are contained in Articles 500 through 504 of NFPA 70–2002, *National Electrical Code.*

New requirements covering wiring methods, sealing, and use of flexible cord were added to Article 505 in the 2002 edition of the *Code.* The Zone 0, Zone 1, and Zone 2 classification concept can be used as a parallel set of requirements to the division concept. The zone classification concept, based on area classification standards used by the International Electrotechnical Commission (IEC), provides an alternative method of classifying Class I hazardous locations.

The IEC classification scheme covers underground mines. In the United States, mines are under the jurisdiction of the Mine Safety and Health Administration (MSHA) and are outside of the scope of the *Code.*

**(B) Threading.** All threaded conduit or fittings referred to herein shall be threaded with a National (American) Standard Pipe Taper (NPT) standard conduit cutting die that provides a taper of 1 in 16 (¾ in. taper per foot). Such conduit shall be made wrenchtight to prevent sparking when fault current flows through the conduit system and to ensure the explosionproof or flameproof integrity of the conduit system where applicable. Equipment provided with threaded entries for field wiring connection shall be installed in accordance with 440.4(B)(1) or 440.4(B)(2).

**(1) Equipment Provided with Threaded Entries for NPT-Threaded Conduit or Fittings.** For equipment provided with threaded entries for NPT-threaded conduit or fittings, listed conduit, conduit fittings, or cable fittings shall be used.

**(2) Equipment Provided with Threaded Entries for Metric Threaded Conduit or Fittings.** For equipment with metric threaded entries, such entries shall be identified as being metric, or listed adapters to permit connection to conduit or NPT-threaded fittings shall be provided with the equipment. Adapters shall be used for connection to conduit or NPT-threaded fittings. Listed cable fittings that have metric threads shall be permitted to be used.

Exhibit 440.1 is an example of an adapter that provides a means of connecting conduit or fitting with NPT threads to an "increased safety" enclosure that has metric threads.

FPN: Threading specifications for metric threaded entries are located in ISO 965/1-1980, *Metric Screw Threads,* and ISO 965/3-1980, *Metric Screw Threads.*

***EXHIBIT 440.1*** *A typical hub providing an NPT threaded entry for conduit or cable into an increased safety enclosure. (Courtesy of Cooper Crouse-Hinds)*

**(C) Special Precaution.** Article 440 requires equipment construction and installation that will ensure safe performance under conditions of proper use and maintenance.

FPN No. 1: It is important that inspection authorities and users exercise more than ordinary care with regard to the installation and maintenance of electrical equipment in hazardous (classified) locations.

FPN No. 2: Low ambient conditions require special consideration. Electrical equipment depending on the protection techniques may not be suitable for use at temperatures lower than –20°C (–4°F) unless they are approved for use at lower temperatures. However, at low ambient temperatures, flammable concentrations of vapors may exist in a location classified Class I, Zones 0, 1, or 2 at normal ambient temperature.

**(1) Supervision of Work.** Classification of areas and selection of equipment and wiring methods shall be under the supervision of a qualified Registered Professional Engineer.

Note that 440.4(C)(1) requires area classification, wiring, and equipment selection to be under the supervision of a qualified Registered Professional Engineer for installations in Class I, Zone 0, 1, and 2 locations.

**(2) Dual Classification.** In instances of areas within the same facility classified separately, Class I, Zone 2 locations shall be permitted to abut, but not overlap, Class I, Division 2 locations. Class I, Zone 0 or Zone 1 locations shall not abut Class I, Division 1 or Division 2 locations.

An installation may use either the Article 500 classification scheme or the Article 505 classification scheme. The classification schemes must not be mixed within the same area. In areas within the same facility, Class I, Zone 2 locations may be adjacent to and share the same border with Class I, Division 2 locations. However, Class I, Zone 0 or Zone 1 locations must not be adjacent to or share the same border with Class I, Division 1 locations.

**(3) Reclassification Permitted.** A Class I, Division 1 or Division 2 location shall be permitted to be reclassified as a Class I, Zone 0, Zone 1, or Zone 2 location, provided all of the space that is classified because of a single flammable gas or vapor source is reclassified under the requirements of the section.

**(D) Class I Temperature.** The temperature marking shall not exceed the ignition temperature of the specific gas or vapor to be encountered.

FPN: For information regarding ignition temperatures of gases and vapors, see NFPA 497-1997, *Recommended Practice for the Classification of Flammable Liquids, Gases, or Vapors and of Hazardous (Classified) Locations for Electrical Installations in Chemical Process Areas,* and IEC 79-20-1996, *Electrical Apparatus for Explosive Gas Atmospheres, Data for Flammable Gases and Vapours, Relating to the Use of Electrical Apparatus.*

**(E) Equipment.**

**(1) Suitability.** Suitability of identified equipment shall be determined by one of the following:

(1) Equipment listing or labeling
(2) Evidence of equipment evaluation for a qualified testing laboratory or inspection agency concerned with product evaluation
(3) Evidence acceptable to the authority having jurisdiction such as a manufacturer's self-evaluation or an owner's engineering judgment

**(2) Listing.**

(a) Equipment that is listed for a Zone 0 location shall be permitted in a Zone 1 or Zone 2 location of the same gas or vapor. Equipment that is listed for a Zone 1 location shall be permitted in a Zone 2 location of the same gas or vapor.

(b) Equipment shall be permitted to be listed for a specific gas or vapor, specific mixtures of gases or vapors, or any specific combination of gases or vapors.

**(3) Marking.** Equipment shall be marked in accordance with 440.3(E)(3)(a) or 440.3(E)(3)(b).

(a) Division Equipment. Equipment approved for Class I, Division 1 or Class I, Division 2 shall, in addition to being marked, be permitted to be marked with all of the following:

(1) Class I, Zone 1 or Class I, Zone 2 (as applicable)
(2) Applicable gas classification group(s)
(3) Temperature classification

(b) Zone Equipment. Equipment meeting one or more of the protection techniques shall be marked with the following in the order shown:

(1) Class
(2) Zone
(3) Symbol "AEx"
(4) Protection technique(s)
(5) Applicable gas classification group(s)
(6) Temperature classification

*Exception: Intrinsically safe associated apparatus shall be required to be marked only with items (4), (5) and (6).*

The symbol AEx identifies the equipment as meeting American standards. In European Common Market countries, the symbol is EEx. In the IEC standards on which American and European standards are based, the symbol is Ex.

(c) Group and Zone Markings for Zone Equipment. Electric equipment of types of protection "e," "m," "p," or "q" shall be marked Group II. Electric equipment of types of protec-

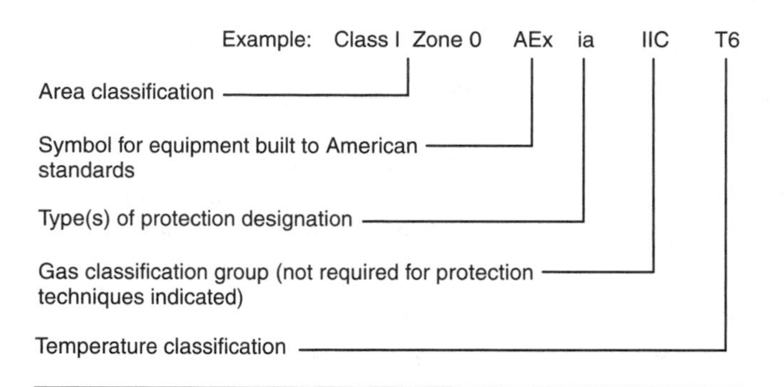


***FIGURE 440.4(E)(3)(C)*** *Class I, Zone 0, AEx ia IIC T6.*

tion "d," "ia," "ib," "[ia]," or "[ib]" shall be marked Group IIA, or IIB, or IIC, or for a specific gas or vapor. Electric equipment of types of protection "n" shall be marked Group II unless it contains enclosed-break devices, nonincendive components, or energy-limited equipment or circuits, in which case it shall be marked Group IIA, IIB, or IIC, or a specific gas or vapor. Electrical equipment of other types of protection shall be marked Group II unless the type of protection utilized by the equipment requires that it be marked Group IIA, IIB, or IIC, or a specific gas or vapor. *[See Figure 440.4(E)(3)(c).]*

FPN: An example of such a required marking is "Class I, Zone 0, AEx ia IIC T6."

**(F) Documentation for Industrial Occupancies.** All areas in industrial occupancies designated as hazardous (classified) locations shall be properly documented. This documentation shall be available to those authorized to design, install, inspect, maintain, or operate electrical equipment at the location.

**(G) Grounding and Bonding.** Grounding and bonding shall comply with the following and other applicable requirements.

**(1) Bonding.** The locknut-bushing and double-locknut type of contacts shall not be depended on for bonding purposes, but bonding jumpers with proper fittings or other approved means of bonding shall be used. Such means of bonding shall apply to all intervening raceways, fittings, boxes, enclosures, and so forth, between Class I locations and the point of grounding for service equipment or point of grounding of a separately derived system.

*Exception: The specific bonding means shall be required only to the nearest point where the grounded circuit conductor and the grounding electrode are connected together on the line side of the building or structure disconnecting means, provided the branch-circuit overcurrent protection is located on the line side of the disconnecting means.*

**(2) Flexible Metal Conduit.** Where flexible metal conduit or liquidtight flexible metal conduit is used and is to be relied on to complete a sole equipment grounding path, it shall be installed with internal or external bonding jumpers in parallel with each conduit.

*Exception: In Class I, Zone 2 locations, the bonding jumper shall be permitted to be deleted where all the following conditions are met:*

*(a) Listed liquidtight flexible metal conduit 1.8 m (6 ft) or less in length, with fittings listed for grounding, is used.*
*(b) Overcurrent protection in the circuit is limited to 10 amperes or less.*
*(c) The load is not a power utilization load.*

# ARTICLE 450
# Special Systems

Articles 400, 410, and 420 contain basic requirements that apply generally to all installations unless modified by the content of Articles 430, 440, and 450. This article defines requirements for special systems that need to be different from the basic installation requirements.

### 450.1 Systems Over 600 Volts, Nominal

Sections 450.1(A) through 450.1(I) cover the general requirements for equipment operating at more than 600 volts, nominal.

**(A) Aboveground Wiring Methods.** Aboveground conductors shall be installed in rigid metal conduit, in intermediate metal conduit, in electrical metallic tubing, in rigid nonmetallic conduit, in cable trays, as busways, as cablebus, in other identified raceways, or as open runs of metal-clad cable suitable for the use and purpose. In locations accessible to qualified persons only, open runs of Type MV cables, bare conductors, and bare busbars shall also be permitted. Busbars shall be permitted to be either copper or aluminum.

Wiring methods are not restricted within the confines of transformer vaults, switch rooms, and similar areas that are restricted to qualified personnel if they are marked and the entry controlled. Open wiring with bare or insulated conductors on insulators is a common wiring method. Any wiring method may be used, including conduit systems.

**(B) Braid-Covered Insulated Conductors—Open Installations.** Open runs of braid-covered insulated conductors shall have a flame-retardant braid. If the conductors used do not have this protection, a flame-retardant saturant shall be applied to the braid covering after installation. This treated braid covering shall be stripped back a safe distance at conductor terminals, according to the operating voltage. This distance shall not be less than 25 mm (1 in.) for each kilovolt of the conductor-to-ground voltage of the circuit, where practicable.

**(C) Insulation Shielding.** Metallic and semiconducting insulation shielding components of shielded cables shall be removed for a distance dependent on the circuit voltage and insulation. Stress reduction means shall be provided at all terminations of factory-applied shielding. Metallic shielding components such as tapes, wires, or braids, or combinations thereof, and their associated conducting and semiconducting components shall be grounded.

**(D) Moisture or Mechanical Protection for Metal-Sheathed Cables.** Where cable conductors emerge from a metal sheath and where protection against moisture or physical damage is necessary, the insulation of the conductors shall be protected by a cable sheath terminating device.

**(E) Circuit-Interrupting Devices.**

**(1) Circuit Breakers: Location.**

(a) Circuit breakers installed indoors shall be mounted either in metal-enclosed units or fire-resistant cell-mounted units, or they shall be permitted to be open-mounted in locations accessible to qualified persons only.

(b) Circuit breakers used to control oil-filled transformers shall either be located outside the transformer vault or be capable of operation from outside the vault.

(c) Oil circuit breakers shall be arranged or located so that adjacent readily combustible structures or materials are safeguarded in an approved manner.

**(2) Power Fuses and Fuseholders—Use.** Where fuses are used to protect conductors and equipment, a fuse shall be placed in each ungrounded conductor. Two power fuses shall be permitted to be used in parallel to protect the same load if both fuses have identical ratings and both fuses are installed in an identified common mounting with electrical connections that will divide the current equally. Power fuses of the vented type shall not be used indoors, underground, or in metal enclosures unless identified for the use.

**(3) Distribution Cutouts and Fuse Links—Expulsion Type.**

(a) Installation. Cutouts shall be located so that they may be readily and safely operated and re-fused, and so that the exhaust of the fuses does not endanger persons. Distribution cutouts shall not be used indoors, underground, or in metal enclosures.

(b) Operation. Where fused cutouts are not suitable to interrupt the circuit manually while carrying full load, an approved means shall be installed to interrupt the entire load. Unless the fused cutouts are interlocked with the switch to prevent opening of the cutouts under load, a conspicuous sign shall be placed at such cutouts identifying that they shall not be operated under load.

**(4) Oil-Filled Cutouts—Enclosure.** Suitable barriers or enclosures shall be provided to prevent contact with nonshielded cables or energized parts of oil-filled cutouts.

**(5) Load Interrupters.** Load interrupter switches shall be permitted if suitable fuses or circuits are used in conjunction with these devices to interrupt fault currents. Where these devices are used in combination, they shall be coordinated electrically so that they will safely withstand the effects of closing, carrying, or interrupting all possible currents up to the assigned maximum short-circuit rating. Where more than one switch is installed with interconnected load terminals to provide for alternate connection to different supply conductors, each switch shall be provided with a conspicuous sign identifying this hazard.

**(F) Isolating Means.** Means shall be provided to completely isolate an item of equipment. The use of isolating switches shall not be required where there are other ways of de-energizing the equipment for inspection and repairs, such as drawout-type metal-enclosed switchgear units and removable truck panels. Isolating switches not interlocked with an approved circuit-interrupting device shall be provided with a sign warning against opening them under load. A fuseholder and fuse, designed for the purpose, shall be permitted as an isolating switch.

**(G) Accessibility of Energized Parts.**

**(1) High-Voltage Equipment.** Doors that would provide unqualified persons access to high-voltage energized parts shall be locked.

**(2) Low-Voltage Control Equipment.** Low-voltage control equipment, relays, motors, and the like shall not be installed in compartments with exposed high-voltage energized parts or high-voltage wiring unless either of the following conditions is met:

(1) The access means is interlocked with the high-voltage switch or disconnecting means to prevent the access means from being opened or removed.
(2) The high-voltage switch or disconnecting means is in the isolating position.

**(3) High-Voltage Instruments or Control Transformers and Space Heaters.** High-voltage instrument or control transformers and space heaters shall be permitted to be installed in the high-voltage compartment without access restrictions beyond those that apply to the high-voltage compartment generally.

**(H) Mobile and Portable Equipment.**

**(1) Enclosures.** All energized switching and control parts shall be enclosed in effectively grounded metal cabinets or enclosures. These cabinets or enclosures shall be marked "DANGER—HIGH VOLTAGE—KEEP OUT" and shall be locked so that only authorized and qualified persons can enter. Circuit breakers and protective equipment shall have the operating means projecting through the metal cabinet or enclosure so these units can be reset without opening locked doors. With doors closed, reasonable safe access for normal operation of these units shall be provided.

FPN: For further information on hazard signs and labels, see ANSI Z535-4, *Product Signs and Safety Labels.*

**(2) Power Cable Connections to Mobile Machines.** A metallic enclosure shall be provided on the mobile machine for enclosing the terminals of the power cable. The enclosure shall include provisions for a solid connection for the ground wire(s) terminal to effectively ground the machine frame. Ungrounded conductors shall be attached to insulators or be terminated in approved high-voltage cable couplers (which include ground wire connectors) of proper voltage and ampere rating. The method of cable termination used shall prevent any strain or pull on the cable from stressing the electrical connections. The enclosure shall have provision for locking so only authorized and qualified persons may open it, and shall be marked

"DANGER—HIGH VOLTAGE—KEEP OUT."

FPN: For further information on hazard signs and labels, see ANSI Z535-4, *Product Signs and Safety Labels.*

**(I) Tunnel Installations.**

**(1) General.** The provisions of 450.1(H) shall apply to the installation and use of high-voltage power distribution and utilization equipment that is portable, mobile, or both, such as substations, trailers, cars, mobile shovels, draglines, hoists, drills, dredges, compressors, pumps, conveyors, and underground excavators, and the like.

**(2) Conductors.** High-voltage conductors in tunnels shall be installed in metal conduit or other metal raceway, Type MC cable, or other approved multiconductor cable. Multiconductor portable cable shall be permitted to supply mobile equipment.

**(3) Protection Against Physical Damage.** Conductors and cables in tunnels shall be located above the tunnel floor and be so placed or guarded to protect them from physical damage.

**(4) Equipment Grounding Conductors.** An equipment grounding conductor shall be run with circuit conductors inside the metal raceway or inside the multiconductor cable jacket. The equipment grounding conductor shall be permitted to be insulated or bare.

**(5) Energized Parts.** Bare terminals of transformers, switches, motor controllers, and other equipment shall be enclosed to prevent accidental contact with energized parts.

**(6) Enclosures.** Enclosures for use in tunnels shall be dripproof, weatherproof, or submersible as required by the environmental conditions. Switch or contactor enclosures shall not be used as junction boxes or raceways for conductors feeding through or tapping off to other switches, unless special designs are used to provide adequate space for this purpose.

**(7) Disconnecting Means.** A switch or circuit breaker that simultaneously opens all ungrounded conductors of the circuit shall be installed within sight of each transformer or motor. The switch or circuit breaker for a transformer shall have an ampere rating not less than the ampacity of the transformer supply conductors.

**(8) Grounded and Bonded.** All non–current-carrying metal parts of electric equipment and all metal raceways and cable sheaths shall be effectively grounded and bonded to all metal pipes and rails at the portal and at intervals not exceeding 300 m (1000 ft) throughout the tunnel.

**450.2 Emergency Systems**

**(A) Scope.** The provisions of this section shall apply to the electrical safety of the installation, operation, and maintenance of emergency systems consisting of circuits and equipment intended to supply, distribute, and control electricity for illumination, power, or both, to required facilities when the normal electrical supply or system is interrupted. Emergency systems are those systems legally required and classified as emergency by municipal, state, federal, or other codes, or by any governmental agency having jurisdiction. These systems are intended to automatically supply illumination, power, or both, to designated areas and equipment in the event of failure of the normal supply or in the event of accident to elements of a system intended to supply, distribute, and control power and illumination essential for safety to human life.

Emergency systems provide limited lighting and power to essential equipment or circuits if the normal electricity supply fails. Emergency systems are those that are essential for safety to human life and are required legally by municipal, state, federal, or other codes or by a governmental agency. Examples of essential equipment include operating room equipment and other life-support equipment in hospitals.

Article 700 of the *National Electrical Code®* (*NEC®*, NFPA 70, or *Code*) defines installation requirements that apply to emergency systems. However, the *NEC* does not require emergency systems or exit lights. NFPA *101®*, *Life Safety Code®*, defines these and other life safety issues. If emergency lights or exit lights are required for safe egress, passageways that lead to the exits must also be lighted.

**(B) Wiring.** Wiring of two or more emergency circuits supplied from the same source shall be permitted in the same raceway, cable, box, or cabinet. Wiring from an emergency source or emergency source distribution overcurrent protection to emergency loads shall be kept entirely independent of all other wiring and equipment, unless otherwise permitted in (1) through (4):

(1) Wiring from the normal power source located in transfer equipment enclosures
(2) Wiring supplied from two sources in exit or emergency luminaires (lighting fixtures)
(3) Wiring from two sources in a common junction box, attached to exit or emergency luminaires (lighting fixtures)
(4) Wiring within a common junction box attached to unit equipment, containing only the branch circuit supplying the unit equipment and the emergency circuit supplied by the unit equipment

To protect wiring for emergency circuits from damage by faults in wiring for general use circuits, emergency circuit wiring must be segregated from general use wiring. Emergency circuit wiring must not be co-mingled with wiring for other circuits, to provide maximum serviceability of the emergency circuit wiring.

Exit or emergency light fixtures that contain two lamps may be wired in such a way that one lamp is connected to the normal supply and the other lamp is connected to the alternate supply. Both lamps may serve as part of the regular building lighting system.

Both the normal energy source and the emergency energy source must be present within a transfer switch enclosure to enable rapid transfer from the normal source to the emergency source, according to 450.2(B)(1) of this standard.

Wiring on the load side of a transfer switch delivers energy from both the emergency circuit wiring and the normal circuit wiring. Using two sets of wiring to supply emergency loads is not intended, either from the transfer switch to the emergency load distribution panel, as shown in Exhibits 450.1 and 450.2, or from the emergency distribution panel to the emergency loads.

At least two sources of power must be provided: one normal supply and one or more emergency systems as described in 700.12 of the *NEC*. The sources may be any of the following:

- Two services, one normal supply and one emergency supply (preferably from separate utility stations)
- One normal service and a storage battery (or unit equipment) system
- One normal service and a generator set

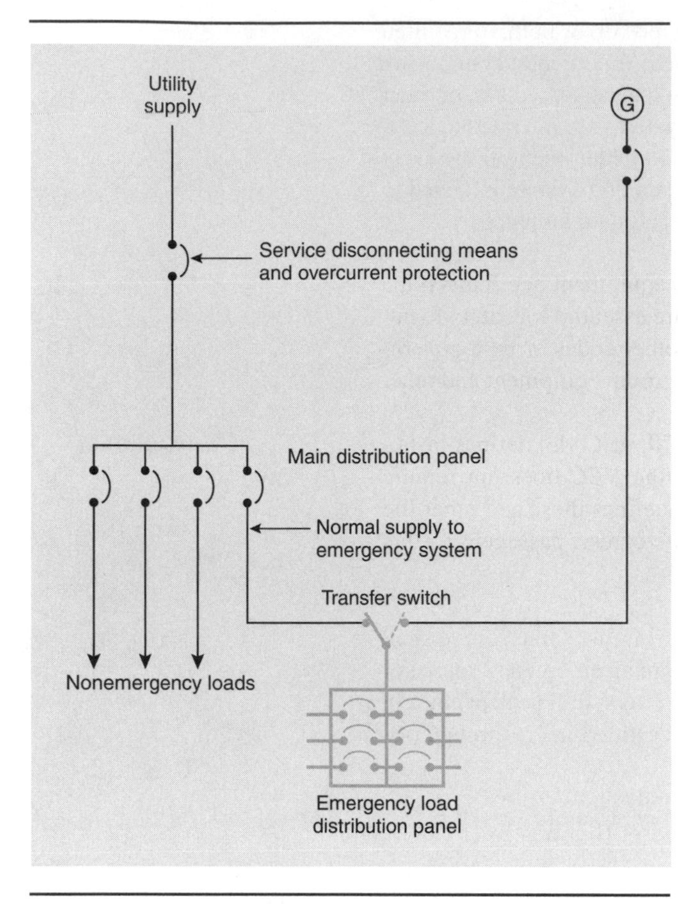


***EXHIBIT 450.1*** *Emergency load arranged to be supplied from a generator, as permitted by 700.12(B) of the NEC.*

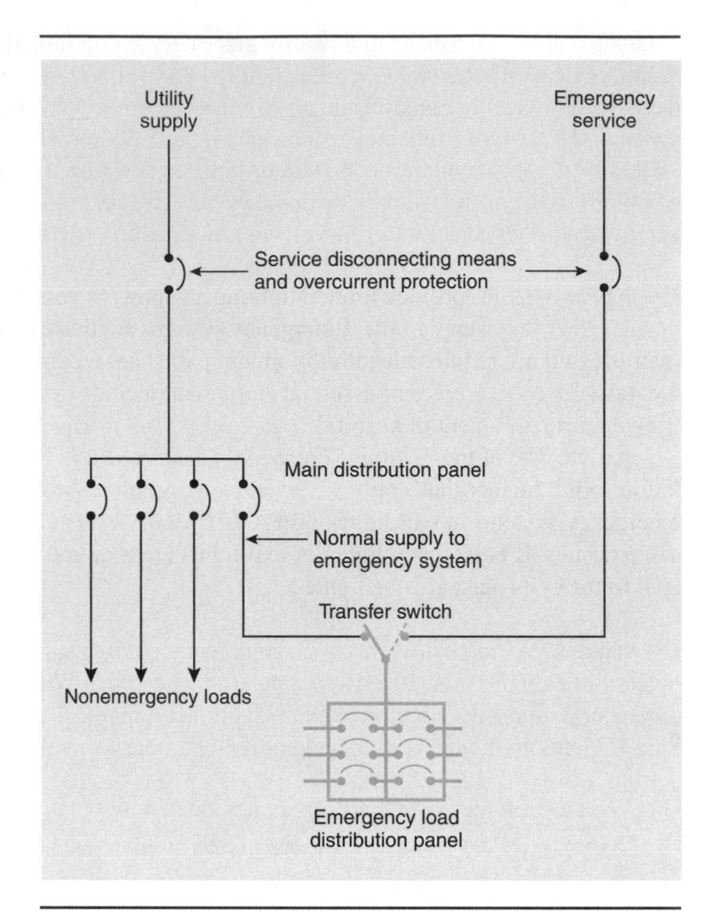


***EXHIBIT 450.2*** *Emergency load arranged to be supplied from two widely separated services, as permitted by 700.12(D) of the NEC. When one service fails, the emergency load will be transferred to the other service.*

A transfer switch (or throw-over switch) must be provided to transfer the emergency equipment supply source from the alternate supply when the normal source of supply is interrupted. If a separate supply is provided, both sources of supply may operate normally. However, all equipment supplying emergency lighting and power must be arranged to be capable of being energized from either service.

If the alternate, or emergency, source of supply is a storage battery or generator set, the single emergency system usually is supplied from the normal service, and the battery (or batteries) or generator operates only if the normal service fails. A generator may be used, however, for peak load shaving and the like, in accordance with 700.5 of the *NEC.*

Two or more separate and complete systems may provide power for emergency lighting, but a means must be provided for energizing emergency loads from the remaining system if the other one fails.

Disconnecting means and overcurrent protection are required for emergency systems as defined by Article 230 of the *NEC.*

**(C) Emergency Illumination.** Emergency illumination shall include all required means of egress lighting, illuminated exit signs, and all other lights specified as necessary to provide required illumination. Emergency lighting systems shall be designed and installed so that the failure of any individual lighting elements, such as the burning out of a light bulb, cannot leave in total darkness any space that requires emergency illumination. Where high-intensity discharge lighting such as high- and low-pressure sodium, mercury vapor, and metal halide is used as the sole source of normal illumination, the emergency lighting system shall be required to operate until normal illumination has been restored.

*Exception: Alternative means that ensure the emergency lighting illumination level is maintained shall be permitted.*

High-intensity discharge (HID) fixtures take some time to reach full intensity once they are energized. Therefore, if HID fixtures are the sole source of normal illumination in an area, the *Code* requires that the emergency lighting system operate not only until the normal system is returned to service but also until the HID fixtures provide adequate illumination. This might require a timing circuit, photoelectric monitoring system, or the equivalent.

**(D) Signs.**

**(1) Emergency Sources.** A sign shall be placed at the service entrance equipment indicating type and location of on-site emergency power sources.

*Exception: A sign shall not be required for individual unit equipment.*

**(2) Grounding.** Where the grounded circuit conductor connected to the emergency source is connected to a grounding electrode conductor at a location remote from the emergency source, there shall be a sign at the grounding location that shall identify all emergency and normal sources connected at that location.

Section 700.8(B) of the *NEC* requires that a sign, indicating type and location of on-site emergency power sources, be placed at the grounding location if the emergency source is a separately derived system and is connected to a grounding electrode conductor at a location that is remote from the emergency source.

### 450.3 Class 1, Class 2, and Class 3 Remote Control, Signaling, and Power-Limited Circuits

**(A) Classification.** Class 1, Class 2, or Class 3 remote control, signaling, or power-limited circuits shall be characterized by usage and electrical power limitations that differentiate them from light and power circuits.

**(1) Class 1 Circuits.**

(a) Class 1 Power-Limited Circuits. These circuits shall be supplied from a source that has a rated output of not more than 30 volts and 1000 volt-amperes.

(b) Class 1 Remote-Control and Signaling Circuits. These circuits shall not exceed 600 volts. The power output of the source shall not be required to be limited.

Remote-control and signaling circuits that do not meet the requirements of a Class 2 or a Class 3 circuit are often classified Class 1 circuits by default. For example, a listed nurses call system might contain a power supply with an output of 500 watts at 24 volts. Because this power supply obviously exceeds the maximum permitted output of a Class 2 power supply and the output terminals are not marked to indicate this equipment as suitable for a Class 2 power supply, the output circuit wiring is classified as Class 1 and subject to all the Class 1 circuit requirements.

**(2) Power Sources for Class 2 and Class 3 Circuits.** The power source for a Class 2 or a Class 3 circuit shall be as specified in (1) through (5):

(1) A listed Class 2 or 3 transformer
(2) A listed Class 2 or 3 power supply
(3) Other listed equipment marked to identify the Class 2 or Class 3 power source

*Exception to (3): Thermocouples shall not require listing as a Class 2 power source.*

(4) Listed information technology (computer) equipment limited power circuits
(5) A dry cell battery shall be considered an inherently limited Class 2 power source, provided the voltage is 30 volts or less and the capacity is equal to or less than that available from series connected No. 6 carbon zinc cells

**(3) Separation from Electric Light, Power, Class 1, Non–Power-Limited Fire Alarm Circuit Conductors, and Medium Power Network-Powered Broadband Communications Cables.** Cables and conductors of Class 2 and Class 3 circuits shall not be placed in any cable, cable tray, compartment, enclosure, manhole, outlet box, device box, raceway, or similar fitting with conductors of electric light, power, Class 1, non–power-limited fire alarm circuits, and medium power network-powered broadband communications cables.

**(4) Class 1, Class 2, and Class 3 Circuit Identification.** Class 1, Class 2, and Class 3 circuits shall be identified at terminal and junction locations in a manner that prevents unintentional interference with other circuits during testing and servicing.

### 450.4 Fire Alarm Systems

**(A) Classifications.** Fire alarm circuits shall be classified as either non–power-limited or power-limited.

**(B) Power Sources.** The power sources for use with fire alarm circuits shall be either power-limited or non–power-limited as required in 450.4(B)(1) and 450.4(B)(2).

**(1) Non–Power-Limited Fire Alarm (NPLFA) Circuits.** The power source of non–power-limited fire alarm circuits shall have an output voltage not more than 600 volts, nominal. These circuits shall not be supplied through ground-fault circuit interrupters.

**(2) Power-Limited Fire Alarm (PLFA) Circuits.** The power source for a power-limited fire alarm circuit shall be as specified in (1) through (3):

(1) Transformers. A listed PLFA for Class 3 transformer
(2) Power Supplies. A listed PLFA for Class 3 power supply
(3) Listed Equipment. Listed equipment marked to identify the PLFA power source

**(C) Conductors of Different Circuits in Same Cable, Enclosure, or Raceway: Class 1 with NPLFA Circuits.** Class 1 and non–power-limited fire alarm circuits shall be permitted to occupy the same cable, enclosure, or raceway without regard to whether the individual circuits are alternating current or direct current, provided that all conductors are insulated for the maximum voltage of any conductor in the enclosure or raceway.

**(D) Separation from Electric Light, Power, Class 1, NPLFA, and Medium Power Network-Powered Broadband Communications Circuit Conductors.** Power-limited fire alarm circuit cables and conductors shall not be placed in any cable, cable tray, compartment, enclosure, manhole, outlet box, device box, raceway, or similar fitting with conductors of electric light, power, Class 1, non–power-limited fire alarm circuits and medium power network-powered broadband communications circuits.

Jackets of listed power-limited fire alarm cables do not have sufficient integrity to be co-mingled with electric light, power, Class 1, non–power-limited fire alarm circuits, and medium power network-powered broadband communications cables. Failure of the cable insulation due to a fault could lead to hazardous voltages being imposed on the power-limited fire alarm circuit conductors.

**(E) Other Applications.** For other applications, power-limited fire alarm circuit conductors shall be separated by at least 50 mm (2 in.) from conductors of any electric light, power, Class 1, non–power-limited fire alarm, or medium power network-powered broadband communications circuits unless one of the following conditions is met:

(1) Either (a) all of the electric light, power, Class 1, non–power-limited fire alarm, and medium power network-powered broadband communications circuit conductors or (b) all of the power-limited fire alarm circuit conductors are in a raceway or in metal-sheathed, metal-clad, nonmetallic-sheathed, or Type UF cables.
(2) All of the electric light, power, Class 1 non–power-limited fire alarm, and medium power network-powered broadband communications circuit conductors are permanently separated from all of the power-limited fire alarm circuit conductors by a continuous and firmly fixed nonconductor, such as porcelain tubes or flexible tubing, in addition to the insulation on the conductors.

**(F) Class 2 Circuits with PLFA Circuits.** Conductors of one or more Class 2 circuits shall be permitted within the same cable, enclosure, or raceway with conductors of power-limited fire alarm circuits, provided that the insulation of the Class 2 circuit conductors in the cable, enclosure, or raceway is at least that required by the power-limited fire alarm circuits.

**(G) Fire Alarm Circuit Identification.** Fire alarm circuits shall be identified at terminal and junction locations, in a manner that prevents unintentional interference with the signaling circuit during testing and servicing.

### 450.5 Communications Systems

**(A) Scope.** These provisions for communications systems shall apply to telephone, telegraph (except radio), outside wiring for fire alarm and burglar alarm, and similar central station systems, and telephone systems not connected to a central station system but using similar types of equipment, methods of installation, and maintenance.

Section 90.3 of the *NEC,* Code Arrangement, states that Chapter 8 (in the *NEC*), comprising Articles 800, 810, 820, and 830, covers communications systems and is not subject to the requirements of Chapters 1 through 7 except where a requirement from these chapters is specifically referenced in Chapter 8. For instance, *NEC* 800.10(A)(3) references 225.14(D), 800.30(C) references Article 500, and 800.52(C) references 300.22(C).

Although information technology equipment systems often are used for or with communications systems, Article 800 of the *NEC* does not cover wiring of this equipment. Instead, Article 645 of the *NEC* provides requirements for wiring contained solely within an information technology equipment (computer) room. See 645.2 of the *NEC* for a description of the type of information technology equipment room to which Article 645 applies. Article 725 of the *NEC* provides requirements for wiring extending beyond a computer room, and Article 760 covers wiring requirements for a fire alarm system.

In some cases, telephone system wiring is used for data transmission; this use is covered by Article 800 of the *NEC.* Telephone company central offices are exempt from the requirements of Article 800 by 90.2(B)(4) of the *NEC.* The format for Article 800 is similar to that for Articles 725, 760, 770, and 820.

**(B) Protective Devices: Application.** A listed primary protector shall be provided on each circuit run partly or entirely in aerial wire or aerial cable not confined within a block. Also, a listed primary protector shall be provided on each circuit, aerial or underground, located within the block containing the building served so as to be exposed to accidental contact with electric light or power conductors operating at over 300 volts to ground. In addition, where there exists a lightning exposure, each interbuilding circuit on a premises shall be protected by a listed primary protector at each end of the interbuilding circuit.

**(C) Lightning Conductors.** Where practicable, a separation of at least 1.8 m (6 ft) shall be maintained between communications wires and cables on buildings and lightning conductors.

**(1) Overhead Communications Wires and Cables: On Poles and In-Span.** Where communications wires and cables and electric light or power conductors are supported by the same pole or run parallel to each other in-span, the following conditions shall be met:

(1) Relative Location. Where practicable, the communications wires and cables shall be located below the electric light or power conductors.
(2) Attachment to Crossarms. Communications wires and cables shall not be attached to a crossarm that carries electric light or power conductors.

**(2) Other Applications.** Communications wires and cables shall be separated at least 50 mm (2 in.) from conductors of any electric light, power, Class 1, non–power-limited fire alarm, or medium power network-powered broadband communications circuits.

**(D) Cable and Primary Protector Grounding.** The metallic member(s) of the cable sheath, where required to be grounded, and primary protectors shall be grounded as specified in 450.5(D)(1) through 450.5(D)(4). The metallic sheath of communications cables entering buildings shall be grounded as close as practicable to the point of entrance or shall be interrupted as close to the point of entrance as practicable by an insulating joint or equivalent device.

**(1) Grounding Conductor.**

(a) Insulation. The grounding conductor shall be insulated and shall be listed as suitable for the purpose.

(b) Material. The grounding conductor shall be copper or other corrosion-resistant conductive material, stranded or solid.

(c) Size. The grounding conductor shall not be smaller than 14 AWG.

(d) Length. The primary protector grounding conductor shall be as short as practicable.

(e) Run in Straight Line. The grounding conductor shall be run to the grounding electrode in as straight a line as practicable.

(f) Physical Damage. Where necessary, the grounding conductor shall be guarded from physical damage. Where the grounding conductor is run in a metal raceway, both ends of the raceway shall be bonded to the grounding conductor or the same terminal or electrode to which the grounding conductor is connected.

**(2) Electrode.** The grounding conductor shall be connected as follows:

(1) To the nearest accessible location on the following:
   a. The building or structure grounding electrode system
   b. The grounded interior metal water piping system, within 1.5 m (5 ft) from its point of entrance to the building
   c. The power service accessible means external to enclosures
   d. The metallic power service raceway

e. The service equipment enclosure
f. The grounding electrode conductor or the grounding electrode conductor metal enclosure
g. To the grounding conductor or the grounding electrode of a building or structure disconnecting means that is grounded to an electrode

(2) If the building or structure served has no grounding means, as described in item (1):
a. To any one of the individual electrodes
b. If the building or structure served has no grounding means, as described in 450(D)(1) or 450(D)(2), to an effectively grounded metal structure or to a ground rod or pipe not less than 1.5 m (5 ft) in length and 12.7 mm (½ in.) in diameter, driven, where practicable, into permanently damp earth and separated from lightning conductors and at least 1.8 m (6 ft) from electrodes of other systems. Steam or hot water pipes or air terminal conductors (lightning-rod conductors) shall not be employed as electrodes for protectors.

**(3) Electrode Connection.** Connectors, clamps, fittings, or lugs used to attach grounding conductors and bonding jumpers to grounding electrodes or to each other that are to be concrete-encased or buried in the earth shall be suitable for its application.

**(4) Bonding of Electrodes.** A bonding jumper not smaller than 6 AWG copper or equivalent shall be connected between the communications grounding electrode and power grounding electrode system at the building or structure served where separate electrodes are used. Bond ing together of all separate electrodes shall be permitted

FPN: Bonding together of all separate electrodes limits potential differences between them and between their associated wiring systems.

### 450.6 Solar Photovoltaic Systems

**(A) General.** The provisions of this section shall apply to solar photovoltaic electrical energy systems, including the array circuit(s), inverter(s), and controller(s) for such systems. Solar photovoltaic systems covered by this section may be interactive with other electrical power production sources or stand-alone, with or without electrical energy storage such as batteries. These systems can have ac or dc output for utilization.

Section 450.6 is applicable to a solar photovoltaic system that is interconnected with a second source of electricity. The second source may be an electric utility or an on-site generating system, such as a wind turbine or a hydroelectric generator.

Typical solar photovoltaic systems are illustrated in Exhibits 450.3 through 450.6. Other circuit arrangements are permissible. Exhibit 450.7 diagrams the system components of such systems.

Integrated electrical systems commonly installed in large industrial establishments are designed, installed, and operated by engineering work forces. Control equipment, including overcurrent devices, is located so as to be accessible to qualified personnel, but that location might not meet the definition of *readily accessible* provided in Article 100.

Orderly shutdown is sometimes required to prevent equipment damage or injury to personnel due to sudden loss of electrical power to the equipment. Orderly shutdown is commonly employed in nuclear power-generating facilities, paper mills, and other areas with hazardous processes.

**(B) Conductors of Different Systems.** Photovoltaic source circuits and photovoltaic output circuits shall not be contained in the same raceway, cable tray, cable, outlet box, junction box,

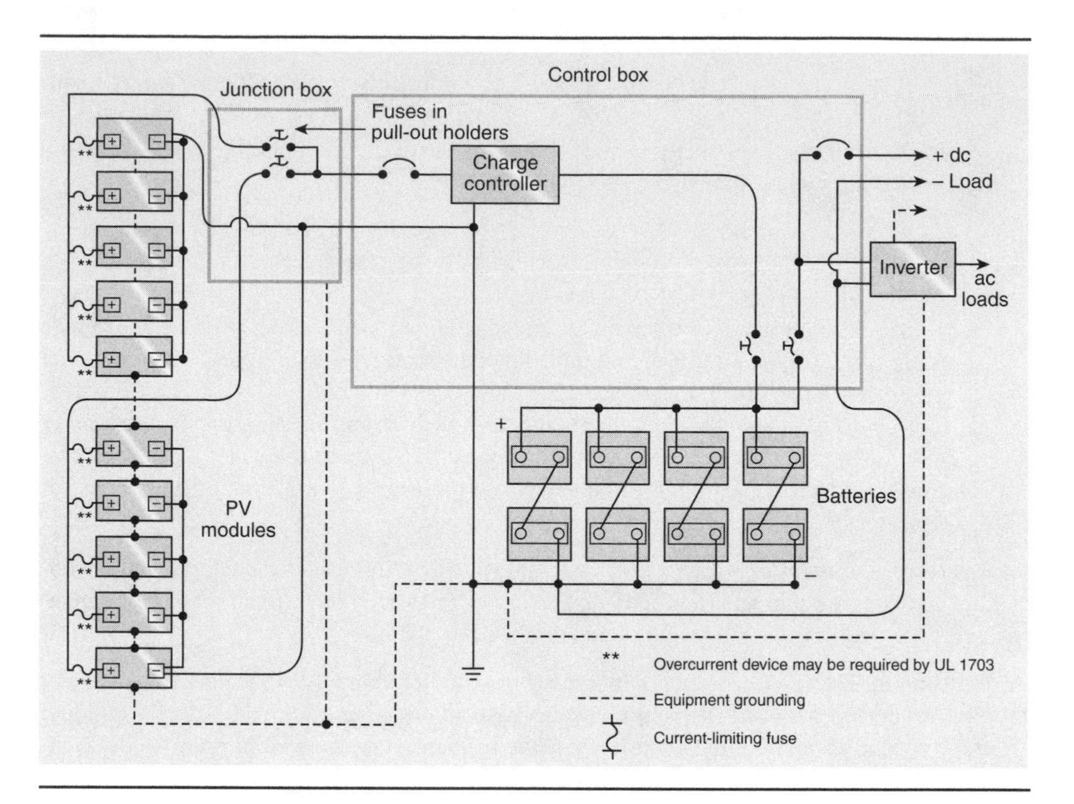


***EXHIBIT 450.3*** *Simplified circuit diagram of a small residential stand-alone system.*

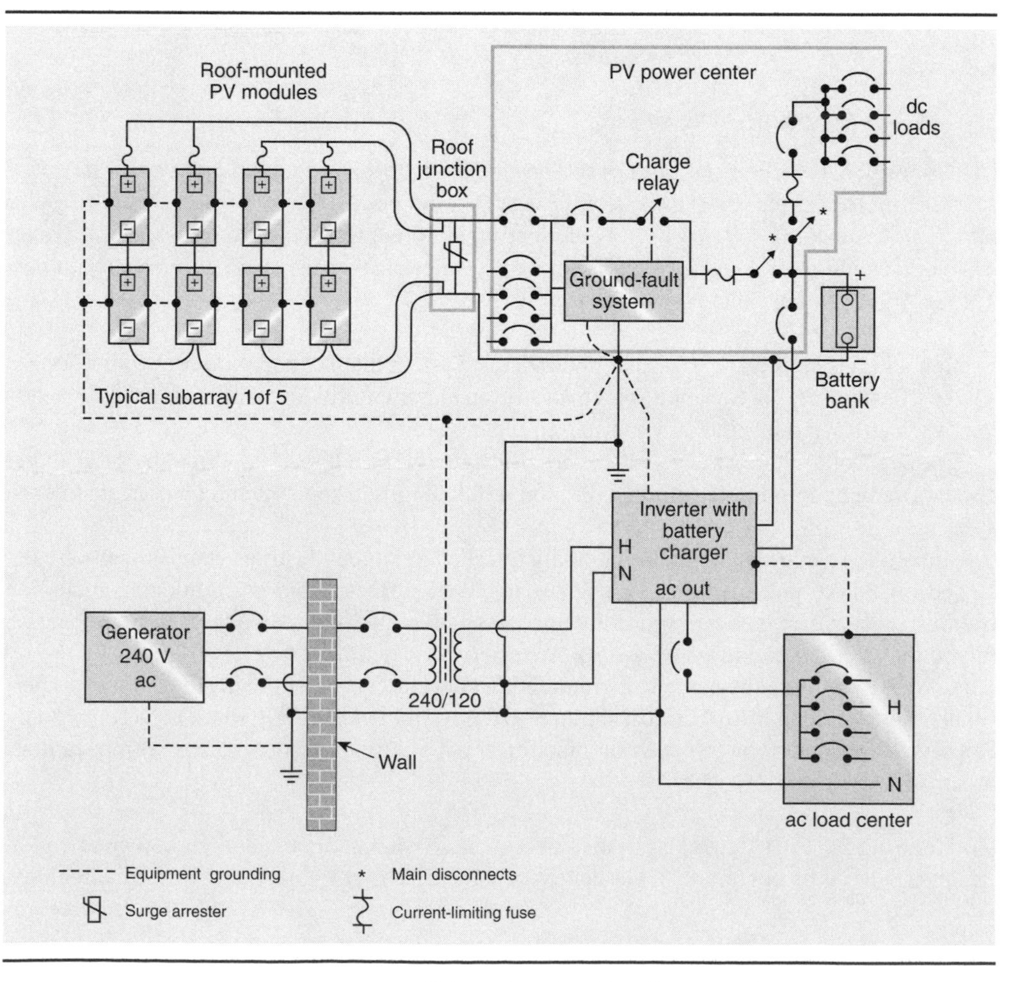


***EXHIBIT 450.4*** *Simplified circuit diagram of a medium-sized residential hybrid system.*

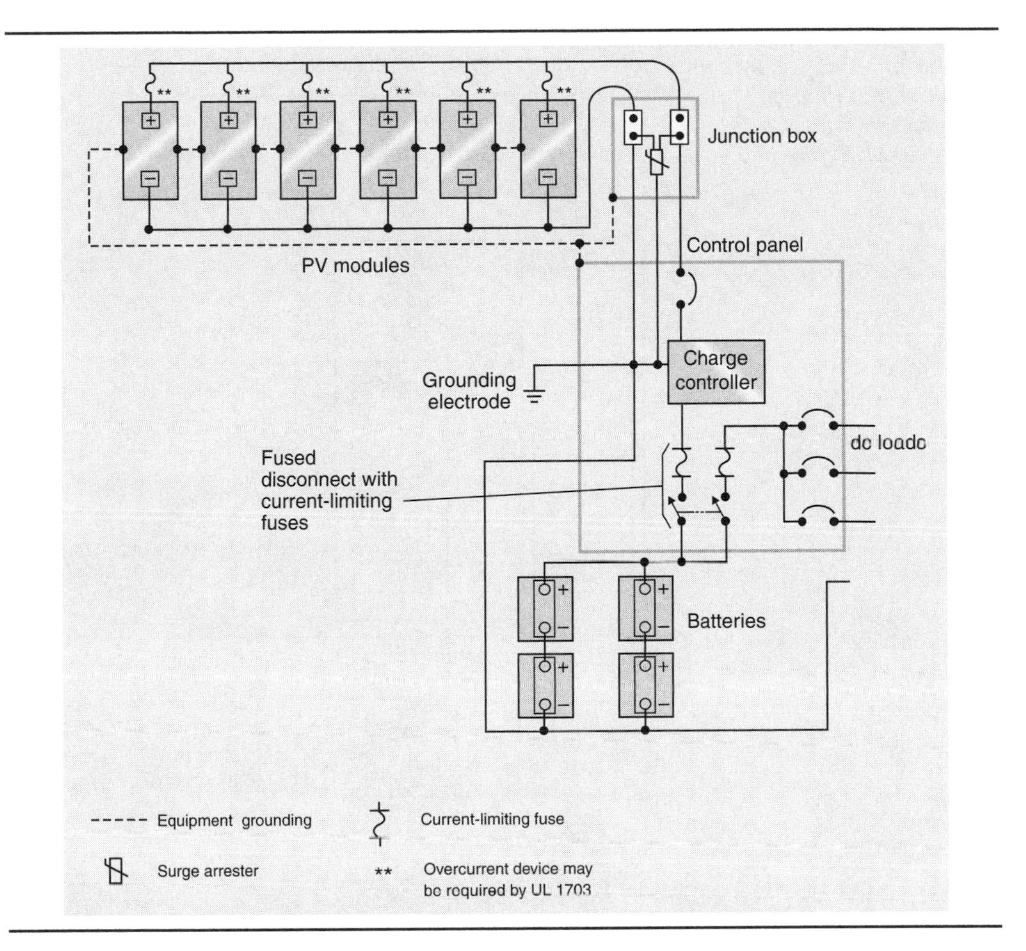


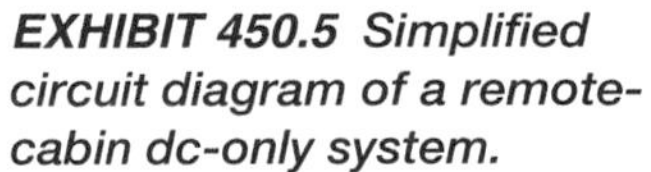

***EXHIBIT 450.5*** *Simplified circuit diagram of a remote-cabin dc-only system.*

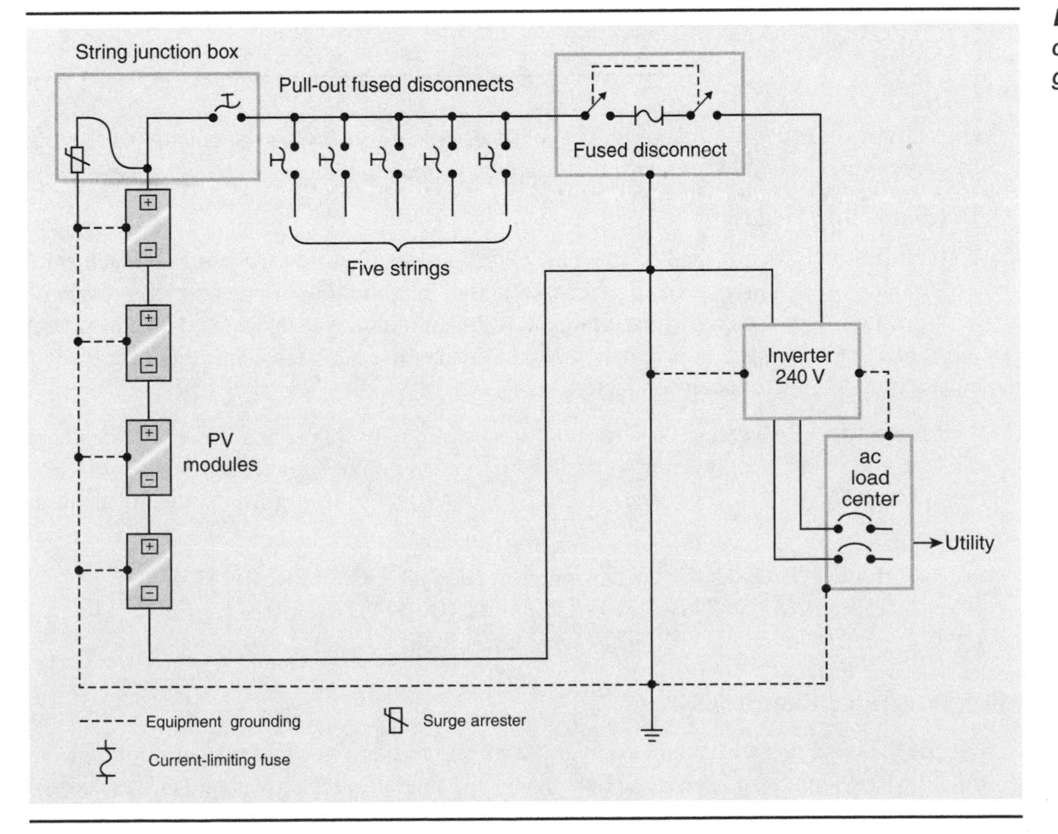


***EXHIBIT 450.6*** *Simplified circuit diagram of a rooftop grid-connected system.*

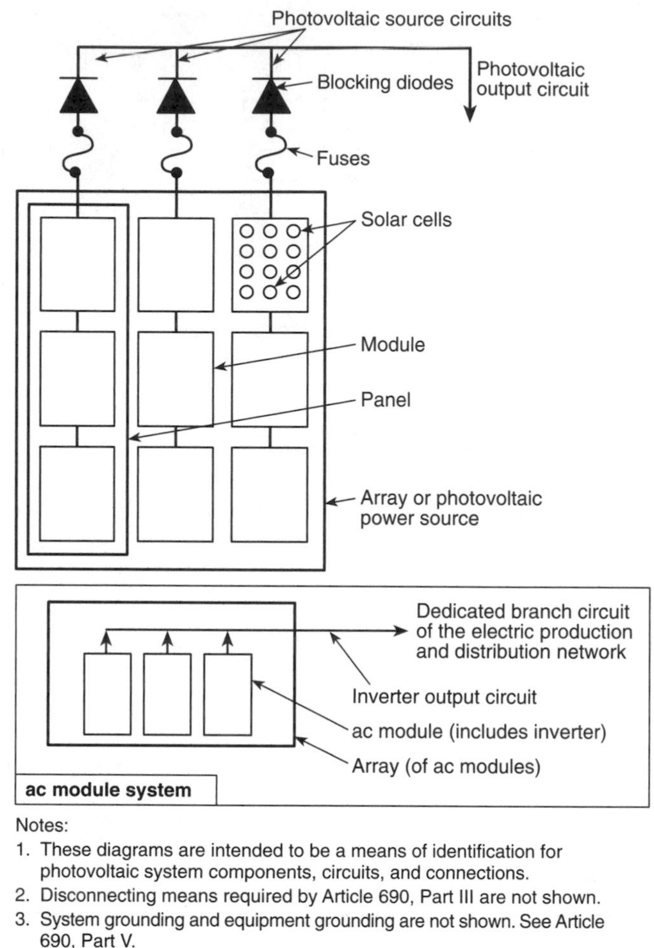


***EXHIBIT 450.7** Identification of solar photovoltaic system components. [Source: National Electrical Code®, NFPA, 2002, Figure 690.1(A)]*

or similar fitting as feeders or branch circuits of other systems, unless the conductors of the different systems are separated by a partition or are connected together.

**(C) Disconnecting Means.**

**(1) All Conductors.** Means shall be provided to disconnect all current-carrying conductors of a photovoltaic power source from all other conductors in a building or other structure. Where a circuit grounding connection is not designed to be automatically interrupted as part of the ground-fault protection system, a switch or circuit breaker used as a disconnecting means shall not have a pole in the grounded conductor.

**(2) Switch or Circuit Breaker.** Where all terminals of the disconnecting means can be energized in the open position, a warning sign shall be mounted on or adjacent to or on the disconnecting means. The sign shall be clearly legible and have the following words or equivalent:

WARNING
ELECTRIC SHOCK HAZARD. DO NOT TOUCH TERMINALS.
TERMINALS ON BOTH THE LINE AND LOAD SIDES MAY BE
ENERGIZED IN OPEN POSITION.

## 450.7 Integrated Electrical Systems

**(A) Scope** This section shall apply to integrated electrical systems, other than unit equipment, in which orderly shutdown is necessary to ensure safe operation. An integrated electrical system as used in this section is a unitized segment of an industrial wiring system where all of the following conditions are met:

(1) An orderly shutdown is required to minimize personnel hazard and equipment damage.
(2) The conditions of maintenance and supervision ensure that qualified persons service the system.
(3) Effective safeguards, acceptable to the authority having jurisdiction, are established and maintained.

**(B) Location of Overcurrent Devices in or on Premises.** Location of overcurrent devices that are critical to integrated electrical systems shall be permitted to be accessible, with mounting heights permitted to ensure security from operation by nonqualified personnel.

## REFERENCES CITED IN COMMENTARY

ANSI C73, *Dimensions of Attachment Plugs and Receptacles,* American National Standards Institute, Inc., New York, NY, 2002.

ANSI/ASME A17.1, *Safety Code for Elevators and Escalators,* American National Standards Institute, Inc., New York, NY, 1998.

Earley, M. W., et al., eds., *National Electrical Code® Handbook,* 2002 edition, National Fire Protection Association, Quincy, MA

ICEA P-32–382, *Short-Circuit Characteristics of Insulated Cable,* Insulated Cable Engineers Association, Carrollton, GA, 1999.

IEEE 141, *IEEE Recommended Practice for Electric Power Distribution for Industrial Plants* ("Red Book"), Institute of Electrical and Electronic Engineers, New York, NY, 1993.

IEEE/ANSI C2, *National Electrical Safety Code,* Institute of Electrical and Electronic Engineers, New York, NY, 2002.

NEMA MG-1, *Motors and Generators,* National Electrical Manufacturers Association, Rosslyn, VA, 2003.

NEMA WD 6, *Wiring, Devices—Dimensional Requirement,* National Electrical Manufacturers Association, Rosslyn, VA, 1997.

NFPA 13, *Standard for the Installation of Sprinkler Systems,* 2002 edition, National Fire Protection Association, Quincy, MA.

NFPA 70, *National Electrical Code®*, 1999 and 2002 editions, National Fire Protection Association, Quincy, MA.

*NFPA 72®, National Fire Alarm Code®*, 2002 edition, National Fire Protection Association, Quincy, MA.

NFPA 79, *Electrical Standard for Industrial Machinery,* 2002 edition, National Fire Protection Association, Quincy, MA.

NFPA *101®, Life Safety Code®*, 2003 edition, National Fire Protection Association, Quincy, MA.

NFPA 780, *Installation of Lightning Protection Systems,* 2000 edition, National Fire Protection Association, Quincy, MA.

OSHA 29 CFR 1926.404(b)(1)(iii), U.S. Government Printing Office, Washington, DC.

Report prepared by the Ad Hoc Committee on Aluminum Terminations prior to the 1975 Code.

UL 486A, *Wire Connections and Solder Lugs for Use with Copper Conductors,* Underwriters Laboratories Inc., Northbrook, IL, 2003.

UL 486B, *Wire Connections for Use with Aluminum Conductors,* Underwriters Laboratories Inc., Northbrook, IL, 2003.

UL 486C, *Standard for Splicing Wire Connectors,* Underwriters Laboratories Inc., Northbrook, IL, 2000.

UL 508, *Standard for Safety of Industrial Control Equipment,* Underwriters Laboratories Inc., Northbrook, IL, 1999.

ANNEX A

# Referenced Publications

## A.1 General

The documents or portions thereof listed in this annex are referenced within this standard and shall be considered part of the requirements of this document.

**A.2 NFPA Publications.** National Fire Protection Association, 1 Batterymarch Park, Quincy, MA 02169-7471.

NFPA 70, *National Electrical Code®*, 2002.

## A.3 Other Publications

**A.3.1 ANSI Publications.** American National Standards Institute, Inc., 25 West 43rd Street, 4th Floor, New York, NY 10036.

ANSI A14.1, *Safety Requirements for Portable Wood Ladders,* 1994.
ANSI A14.3, *Safety Requirements for Fixed Ladders,* 2002.
ANSI A14.4, *Safety Requirements for Job-Made Ladders,* 1992.
ANSI A14.5, *Safety Requirement for Portable Reinforced Plastic Ladders,* 2000.
ANSI Z41, *Standard for Personnel Protection, Protective Footwear,* 1998.
ANSI Z87.1, *Practice for Occupational and Educational Eye and Face Protection,* 1998.
ANSI Z89.1, *Requirements for Protective Headwear for Industrial Workers,* 1997.
ANSI Z535, *Series of Standards for Safety Signs and Tags,* 1998.

**A.3.2 ASTM Publications.** American Society for Testing and Materials, 100 Barr Harbor Drive, West Conshohocken, PA 19428-2959.

ASTM D 120-02, *Standard Specification for Rubber Insulating Gloves,* 2002.
ASTM D 1048, *Standard Specification for Rubber Insulating Blankets,* 1999.
ASTM D 1049, *Standard Specification for Rubber Covers,* 1998.
ASTM D 1050, *Standard Specification for Rubber Insulating Line Hoses,* 1990.
ASTM D 1051-02, *Standard Specification for Rubber Insulating Sleeves,* 2002.
ASTM F 478, *Standard Specification for In-Service Care of Insulating Line Hose and Covers,* 1999.
ASTM F 479, *Standard Specification for In-Service Care of Insulating Blankets,* 1999.
ASTM F 496-02, *Standard Specification for In-Service Care of Insulating Gloves and Sleeves,* 2002.
ASTM F 696-02, *Standard Specification for Leather Protectors for Rubber Insulating Gloves and Mittens,* 2002.
ASTM F 711, *Standard Specification for Fiberglass-Reinforced Plastic (FRP) Rod and Tube Used; in Line Tools,* 1989 (R 1997).
ASTM F 712, *Standard Test Methods for Electrically Insulating Plastic Guard Equipment for Protection of Workers,* 1995.

ASTM F 855, *Standard Specification for Temporary Protective Grounds to Be Used on De-energized Electric Power Lines and Equipment,* 1997.
ASTM F 1117-98, *Standard Specification for Dielectric Overshoe Footwear,* 1998.
ASTM F 1236-01, *Standard Guide for Visual Inspection of Electrical Protective Rubber Products,* 2001.
ASTM F 1505, *Standard Specification for Insulated and Insulating Hand Tools,* 2001.
ASTM F 1506-02, *Standard Performance Specification for Textile Material for Wearing Apparel for Use by Electrical Workers Exposed to Momentary Electric Arc and Related Thermal Hazards,* 2002a.
ASTM F 1891-02, *Standard Specification for Arc and Flame Resistant Rainwear,* 2002a.
ASTM F 1958, *Standard Test Method for Determining the Ignitability of Non-Flame Resistant Materials for Clothing by Electric Arc Exposure Method Using Mannequins,* 1999.
ASTM F 1959, *Standard Test Method for Determining the Arc Thermal Performance Value of Materials for Clothing,* 1999.
ASTM F 2178-02, *Standard Test Method for Determining the Arc Rating of Face Protective Products,* 2002.

**A.3.3 IEC Publications.** International Electrotechnical Commission, 3, rue de Varembé, P.O. Box 131, CH-1211 Geneva 20, Switzerland.

479-1 Part 1 General aspects
479-1-1 Chapter 1: Electrical impedance of the human body
479-1-2 Chapter 2: Effects of ac in the range of 15 Hz to 100 Hz
479-2 Part 2: Special aspects
479-2-4: Chapter 4: Effects of ac with frequencies above 100 Hz
479-2-5 Chapter 5: Effects of special waveforms of current
479-2-6 Chapter 6: Effects of unidirectional single impulse currents of short duration

**A.3.4 IRCP Publications.** International Commission for Radiological Protection.

IRCP 15, *Protection Against Ionizing Radiation from External Sources.*

# ANNEX B

# Informational References

## B.1 Referenced Publications

The following documents or portions thereof are referenced within this standard for informational purposes only and are thus not part of the requirements of this document unless also listed in Annex A.

**B.1.1 NFPA Publications.** National Fire Protection Association, 1 Batterymarch Park, Quincy, MA 02169-7471.


NFPA 30, *Flammable and Combustible Liquids Code,* 2000 edition.
NFPA 58, *Liquefied Petroleum Gas Code,* 2001 edition.
NFPA 70, *National Electrical Code*®, 2002 edition.
NFPA 70B, *Recommended Practice for Electrical Equipment Maintenance,* 2002 edition.
NFPA 77, *Recommended Practice on Static Electricity,* 1993 edition.
NFPA 325, *Guide to Fire Hazard Properties of Flammable Liquids, Gases, and Volatile Solids,* 1994 edition (available in NFPA's *Fire Protection Guide to Hazardous Materials*).
NFPA 496, *Standard for Purged and Pressurized Enclosures for Electrical Equipment,* 1998 edition.
NFPA 497, *Recommended Practice for the Classification of Flammable Liquids, Gases or Vapors and of Hazardous (Classified) Locations for Electrical Installations in Chemical Process Areas,* 1997 edition.
NFPA 780, *Standard for the Installation of Lightning Protection Systems,* 2000 edition.


**B.1.2 ANSI Publications.** American National Standards Institute, 25 West 43rd Street, 4th Floor, New York, NY 10036.


ANSI/API RP 14F, *Design and Installation of Electrical Systems for Offshore Production Platforms,* 1991.
ANSI C2, *National Electrical Safety Code,* 2002.
ANSI/UL 2279, *Electrical Equipment for Use in Class I, Zone 0, 1, and 2 Hazardous (Classified) Locations,* 1997.
ANSI C84.1, *Electric Power Systems and Equipment—Voltage Ratings (60 Hz),* 1995.
ANSI Z535-4, *Product Signs and Safety Labels.*


**B.1.3**


Lee, Ralph, Life Fellow IEEE, "The Other Electrical Hazard: Electrical Arc Blast Burns," 1982.


**B.1.4 U.S. Government Publication.** Superintendent of Documents, U.S. Government Printing Office, Washington, DC 20402.


Title 29, *Code of Federal Regulations,* Parts 1926 and 1910.
OSHA 1926.403, *Battery Rooms and Battery Charging*

OSHA 1910.178(g), *Changing and Charging Batteries*
OSHA 1910.305(j)(7), *Storage Batteries*

**B.1.5 ISA Publications.**

Electrical Apparatus for Use in Class I, Zone 1 Hazardous (Classified) Locations, Type of Protection—Oil-Immersion "o," ISA S12.26.01-1996
Electrical Apparatus for Use in Class I, Zone 1 Hazardous (Classified) Locations, Type of Protection—Increased Safety "e," ISA S12.16.01-1996
Electrical Apparatus for Use in Class I, Zone 1 Hazardous (classified) Locations, Type of Protection—Encapsulation "m," ISA S12.23.01-1996
Electrical Apparatus for Use in Class I, Zone 0, 1 Hazardous (Classified) Locations—General Requirements, ISA S12.0.01-1997
Electrical Apparatus for Use in Class I, Zone 1 and 2 Hazardous (Classified) Locations, Type of Protection—Flameproof "d," ISA S 12.22.01-1996
Electrical Apparatus for Explosive Gas Atmospheres, Classification of Hazardous Areas, IEC 79-10-1995
Recommended Practice for Classification of Locations for Electrical Installations at Petroleum Facilities Classified as Class I, Zone 0, Zone 1, or Zone 2, API RP 505-1997
Electrical Apparatus for Explosive Gas Atmospheres, Classifications of Hazardous (Classified) Locations, ISA S12.24.01-1997
Electrical Apparatus for Explosive Gas Atmospheres, Classifications of Hazardous (Classified) Locations, ISA S12.24.01-1997
Electrical Apparatus for Explosive Gas Atmospheres, Classifications of Hazardous (Classified) Locations, ISA S12.24.01-1997

**B.1.6 IEC Publications.** International Electrotechnical Commission, 3, rue de Varembé, P.O. Box 131, CH-1211 Geneva 20, Switzerland.

Electrical apparatus for explosive gas atmospheres, part 6—oil immersion "o," IEC 79-6-1995
Electrical apparatus for explosive gas atmospheres, part 7—increased safety "e," IEC 79-7-1990, Amendment No. 1 (1991) and Amendment No. 2 (1993)
Electrical apparatus for explosive gas atmospheres, part 18—encapsulation "m," IEC 79-18-1992
Construction and verification test of flameproof enclosures of electrical apparatus, IEC 79-1A-1975, Amendment No. 1 (1993)
Spark-test apparatus for intinsically safe circuits, IEC 79-3-1990
Classification of mixtures of gases or vapors with air according to their maximum experimental safe gaps and minimum igniting currents, IEC 79-12-1978
Electrical apparatus for explosive gas atmospheres—part O—general requirements, IEC 79-0-1983, Amendment No. 1 (1987) and Amendment No. 2 (1991)
Electrical apparatus for explosive gas atmospheres, part 1—construction and verification test of flameproof enclosures of electrical apparatus, IEC 79-1-1990, Amendment No. 1 (1993).
Electrical apparatus for explosive gas atmospheres, part 15—electrical apparatus with type of protection "n," IEC 79-15-1987
Electrical apparatus for Explosive Gas Atmospheres—Part 2: Electrical Apparatus, Type of Protection "p," IEC 79-2-1983; and electrical apparatus for gas atmospheres, Part 13—construction and use of rooms or buildings protected by pressurization, IEC 79-13-1982
Intrinsically Safe Apparatus and Associated Apparatus for Use in Class I, II, and III, Hazardous Locations, ANSI/UL 913-1997; Electrical apparatus for explosive gas atmospheres, part II—instrinsic safety "i," IEC 79-11-1991; and electrical apparatus for intrinsically safe circuits, IEC 79-3-1990

Electrical Apparatus for Use in Class I, Zone I Hazardous (Classified) Locations Type of Protection—Powder Filling "q," ISA S12.25.01-1996, and Electrical Apparatus for Explosive Gas Atmospheres—Part 5: Powder Filling, Type of Protection "q," IEC 79-5-1967

Electrical Apparatus for Explosive Gas Atmospheres—Part 14: Electrical Installations in Explosive Gas Atmospheres (Other than Mines), IEC 79-14-1996

Electrical Apparatus for Explosive Gas Atmospheres—Part 16: Artificial Ventilation for the Protection of Analyzer(s) Houses, IEC 79-16-1990

Electrical Apparatus for Explosive Gas Atmospheres, Data for Flammable Gases and Vapours, Relating to the Use of Electrical Apparatus, IEC 79-20-1996

Electrical Apparatus for Explosive Gas Atmospheres, Classification of Hazardous Areas, IEC 600 79-10-1995

**B.1.7 IEEE Publications.** Institute of Electrical and Electronic Engineers, 445 Hoes Lane, P.O. Box 1331, Piscataway, NJ 08855-1331.

IEEE Std 4, *Standard Techniques for High Voltage Testing,* 1978 (4th printing)

IEEE Std 463, *Standard for Electrical Safety Practices in Electrolytic Cell Line Work Zones,* 1993

IEEE Std 484, *Recommended Practice for Installation Design and Installation of Vented Lead-Acid Batteries for Stationary Applications,* 1996

IEEE Std 516, *Guide for Maintenance Methods on Energized Power Lines,* 2003

IEEE Std 937, *Recommended Practice for Installation and Maintenance of Lead-Acid Batteries for Photovoltaic Systems,* 1987 (R 1993)

IEEE Std 1187, *Recommended Practice for Installation Design and Installation of Valve-Regulated Lead-Acid Storage Batteries for Stationary Applications,* 1996

IEEE Std 1584, *Guide for Performing Arc Flash Calculations,* 2002

See also R. H. Lee, "The Other Electrical Hazard: Electric Arc Blast Burns," IEEE Trans. Industrial Applications, Vol 1A-18. No. 3, Page 246, May/June 1982.

**B.1.8 ISO Publications.** International Standards Organization, 1, rue Varembé, Case postal 56, CH-1211 Geneva 20, Switzerland.

*Metric Screw Threads,* ISO 965/1:1980

*Metric Screw Threads,* ISO 965/3:1980

UL 943, *Standard for Ground-Fault Circuit Interrupters.*

**B.1.9 Other Publications.**

UL Technical Report No. 58 (1993)

API RP 500, *Recommended Practice for Classification of Locations for Electrical Installations at Petroleum Facilities Classified as Class I, Division 1 or Division 2,* 1997

API RP 505, *Classification of Locations for Electrical Installations at Petroleum Facilities Classified as Class I, Zone 0, Zone 1, or Zone 2,* 1997

API RP 2003, *Protection Against Ignitions Arising Out of Static Lightning and Stray Currents,* 1998

*Area Classification Code for Petroleum Installations,* Model Code—Part 15, Institute for Petroleum

*Model Code of Safe Practice in the Petroleum Industry,* Part 15—Area Classification Code for Petroleum Installations, IP 15, The Institute of Petroleum, London

ANNEX C

# Limits of Approach

*This annex is not a part of the requirements of this NFPA document but is included for informational purposes only.*

The NFPA 70E Technical Committee has provided the information in Annex C to illustrate the approach boundaries. The information is intended to provide users of the standard with suggestions regarding worker approach to each safe approach limit. The approach limits are intended to trigger the need for greater control of work performed closer than the approach limit.

## C.1 Preparation for Approach

Observing a safe approach distance from exposed energized electrical conductors or circuit parts is an effective means of maintaining electrical safety. As the distance between a person and the exposed energized conductors or circuit parts decreases, the potential for electrical accident increases.

**C.1.1 Unqualified Persons, Safe Approach Distance.** Unqualified persons are safe when they maintain a distance from the exposed energized conductors or circuit parts, including the longest conductive object being handled, so that they cannot contact or enter a specified air insulation distance to the exposed energized electrical conductors or circuit parts. This safe approach distance is the Limited Approach Boundary. Further, persons must not cross the Flash Protection Boundary unless they are wearing appropriate personal protective clothing and are under the close supervision of a qualified person. Only when continuously escorted by a qualified person may an unqualified person cross the Limited Approach Boundary. Under no circumstance may an unqualified person cross the Restricted Approach Boundary, where special shock protection techniques and equipment are required.

**C.1.2 Qualified Persons, Safe Approach Distance.**

**C.1.2.1** Determine the Flash Protection Boundary and, if the boundary is to be crossed, appropriate flash-flame protection equipment must be utilized.

**C.1.2.2** For a person to cross the Limited Approach Boundary and enter the limited space, he or she must be qualified to perform the job/task.

**C.1.2.3** To cross the Restricted Approach Boundary and enter the restricted space, qualified persons must do the following:

(1) Have a plan that is documented and approved by authorized management
(2) Use personal protective equipment that is appropriate for working near exposed energized conductors or circuit parts and is rated for the voltage and energy level involved
(3) Be certain that no part of the body enters the prohibited space
(4) Minimize the risk from inadvertent movement by keeping as much of the body out of the restricted space as possible, using only protected body parts in the space as necessary to accomplish the work

**C.1.2.4** Crossing the Prohibited Approach Boundary and entering the prohibited space is considered the same as making contact with exposed energized conductors or circuit parts. See Figure C.1.2.4. Therefore, qualified persons must do the following:

(1) Have specified training to work on energized conductors or circuit parts
(2) Have a documented plan justifying the need to work that close
(3) Perform a risk analysis
(4) Have (2) and (3) approved by authorized management
(5) Use personal protective equipment that is appropriate for working on exposed energized conductors or circuit parts and is rated for the voltage and energy level involved

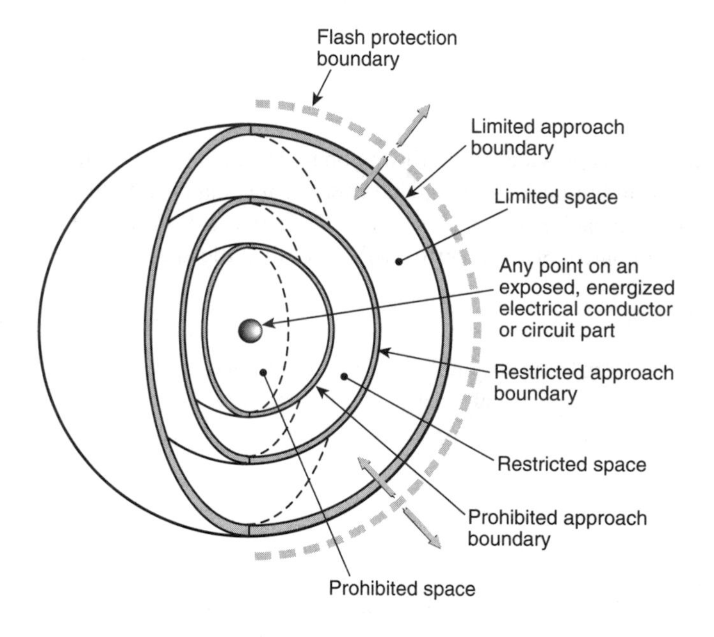


***FIGURE C.1.2.4*** *Limits of Approach.*

## C.2 Basis for Distance Values in Table 130.2(B)

This section of Annex C is provided to illustrate how the information contained in Table 130.2(B) was derived.

**C.2.1 General Statement.** Columns 1 through 5 of Table 130.2(C) show various distances from the exposed energized electrical conductors or circuit parts. They include dimensions that are added to a basic minimum air insulation distance. Those basic minimum air insulation distances for voltages 72.5 kV and under are based on ANSI/IEEE 4-1978 4th Printing, *Standard Techniques for High-Voltage Testing,* Appendix 2B; and voltages over 72.5 kV are based on ANSI/IEEE 516-2003, *Guide for Maintenance Methods on Energized Power Lines.* The minimum air insulation distances that are required to avoid flashover are as follows:

(1) ≤300 V: 1 mm (0 ft 0.03 in.)
(2) >300 V ≤ 750 V: 2 mm (0 ft 0.07 in.)
(3) >750 V ≤ 2 kV: 5 mm (0 ft 0.19 in.)
(4) >2 kV ≤ 15 kV: 39 mm (0 ft 1.5 in.)
(5) >15 kV ≤ 36 kV: 161 mm (0 ft 6.3 in.)
(6) >36 kV ≤ 48.3 kV: 254 mm (0 ft 10.0 in.)

(7) >48.3 kV ≤ 72.5 kV: 381 mm (1 ft 3.0 in.)
(8) >72.5 kV ≤ 121 kV: 640 mm (2 ft 1.2 in.)
(9) >138 kV ≤ 145 kV: 778 mm (2 ft 6.6 in.)
(10) >161 kV ≤ 169 kV: 915 mm (3 ft 0.0 in.)
(11) >230 kV ≤ 242 kV: 1.281 m (4 ft 2.4 in.)
(12) >345 kV ≤ 362 kV: 2.282 m (7 ft 5.8 in.)
(13) >500 kV ≤ 550 kV: 3.112 m (10 ft 2.5 in.)
(14) >765 kV ≤ 800 kV: 4.225 m (13 ft 10.3 in.)

**C.2.1.1 Column 1.** The voltage ranges have been selected to group voltages that require similar approach distances based on the sum of the electrical withstand distance and an inadvertent movement factor. The value of the upper limit for a range is the maximum voltage for highest nominal voltage in the range, based on ANSI C84.1-1995, *Electric Power Systems and Equipment (60 Hz)*. For single-phase systems, select the range that is equal to the system's maximum phase-to-ground voltage times 1.732.

**C.2.1.2 Column 2.** The distances in this column are based on OSHA's rule for unqualified persons to maintain a 3.05-m (10-ft) clearance for all voltages up to 50 kV (voltage-to-ground), plus 102 mm (4.0 in.) for each 1 kV over 50 kV.

**C.2.1.3 Column 3.** The distances are based on the following:

(1) ≤750 V: Use *NEC* Table 110.26(A)(1), Working Spaces, Condition 2 for 151 V–600 V range.
(2) >750 V ≤145 kV: Use *NEC* Table 110.34(A), Working Space, Condition 2.
(3) >145 kV: Use OSHA's 3.05-m (10-ft) rules as used in Column 2.

**C.2.1.4 Column 4.** The distances are based on adding to the flashover dimensions shown above the following inadvertent movement distance:

≤300 V: Avoid contact.

Based on experience and precautions for household 120/240 V systems:

>300 V ≤ 750 V: Add 304.8 mm (1 ft 0 in.) inadvertent movement.

These values have been found to be adequate over years of use in ANSI C2, *National Electrical Safety Code,* in the approach distances for communication workers.

>72.5 kV: Add 304.8 mm (1 ft 0 in.) for inadvertent movement.

These values have been found to be adequate over years of use in the *National Electrical Safety Code* in the approach distances for supply workers.

**C.2.1.5 Column 5.** The distances are based on the following:

(1) ≤300 V: Avoid contact.
(2) >300 ≤ 750 V: Use *NEC* Table 230.51(C), Clearances.

Between open conductors and surfaces, 600 V not exposed to weather.

(1) >750 V ≤ 2.0 kV: Select value that fits in with adjacent values.
(2) >2 kV ≤ 72.5 kV: Use *NEC* Table 490.24, Minimum Clearance of Live Parts, outdoor phase-to-ground values.
(3) >72.5 kV: Add 152.4 mm (0 ft 6 in.) for inadvertent movement.

These values have been found to be adequate over years of use where there has been a hazard/risk analysis, either formal or informal, of a special work procedure that allows a closer approach than that permitted by the Restricted Approach Boundary distance.

ANNEX D

# Sample Calculation of Flash Protection Boundary

*This annex is not a part of the requirements of this NFPA document but is included for informational purposes only.*

The NFPA 70E Technical Committee has provided Annex D to illustrate how the Flash Protection Boundary might be calculated. These illustrations are not intended to limit the choices of calculation methods. All publicly known methods of calculating the Flash Protection Boundary produce results that are not necessarily accurate. The thermal hazard associated with an arcing fault is very complex, with many variable attributes having an impact on the calculation. An employer should choose one method and apply that methodology uniformly across all facilities.

Workers who might be exposed to an arcing fault must wear flame-resistant clothing or other equipment to avoid a thermal injury. Protecting a worker from the thermal effects of an arcing fault does not necessarily protect the worker from injury. An arcing fault exhibits characteristics of other hazards. For instance, the arc generates a significant pressure wave. A worker could be injured by the pressure differential that could develop between the outside and inside of a person's body. The calculations illustrated in this annex do not offer protection from the effects of any pressure wave.

If the calculations illustrated in this annex (or from any other method) indicate that incident energy is 40 calories per square centimeter or greater, the pressure wave might be hazardous. That work task should not be performed unless an electrically safe work condition is established.

### D.1 Introduction

Existing knowledge about arc flash exposure at voltage levels above 600 volts is limited. Other methods of calculating such exposure exist and may be used. Commercial and shareware programs are available for calculating these values. It is important to investigate the limitations of any programs to be used. For example, some programs only calculate for single phase conditions, while others have current limitations.

The following example is conservative at voltage levels above 600 volts. Experience suggests that the example is conservative and becomes more conservative as the voltage increases. It should be noted that all present methods of calculating incident energy at higher voltage levels have limitations.

See 130.3 for requirements for flash hazard analysis.

### D.2 Development of Arc Energy and Temperature Rise on a Person's Exposed Skin

The following sections provide an explanation of the development of the arc energy and temperature rise on a person's exposed skin due to various strengths of electrical arc blasts at various distances from the involved person. The formulas used in this explanation are from Ralph

Lee's paper "The Other Electrical Hazard: Electrical Arc Blast Burns," *IEEE Trans. Industrial Applications,* Vol 1A-18. No. 3, Page 246, May/June 1982. The calculations are based on the worst-case arc impedance. See Table D.2.

***TABLE D.2** Flash Burn Hazards at Various Levels in a Large Petrochemical Plant*

| (1) | (2) | (3) | (4) | (5) | (6) | (7) | (8) |
|---|---|---|---|---|---|---|---|
| **Bus Nominal Voltage Levels** | **System (MVA)** | **Transformer (MVA)** | **System or Transformer (% Z)** | **Short Circuit Symmetrical (Amperes)** | **Arc (MW)** | **Clearing Time of Fault (Cycles)** | **Flash Protection Boundary Typical Distance[1]** |
| 230 kV | 9000 | | 1.11 | 23,000 | 4000 | 6.0 | 14.03 m (46.0 ft) |
| 13.8 kV | 750 | | 9.4 | 31,300 | 374 | 6.0 | 4.3 m (14.1 ft) |
| Load side of all 13.8 kV fuses | 750 | | 9.4 | 31,300 | 374 | 1.0 | 1.77 m (5.8 ft) |
| 4.16 kV | | 10.0 | 5.5 | 25,000 | 91 | 6.0 | 2.23 m (7.3 ft) |
| 4.16 kV | | 5.0 | 5.5 | 12,600 | 45 | 6.0 | 1.68 m (5.5 ft) |
| Line side of incoming 600 V fuse | | 2.5 | 5.5 | 44,000 | 23 | 6.0 | 1.13 m (3.7 ft) |
| 600 V bus | | 2.5 | 5.5 | 44,000 | 23 | 0.25 | 225.6 mm (0.74 ft) |
| 600 V bus | | 1.5 | 5.5 | 26,000 | 27 | 6.0 | 853.5 mm (2.8 ft) |
| 600 V bus | | 1.0 | 5.75 | 17,000 | 17 | 6.0 | 702 mm (2.3 ft) |

[1]Distance from an open arc to limit skin damage to a curable second-degree skin burn [less than 80°C (176°F) on skin] in free air.

### D.3 Basic Equations for Calculating Flash Protection Boundary Distances

The short-circuit symmetrical ampacity from a bolted 3-phase fault at the transformer terminals is calculated with the following formula:

$$I_{sc} = \{[MVA \text{ Base} \times 10^6] \div [1.732 \times V]\} \times \{100 \div \%Z\} \quad \textbf{[D.3(a)]}$$

where $I_{sc}$ is in amperes, $V$ is in volts, and $\%Z$ is based on the transformer $MVA$.

A typical value for the maximum power (in MW) in a 3-phase arc can be calculated using the following formula:

$$P = [\text{maximum bolted fault in } MVA_{bf}] \times 0.707^2 \quad \textbf{[D.3(b)]}$$

The Flash Protection Boundary distance is calculated in accordance with the following formulae:

$$P = 1.732 \times V \times I_{sc} \times 10^{-6} \times 0.707^2 \quad \textbf{[D.3(c)]}$$

$$D_c = [2.65 \times MVA_{bf} \times t]^{1/2} \quad \textbf{[D.3(d)]}$$

$$D_c = [53 \times MVA \times t]^{1/2} \quad \textbf{[D.3(e)]}$$

where:

$D_c$ = distance in feet of person from arc source for a just curable burn (i.e., skin temperature remains less than 80 degrees)

$MVA_{bf}$ = bolted fault $MVA$ at point involved

$MVA$ = $MVA$ rating of transformer. For transformers with $MVA$ ratings below 0.75 $MVA$, multiply the transformer $MVA$ rating by 1.25.

$t$ = time of arc exposure in seconds

The clearing time for a current limiting fuse is approximately ¼ cycle or 0.004 second. The clearing time of a 5 kV and 15 kV circuit breaker is approximately 0.1 second or 6 cycles. This can be broken down as follows: actual breaker time (approximately 2.0 cycles), plus relay operating time of approximately 1.74 cycles, plus an additional safety margin of 2 cycles, giving a total time of approximately 6 cycles.

## D.4 Single Line Diagram of a Typical Petrochemical Complex

The single line diagram (*see Figure D.4*) illustrates the complexity of a distribution system in a typical petrochemical plant.

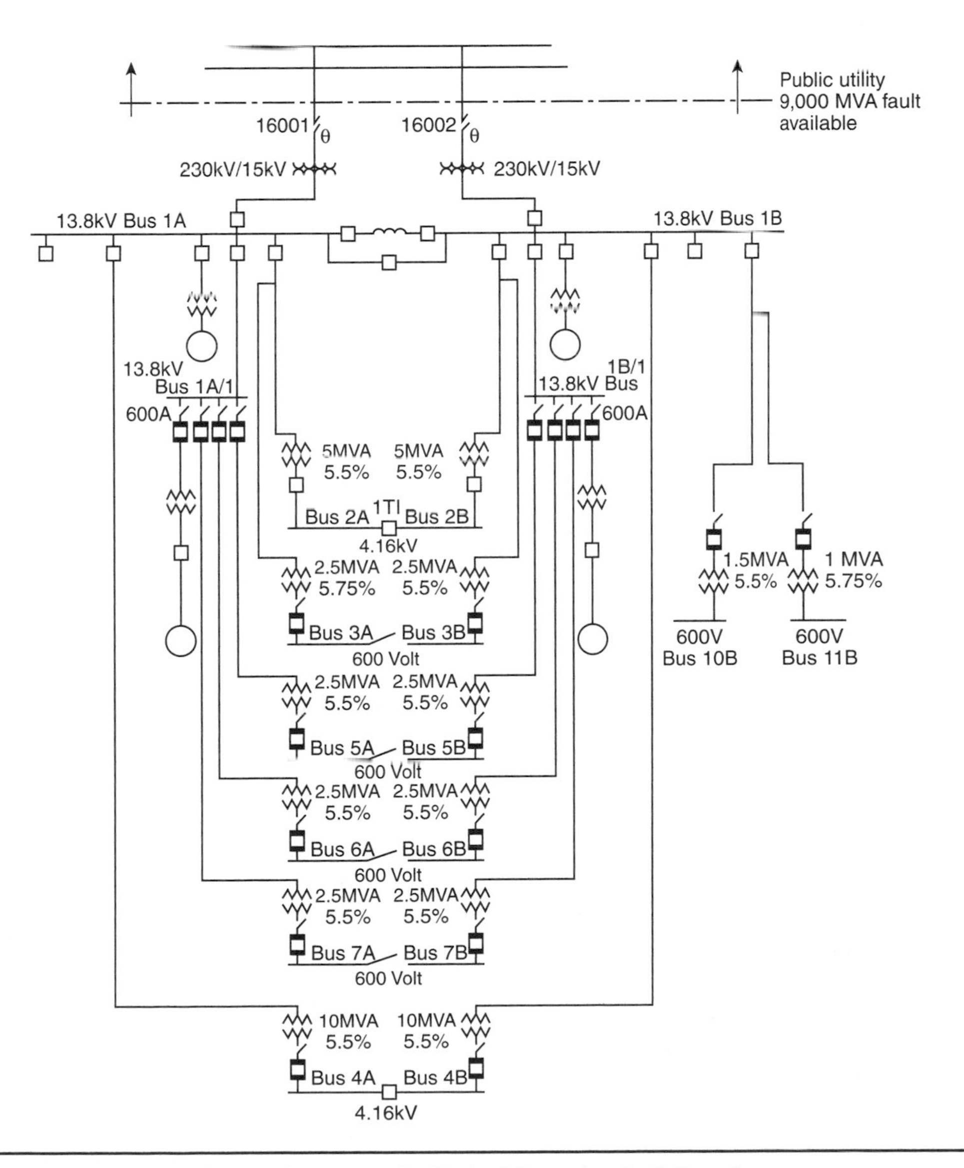


*FIGURE D.4 Single Line Diagram of a Typical Petrochemical Complex.*

## D.5 Sample Calculation

Many of the electrical characteristics of the systems and equipment are provided in Table D.2. The sample calculation is made on the 4160-volt bus 4A or 4B. Table D.2 tabulates the results of calculating the Flash Protection Boundary for each part of the system. For this calculation, based on Table D.2, the following results are obtained:

(1) Calculation is made on a 4160-volt bus.
(2) Transformer MVA (and base *MVA*) = 10 MVA.
(3) Transformer impedance on 10 *MVA* base = 5.5 percent.
(4) Circuit breaker clearing time = 6 cycles.

Using Equation D.3(a), calculate the short-circuit current:

$$I_{sc} = \{[MVA \text{ Base} \times 10^6] \div [1.732 \times V]\} \times \{100 \div \%Z\}$$
$$= \{[10 \times 10^6] \div [1.732 \times 4160]\} \times \{100 \div 5.5\}$$
$$= 25{,}000 \text{ amperes}$$

Using Equation D.3(b), calculate the power in the arc:

$$P = 1.732 \times 4160 \times 25{,}000 \times 10^{-6} \times 0.707^2$$
$$= 91 \text{ MW}$$

Using the Equation D.3(d), calculate the second-degree burn distance:

$$D_c = \{2.65 \times [1.732 \times 25{,}000 \times 4160 \times 10^{-6}] \times 0.1\}^{1/2}$$
$$= 6.9 \text{ or } 7.00 \text{ ft}$$

Or, using Equation D.3(e), calculate the second-degree burn distance using an alternative method:

$$D_c = [53 \times 10 \times 0.1]^{1/2}$$
$$= 7.28 \text{ ft}$$

## D.6 Calculation of Incident Energy Exposure for a Flash Hazard Analysis

The following equations can be used to predict the incident energy produced by a three-phase arc on systems rated 600 volts and below. The results of these equations might not represent the worst case in all situations. It is essential that the equations be used only within the limitations indicated in the definitions of the variables shown under the equations. The equations must be used only under qualified engineering supervision. (Note: Experimental testing continues to be performed to validate existing incident energy calculations and to determine new formulas.)

The parameters required to make the calculations follow:

(1) The maximum "bolted fault" three-phase short-circuit current available at the equipment and the minimum fault level at which the arc will self-sustain (Calculations should be made using the maximum value, and then at lowest fault level at which the arc is self-sustaining. For 480-volt systems, the industry accepted minimum level for a sustaining arcing fault is 38 percent of the available "bolted fault" three-phase short-circuit current. The highest incident energy exposure could occur at these lower levels where the overcurrent device could take seconds or minutes to open.)
(2) The total protective device clearing time (upstream of the prospective arc location) at the maximum short-circuit current, and at the minimum fault level at which the arc will sustain itself.
(3) The distance of the worker from the prospective arc for the task to be performed.

**D.6.1 Arc in Open Air.** The estimated incident energy for an arc in open air is

$$E_{MA} = 5271 D_A^{-1.9593} t_A [0.0016F^2 - 0.0076F + 0.8938] \quad \textbf{[D.6.1(a)]}$$

where:

$E_{MA}$ = maximum open arc incident energy, cal/cm$^2$
$D_A$ = distance from arc electrodes, in. (for distances 18 in. and greater)
$t_A$ = arc duration, seconds
$F$ = short-circuit current, kA (for the range of 16 kA to 50 kA)

Using Equation D.6.1(a), calculate the maximum open arc incident energy, cal/cm$^2$, where $D_A$ = 18 in., $t_A$ = 0.2 second, and $F$ = 20 kA.

$$\begin{aligned} E_{MA} &= 5271 D_A^{-1.9593} t_A [0.0016F^2 - 0.0076F + 0.8938] \quad \textbf{[D.6.1(b)]} \\ &= 5271 \times .0035 \times 0.2[0.0016 \times 400 - 0.0076 \times 20 + 0.8938] \\ &= 3.69 \times [1.381] \\ &= 21.33 \text{ J/cm}^2 \ (5.098 \text{ cal/cm}^2) \end{aligned}$$

**D.6.2 Arc in a Cubic Box.** The estimated incident energy for an arc in a cubic box (20 in. on each side, open on one end) is given in the following equation. This equation is applicable to arc flashes emanating from within switchgear, motor control centers, or other electrical equipment enclosures.

$$E_{MB} = 1038.7 D_B^{-1.4738} t_A [0.0093F^2 - 0.3453F + 5.9675] \quad \textbf{[D.6.2(a)]}$$

where:

$E_{MB}$ = maximum 20 in. cubic box incident energy, cal/cm$^2$
$D_B$ = distance from arc electrodes, inches (for distances 18 in. and greater)
$t_A$ = arc duration, seconds
$F$ = short circuit current, kA (for the range of 16 kA to 50 kA)

*Sample Calculation:* Using Equation D.6.2, calculate the maximum 20 in. cubic box incident energy, cal/cm$^2$, using the following:

(1) $D_A$ = 18 in.
(2) $t_A$ = 0.2 second
(3) $F$ = 20 kA

$$\begin{aligned} E_{MB} &= 1038.7 D_B^{-1.4738} t_A [0.0093F^2 - 0.3453F + 5.9675] \quad \textbf{[D.6.2(b)]} \\ &= 1038 \times 0.0141 \times 0.2[0.0093 \times 400 - 0.3453 \times 20 + 5.9675] \\ &= 2.928 \times [2.7815] \\ &= 34.1 \text{ J/cm}^2 \ (8.144 \text{ cal/cm}^2) \end{aligned}$$

**D.6.3 Reference.** The equations for this section were derived in the IEEE paper by R. L. Doughty, T. E. Neal, and H. L. Floyd, II, "Predicting Incident Energy to Better Manage the

Electric Arc Hazard on 600 V Power Distribution Systems," *Record of Conference Papers IEEE IAS 45th Annual Petroleum and Chemical Industry Conference,* September 28–30, 1998.

**D.7 Calculation of Incident Energy Exposure Greater Than 600 V for a Flash Hazard Analysis**

The following equation can be used to predict the incident energy produced by a three-phase arc in open air on systems rated above 600 V. The parameters required to make the calculations are as follows:

(1) The maximum "bolted fault" three-phase short circuit current available at the equipment
(2) The total protective device clearing time (upstream of the prospective arc location) at the maximum short circuit current
(3) The distance from the arc source
(4) Rated phase-to-phase voltage of the system:

$$E = \frac{793 \times F \times V \times t_A}{D^2}$$

where:

$E$ = incident energy, cal/cm$^2$
$F$ = bolted fault short circuit current, kA
$V$ = system phase-to-phase voltage, kV
$t_A$ = arc duration, seconds
$D$ = distance from the arc source, inches

**D.8 Basic Equations for Calculating Incident Energy and Flash Protection Boundary**

This section offers equations for estimating incident energy and Flash Protection Boundaries based on statistical analysis and curve fitting of available test data. An IEEE working group produced the data from tests it performed to produce models of incident energy. Based on the selection of standard personal protective equipment (PPE) levels (1.2, 8, 25, and 40 cal/cm$^2$), it is estimated that the PPE is adequate or more than adequate to protect employees from second-degree burns in 95 percent of the cases.

FPN: When incident energy exceeds 40 cal/cm$^2$ at the working distance, greater emphasis than normal should be placed on de-energizing before working on or near the exposed electrical conductors or circuit parts.

The complete data, including a spreadsheet calculator to solve the equations, can be found in the *IEEE Guide for Performing Arc Flash Hazard Calculations* (IEEE Std 1584™-2002). It can be ordered from the Institute of Electrical and Electronics Engineers, Inc., 3 Park Avenue, New York, NY 10016-5997.

**D.8.1 System Limits.** An equation for calculating incident energy can be empirically derived using statistical analysis of raw data along with a curve-fitting algorithm. It can be used for systems with the following limits:

(1) 0.208 kV to 15 kV, three-phase
(2) 50 Hz to 60 Hz
(3) 700 A to 106,000 A available short-circuit current
(4) 13 mm to 152 mm conductor gaps

For three-phase systems in open-air substations, open-air transmission systems, and distribution systems, a theoretically derived model is available. This theoretically derived model is intended for use with applications where faults escalate to three-phase faults. Where such an escalation is not possible or likely or where single-phase systems are encountered, this equation will likely provide conservative results.

**D.8.2 Arcing Current.** To determine the operating time for protective devices, find the predicted three-phase arcing current.

For applications with a system voltage under 1 kV, solve Equation D.8.2(a):

$$\lg I_a = K + 0.662 \lg I_{bf} + 0.0966V + 0.000526G + 0.5588V(\lg I_{bf}) - 0.00304G(\lg I_{bf}) \quad \textbf{[D.8.2(a)]}$$

where:

$\lg$ = the $\log_{10}$

$I_a$ = arcing current in kA

$K$ = −0.153 for open air arcs; −0.097 for arcs-in-a-box

$I_{bf}$ = bolted three-phase available short-circuit current (symmetrical rms) (kA)

$V$ = system voltage in kV

$G$ = conductor gap (mm) (See Table D.8.2.)

For systems greater than or equal to 1 kV, use Equation D.8.2(b):

$$\lg I_a = 0.00402 + 0.983 \lg I_{bf} \quad \textbf{[D.8.2(b)]}$$

This higher voltage formula is utilized for both open-air arcs and for arcs-in-a-box.

Convert from lg:

$$I_a = 10^{\lg I_a} \quad \textbf{[D.8.2(c)]}$$

Use $0.85I_a$ to find a second arcing time. This second arc current accounts for variations in the arcing current and the time for the overcurrent device to open. Calculate the incident energy using both values ($I_a$ and $0.85\ I_a$), and use the higher value.

*TABLE D.8.2 Factors for Equipment and Voltage Classes*

| System Voltage (kV) | Type of Equipment | Typical Conductor Gap (mm) | Distance X-Factor |
|---|---|---|---|
| 0.208–1 | Open-air | 10–40 | 2.000 |
| | Switchgear | 32 | 1.473 |
| | MCCs and panels | 25 | 1.641 |
| | Cables | 13 | 2.000 |
| >1–5 | Open-air | 102 | 2.000 |
| | Switchgear | 13–102 | 0.973 |
| | Cables | 13 | 2.000 |
| >5–15 | Open-air | 13–153 | 2.000 |
| | Switchgear | 153 | 0.973 |
| | Cables | 13 | 2.000 |

**D.8.3 Incident Energy at Working Distance—Empirically Derived Equation.** To determine the incident energy using the empirically derived equation, determine the $\log_{10}$ of the normalized incident energy. This equation is based on data normalized for an arc time of 0.2 second and a distance from the possible arc point to the person of 610 mm:

$$\lg E_n = k_1 + k_2 + 1.081 \lg I_a + 0.0011G \quad \textbf{[D.8.3(a)]}$$

where:

$E_n$ = incident energy (J/cm$^2$) normalized for time and distance
$k_1$ = –0.792 for open air arcs; –0.555 for arcs-in-a-box
$k_2$ = 0 for ungrounded and high-resistance grounded systems
= –0.113 for grounded systems
$G$ = the conductor gap (mm) (See Table D.8.2.)

Then,

$$E_n = 10^{\lg E_n} \quad \textbf{[D.8.3(b)]}$$

Converting from normalized:

$$E = 4.184 C_f E_n \left(\frac{t}{0.2}\right)\left(\frac{610^x}{D^x}\right) \quad \textbf{[D.8.3(c)]}$$

where:

$E$ = incident energy in J/cm$^2$
$C_f$ = calculation factor
= 1.0 for voltages above 1 kV
= 1.5 for voltages at or below 1 kV
$E_n$ = incident energy normalized
$t$ = arcing time (seconds)
$D$ = distance (mm) from the arc to the person (working distance)
$X$ = the distance exponent from Table D.8.2

**D.8.4 Incident Energy at Working Distance—Theoretical Equation.** The theoretically derived equation can be applied in cases where the voltage is over 15 kV or the gap is outside the range:

$$E = 2.142 \times 10^6 V I_{bf} \left(\frac{t}{D^2}\right) \quad \textbf{(D.8.4)}$$

where:

$E$ = incident energy (J/cm$^2$)
$V$ = system voltage (kV)
$t$ = arcing time (seconds)
$D$ = distance (mm) from the arc to the person (working distance)
$I_{bf}$ = available three-phase bolted-fault current

For voltages over 15 kV, arcing-fault current and bolted-fault current are considered equal.

**D.8.5 Flash Protection Boundary.** The Flash Protection Boundary is the distance at which a person is likely to receive a second-degree burn. The onset of a second-degree burn is assumed to be when the skin receives 5.0 J/cm$^2$ of incident energy.

For the empirically derived equation,

$$D_B = \left[4.184 C_f E_n \left(\frac{t}{0.2}\right)\left(\frac{610^x}{E_B}\right)\right]^{\frac{1}{x}} \quad \textbf{[D.8.5(a)]}$$

For the theoretically derived equation:

$$D_B = \sqrt{2.142 \times 10^6 V I_{bf} \left(\frac{t}{E_B}\right)} \qquad \textbf{[D.8.5(b)]}$$

where:

$D_B$ = the distance (mm) of the Flash Protection Boundary from the arcing point
$C_f$ = a calculation factor
= 1.0 for voltages above 1 kV
= 1.5 for voltages at or below 1 kV
$E_n$ = incident energy normalized
$E_B$ = incident energy in J/cm² at the distance of the Flash Protection Boundary
$t$ = time (seconds)
$X$ = the distance exponent from Table 10.8.2
$I_{bf}$ = bolted three phase available short-circuit current
$V$ = system voltage in kV

FPN: These equations could be used to determine whether selected PPE is adequate to prevent thermal injury at a specified distance in event of an arc flash.

**D.8.6 Current-Limiting Fuses.** The formulas in this section were developed for calculating arc-flash energies for use with current-limiting Class L and Class RK1 fuses. The testing was done at 600 volts and at a distance of 455 mm, using commercially available fuses from one manufacturer. The following variables are noted:

$I_{bf}$ = available three-phase bolted-fault current (symmetrical rms) (kA)
$E$ = incident energy (J/cm²)

**(A) Class L Fuses 1,601 A–2,000 A.** Where $I_{bf} < 22.6$ kA, calculate the arcing current using Equation D.8.2(a), and use time-current curves to determine the incident energy using Equations D.8.3(a), D.8.3(b), and D.8.3(c).

Where 22.6 kA $\leq I_{bf} \leq$ 65.9 kA.

$$E = 4.184(-0.1284 I_{bf} + 32.262) \qquad \textbf{[D.8.6(a)]}$$

Where 65.9 kA $< I_{bf} \leq$ 106 kA.

$$E = 4.184(-0.5177 I_{bf} + 57.917) \qquad \textbf{[D.8.6(b)]}$$

Where $I_{bf} >$ 106 kA, contact manufacturer.

**(B) Class L Fuses 1,201 A–1,600 A.** Where $I_{bf} < 15.7$ kA, calculate the arcing current using Equation D.8.2(a), and use time-current curves to determine the incident energy using Equations D.8.3(a), D.8.3(b), and D.8.3(c).

Where 15.7 kA $\leq I_{bf} \leq$ 31.8 kA.

$$E = 4.184(-0.1863 I_{bf} + 27.926) \qquad \textbf{[D.8.6(c)]}$$

Where 31.8 kA $< I_{bf} <$ 44.1 kA.

$$E = 4.184(-1.5504 I_{bf} + 71.303) \qquad \textbf{[D.8.6(d)]}$$

Where 44.1 kA $\leq I_{bf}$ 65.9 $\leq$ kA.

$$E = 12.3\text{J/cm}^2(2.94\text{ cal/cm}^2) \qquad \textbf{[D.8.6(e)]}$$

Where 65.9 kA < $I_{bf} \leq$ 106 kA.

$$E = 4.184(-0.0631 I_{bf} + 7.0878) \quad \textbf{[D.8.6(f)]}$$

Where $I_{bf}$ >106 kA, contact manufacturer.

**(C) Class L Fuses 801 A–1,200 A.** Where $I_{bf}$ < 15.7 kA, calculate the arcing current per Equation D.8.2(a), and use time-current curves to determine the incident energy per Equations D.8.3(a), D.8.3(b), and D.8.3(c).

Where 15.7 kA ≤ $I_{bf} \leq$ 22.6 kA.

$$E = 4.184(-0.1928 I_{bf} + 14.226) \quad \textbf{[D.8.6(g)]}$$

Where 22.6 kA < $I_{bf} \leq$ 44.1 kA.

$$E = 4.184(0.0143 I_{bf}^2 - 1.3919 I_{bf} + 34.045) \quad \textbf{[D.8.6(h)]}$$

Where 44.1 kA < $I_{bf} \leq$ 106 kA.

$$E = 1.63 \quad \textbf{[D.8.6(i)]}$$

Where $I_{bf}$ > 106 kA, contact manufacturer.

**(D) Class L Fuses 601 A–800 A.** Where $I_{bf}$ < 15.7 kA, calculate the arcing current per Equation D.8.2(a), and use time-current curves to determine the incident energy using Equations D.8.3(a), D.8.3(b), and D.8.3(c).

Where 15.7 kA ≤ $I_{bf} \leq$ 44.1 kA.

$$E = 4.184(-0.0601 I_{bf} + 2.8992) \quad \textbf{[D.8.6(j)]}$$

Where 44.1 kA < $I_{bf} \leq$ 106 kA.

$$E = 1.046 \quad \textbf{[D.8.6(k)]}$$

Where $I_{bf}$ > 106 kA, contact manufacturer.

**(E) Class RK1 Fuses 401 A–600 A.** Where $I_{bf}$ < 8.5 kA, calculate the arcing current using Equation D.8.2(a), and use time-current curves to determine the incident energy using Equations D.8.3(a), D.8.3(b), and D.8.3(c).

Where 8.5 kA ≤ $I_{bf} \leq$ 14 kA.

$$E = 4.184(-3.0545 I_{bf} + 43.364) \quad \textbf{[D.8.6(l)]}$$

Where 14 kA < $I_{bf} \leq$ 15.7 kA.

$$E = 2.510 \quad \textbf{[D.8.6(m)]}$$

Where 15.7 kA < $I_{bf} \leq$ 22.6 kA.

$$E = 4.184(-0.0507 I_{bf} + 1.3964) \quad \textbf{[D.8.6(n)]}$$

Where 22.6 kA < $I_{bf} \leq$ 106 kA.

$$E = 1.046 \quad \textbf{[D.8.6(o)]}$$

Where $I_{bf}$ > 106 kA, contact manufacturer.

**(F) Class RK1 Fuses 201 A–400 A.** Where $I_{bf}$ < 3.16 kA, calculate the arcing current using Equation D.8.2(a), and use time-current curves to determine the incident energy using Equations D.8.3(a), D.8.3(b), and D.8.3(c).

Where 3.16 kA ≤ $I_{bf} \leq$ 5.04 kA.

$$E = 4.184(-19.053 I_{bf} + 96.808) \quad \textbf{[D.8.6(p)]}$$

Where 5.04 kA < $I_{bf} \leq$ 22.6 kA.

$$E = 4.184(-0.0302 I_{bf} + 0.9321) \quad \textbf{[D.8.6(q)]}$$

Where 22.6 kA < $I_{bf} \leq$ 106 kA.

$$E = 1.046 \quad \textbf{[D.8.6(r)]}$$

Where $I_{bf} >$ 106 kA, contact manufacturer.

**(G) Class RK1 Fuses 101 A–200 A.** Where $I_{bf} <$ 1.16 kA, calculate the arcing current using Equation D.8.2(a), and use time-current curves to determine the incident energy using Equations D.8.3(a), D.8.3(b), and D.8.3(c).

Where 1.16 kA $\leq I_{bf} \leq$ 1.6 kA.

$$E = 4.184(-18.409I_{bf} + 36.355) \quad \textbf{[D.8.6(s)]}$$

Where 1.6 kA < $I_{bf} \leq$ 3.16 kA.

$$E = 4.184(-4.2628I_{bf} + 13.721) \quad \textbf{[D.8.6(t)]}$$

Where 3.16 kA < $I_{bf} \leq$ 106 kA.

$$E = 1.046 \quad \textbf{[D.8.6(u)]}$$

Where $I_{bf} >$ 106 kA, contact manufacturer.

**(H) Class RK1 Fuses 1 A–100 A.** Where $I_{bf} <$ 0.65 kA, calculate the arcing current per Equation D.8.2(a), and use time-current curves to determine the incident energy using Equations D.8.3(a), D.8.3(b), and D.8.3(c).

Where 0.65 kA $\leq I_{bf} \leq$ 1.16 kA.

$$E = 4.184(-11.176I_{bf} + 13.565) \quad \textbf{[D.8.6(v)]}$$

Where 1.16 kA < $I_{bf} \leq$ 1.4 kA.

$$E = 4.184(-1.4583I_{bf} + 2.2917) \quad \textbf{[D.8.6(w)]}$$

Where 1.4 kA < $I_{bf} \leq$ 106 kA.

$$E = 1.046 \quad \textbf{[D.8.6(x)]}$$

Where $I_{bf} >$ 106 kA, contact manufacturer.

**D.8.7 Low-Voltage Circuit Breakers.** The equations in Table D.8.7 can be used for systems with low-voltage circuit breakers. The results of the equations will determine the incident energy and Flash Protection Boundary when $I_{bf}$ is within the range as described. Time-current curves for the circuit breaker are not necessary within the appropriate range.

When the bolted-fault current is below the range indicated, calculate the arcing current per Equation D.8.2(a), and use time-current curves to determine the incident energy using Equations D.8.3(a), D.8.3(b), and D.8.3(c).

The range of available three-phase bolted-fault currents is from 700 A to 106,000 A. Each equation is applicable for the range

$$I_1 < I_{bf} < I_2$$

where:

$I_2$ is the interrupting rating of the CB at the voltage of interest.

$I_1$ is the minimum available three-phase, bolted, short-circuit current at which this method can be applied. $I_1$ is the lowest available three-phase, bolted, short-circuit current level that causes enough arcing current for instantaneous tripping to occur or for circuit breakers with no instantaneous trip, that causes short-time tripping to occur.

*TABLE D.8.7 Incident Energy and Flash-Protection Boundary by Circuit Breaker Type and Rating*

| Rating (A) | Breaker Type | Trip-Unit Type | 480 V and Lower: Incident Energy (J/cm$^2$)[a] | 480 V and Lower: Flash Boundary (mm) | 575–600 V: Incident Energy (J/cm$^2$) | 575–600 V: Flash Boundary (mm) |
|---|---|---|---|---|---|---|
| 100–400 | MCCB | TM or M | $0.189\ I_{bf} + 0.548$ | $9.16\ I_{bf} + 194$ | $0.271\ I_{bf} + 0.180$ | $11.8\ I_{bf} + 196$ |
| 600–1,200 | MCCB | TM or M | $0.223\ I_{bf} + 1.590$ | $8.45\ I_{bf} + 364$ | $0.335\ I_{bf} + 0.380$ | $11.4\ I_{bf} + 369$ |
| 600–1,200 | MCCB | E, LI | $0.377\ I_{bf} + 1.360$ | $12.50\ I_{bf} + 428$ | $0.468\ I_{bf} + 4.600$ | $14.3\ I_{bf} + 568$ |
| 1,600–6,000 | MCCB or ICCB | TM or E, LI | $0.448\ I_{bf} + 3.000$ | $11.10\ I_{bf} + 696$ | $0.686\ I_{bf} + 0.165$ | $16.7\ I_{bf} + 606$ |
| 800–6,300 | LVPCB | E, LI | $0.636\ I_{bf} + 3.670$ | $14.50\ I_{bf} + 786$ | $0.958\ I_{bf} + 0.292$ | $19.1\ I_{bf} + 864$ |
| 800–6,300 | LVPCB | E, LS[b] | $4.560\ I_{bf} + 27.230$ | $47.20\ I_{bf} + 2660$ | $6.860\ I_{bf} + 2.170$ | $62.4\ I_{bf} + 2930$ |

[a]$I_{bf}$ is in kA; working distance is 455 mm (18 in.).

[b]Short-time delay is assumed to be set at maximum.

MCCB = Molded-case circuit breaker

ICCB = Insulated-case circuit breaker

LVPC = Low-voltage power circuit breaker

TM = Thermal-magnetic trip units

M = Magnetic (instantaneous only) trip units

E = Electronic trip units have three characteristics that may be used separately or in combination:

L = Long-time, S = Short-time, I = Instantaneous.

To find $I_1$, the instantaneous trip ($I_t$) of the circuit breaker must be found. This can be determined from the time-current curve, or it can be assumed to be 10 times the rating of the circuit breaker for circuit breakers rated above 100 amperes. For circuit breakers rated 100 amperes and below, a value of $I_t$ = 1,300 A can be used. When short-time delay is utilized, $I_t$ is the short-time pick-up current.

The corresponding bolted-fault current, $I_{bf}$, is found by solving the equation for arc current for box configurations by substituting $I_t$ for arcing current. The 1.3 factor in Equation D.8.7(b) adjusts current to the top of the tripping band.

$$\lg(1.3I_t) = 0.084 + 0.096V + 0.586(\lg I_{bf}) + 0.559V(\lg I_{bf}) \qquad \text{[D.8.7(a)]}$$

At 600 V,

$$\lg I_1 = 0.0281 + 1.091\mathrm{g}(1.3I_t) \qquad \text{[D.8.7(b)]}$$

At 480 V and lower,

$$\lg I_1 = 0.0407 + 1.17\lg(1.3I_t) \qquad \text{[D.8.7(c)]}$$

$$I_{bf} = I_1 = 10^{\lg I_1} \qquad \text{[D.8.7(d)]}$$

**D.8.8 References.** The complete data, including a spreadsheet calculator to solve the equations, may be found in the IEEE *Guide for Performing Arc-Flash Hazard Calculations* (IEEE Std 1584™-2002). IEEE publications are available from the Institute of Electrical and Electronic Engineers, 445 Hoes Lane, P.O. Box 1331, Piscataway, NJ 08855-1331, USA (http://standards.ieee.org/).

ANNEX E

# Electrical Safety Program

*This annex is not a part of the requirements of this NFPA document but is included for informational purposes only.*

Annex E provides information that could be used as the foundation for an electrical safety program.

See 110.7, Electrical Safety Program.

## E.1 Typical Electrical Safety Program Principles

Electrical safety program principles include, but are not limited to, the following:

(1) Inspect/evaluate the electrical equipment
(2) Maintain the electrical equipment's insulation and enclosure integrity
(3) Plan every job and document first-time procedures
(4) Deenergize, if possible *(see 120.1)*
(5) Anticipate unexpected events
(6) Identify and minimize the hazard
(7) Protect the employee from shock, burn, blast, and other hazards due to the working environment
(8) Use the right tools for the job
(9) Assess people's abilities
(10) Audit these principles

## E.2 Typical Electrical Safety Program Controls

Electrical safety program controls can include, but are not limited to, the following:

(1) Every electrical conductor or circuit part is considered energized until proven otherwise.
(2) No bare-hand contact is to be made with exposed energized electrical conductors or circuit parts above 50 volts to ground, unless the "bare-hand method" is properly used.
(3) Deenergizing an electrical conductor or circuit part and making it safe to work on is in itself a potentially hazardous task.
(4) Employer develops programs, including training, and employees apply them.
(5) Use procedures as "tools" to identify the hazards and develop plans to eliminate/control the hazards.
(6) Train employees to qualify them for working in an environment influenced by the presence of electrical energy.

(7) Identify/categorize tasks to be performed on or near exposed energized electrical conductors and circuit parts.
(8) Use a logical approach to determine potential hazard of task.
(9) Identify and use precautions appropriate to the working environment.

### E.3 Typical Electrical Safety Program Procedures

Electrical safety program procedures can include, but are not limited to, the following:

(1) Purpose of task
(2) Qualifications and number of employees to be involved
(3) Hazardous nature and extent of task
(4) Limits of approach
(5) Safe work practices to be utilized
(6) Personal protective equipment involved
(7) Insulating materials and tools involved
(8) Special precautionary techniques
(9) Electrical diagrams
(10) Equipment details
(11) Sketches/pictures of unique features
(12) Reference data

ANNEX F

# Hazard/Risk Evaluation Procedure

*This annex is not a part of the requirements of this NFPA document but is included for informational purposes only.*

Before any task is performed on or near exposed live parts, the worker should determine whether he or she might be exposed to a safety hazard. If the worker might be exposed to a hazard, the worker should determine whether the risk of injury is significant.

Annex F illustrates a series of questions in the form of a flow chart that is intended to identify electrical hazards and the degree of risk. Once the hazard and degree of risk are identified, personal protective equipment (PPE) must be selected. The worker and his or her supervisor then can use the degree of risk to evaluate whether the risk of injury is sufficiently low to accept the risk.

**F.1.** See 110.7(F), Hazard/Risk Evaluation Procedure. Figure F.1 illustrates the steps of a hazard/risk analysis evaluation procedure flow chart.

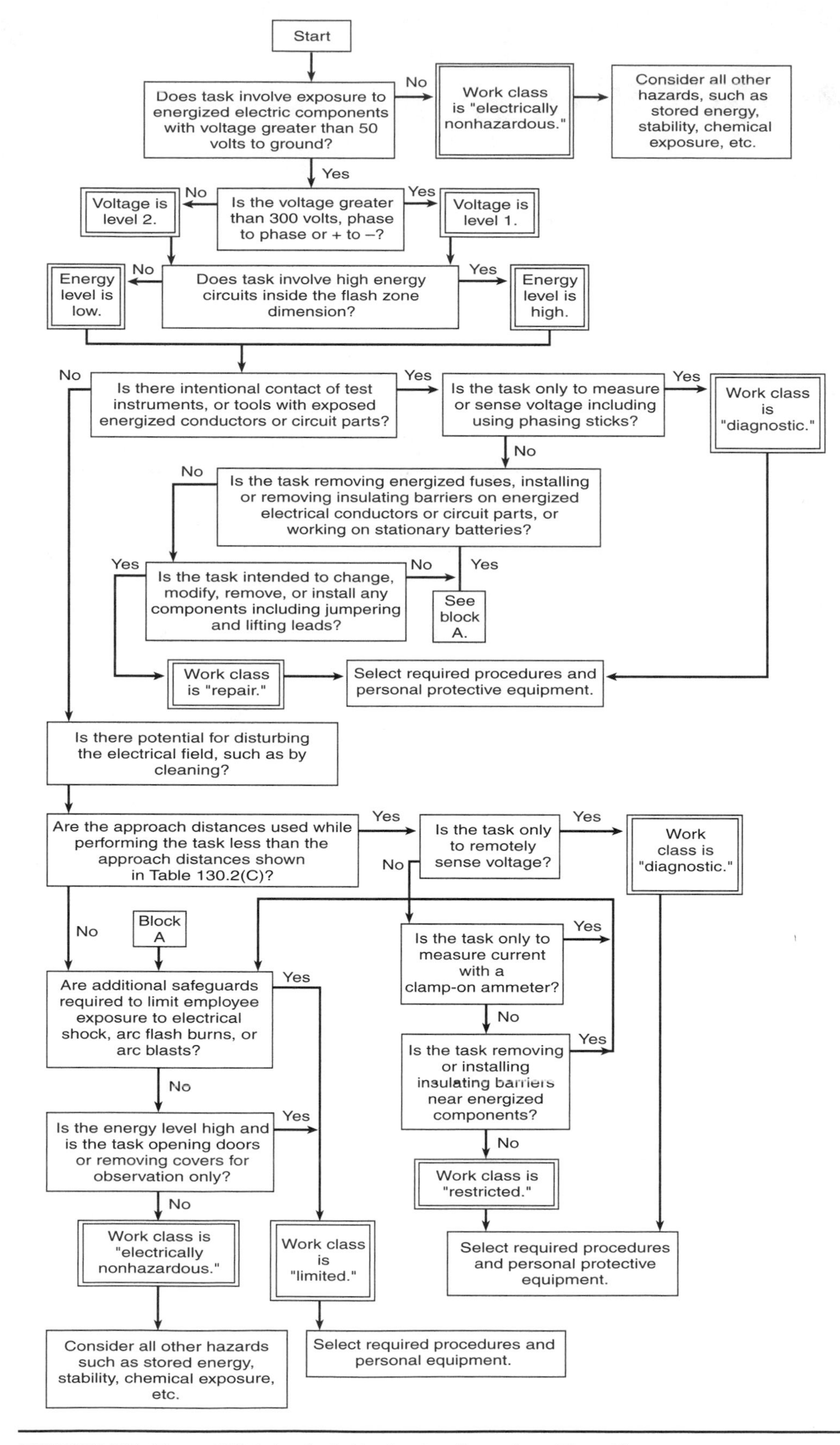


*FIGURE F.1 Hazard/Risk Analysis Evaluation Procedure Flow Chart.*

ANNEX G

# Sample Lockout/ Tagout Procedure

*This annex is not a part of the requirements of this NFPA document but is included for informational purposes only.*

Annex G is provided to illustrate how a lockout/tagout procedure might be published. An employer could "fill in the blanks" and publish the resulting procedure for his or her organization.

Lockout is the preferred method of controlling personnel exposure to electrical energy hazards. Tagout is an alternative method that is available to employers. To assist employers in develop ing a procedure that meets the requirement of 120.3 of NFPA 70E, the following sample procedure is provided for use in lockout or tagout programs. This procedure can be used for an individual employee control, a simple lockout/tagout, or as part of a complex lockout/tagout. Where a job/task is under the control of one person, the individual employee control procedure can be used in lieu of a lockout/tagout procedure. A more comprehensive plan will need to be developed, documented, and utilized for the complex lockout/tagout.

LOCKOUT (TAGOUT) PROCEDURE FOR ABC COMPANY
OR
TAGOUT PROCEDURE FOR _____ COMPANY

**1.0 Purpose.** This procedure establishes the minimum requirements for lockout (tagout) of electrical energy sources. It is to be used to ensure that conductors and circuit parts are disconnected from sources of electrical energy, locked (tagged), and tested before work begins where employees could be exposed to dangerous conditions. Sources of stored energy, such as capacitors or springs, shall be relieved of their energy, and a mechanism shall be engaged to prevent the re-accumulation of energy.

**2.0 Responsibility.** All employees shall be instructed in the safety significance of the lockout (tagout) procedure. All new or transferred employees and all other persons whose work operations are or might be in the area shall be instructed in the purpose and use of this procedure [include the name(s) of person(s) or job title(s) of employees with responsibility] shall ensure that appropriate personnel receive instructions on their roles and responsibilities. All persons installing a lockout (tagout) device shall sign their names and the date on the tag (or state how the name of the individual or person in charge will be available).

**3.0 Preparation for Lockout (Tagout).**

**3.1.** Review current diagrammatic drawings (or other equally effective means), tags, labels, and signs to identify and locate all disconnecting means to determine that power is interrupted by a physical break and not deenergized by a circuit interlock. Make a list of disconnecting means to be locked (tagged).

**3.2.** Review disconnecting means to determine adequacy of their interrupting ability. Determine if it will be possible to verify a visible open point, or if other precautions will be necessary.

**3.3.** Review other work activity to identify where and how other personnel might be exposed to sources of electrical energy hazards. Review other energy sources in the physical area to determine employee exposure to sources of other types of energy. Establish energy control methods for control of other hazardous energy sources in the area.

**3.4.** Provide an adequately rated voltage detector to test each phase conductor or circuit part to verify that they are deenergized. *(See 12.3.)* Provide a method to determine that the voltage detector is operating satisfactorily.

**3.5.** Where the possibility of induced voltages or stored electrical energy exists, call for grounding the phase conductors or circuit parts before touching them. Where it could be reasonably anticipated that contact with other exposed energized conductors or circuit parts is possible, call for applying ground connecting devices.

**4.0 Individual Employee Control Procedure.** The individual employee control procedure can be used when equipment with exposed conductors and circuit parts are deenergized for minor maintenance, servicing, adjusting, cleaning, inspection operating corrections, and the like, and the work shall be permitted to be performed without the placement of lockout/tagout devices on the disconnecting means, provided the disconnecting means is adjacent to the conductor, circuit parts, and equipment on which the work is performed, the disconnecting means is clearly visible to all employees involved in the work, and the work does not extend beyond the work shift.

**5.0 Simple Lockout/Tagout.** The simple lockout/tagout procedure will involve paragraphs 1.0 through 3.0, 5.0 through 9.0, and 11.0 through 13.0.

**6.0 Sequence of Lockout (Tagout) System Procedures.**

**6.1.** The employees shall be notified that a lockout (tagout) system is going to be implemented and the reason therefore. The qualified employee implementing the lockout (tagout) shall know the disconnecting means location for all sources of electrical energy and the location of all sources of stored energy. The qualified person shall be knowledgeable of hazards associated with electrical energy.

**6.2.** If the electrical supply is energized, the qualified person shall deenergize and disconnect the electric supply and relieve all stored energy.

**6.3.** Lockout (tagout) all disconnecting means with lockout (tagout) devices.

FPN: For tagout, one additional safety measure must be employed, such as opening, blocking, or removing an additional circuit element.

**6.4.** Attempt to operate the disconnecting means to determine that operation is prohibited.

**6.5.** A voltage-detecting instrument shall be used. *(See 12.3.)* Inspect the instrument for visible damage. Do not proceed if there is an indication of damage to the instrument until an undamaged device is available.

**6.6.** Verify proper instrument operation and then test for absence of voltage.

**6.7.** Verify proper instrument operation after testing for absence of voltage.

**6.8.** Where required, install grounding equipment/conductor device on the phase conductors or circuit parts, to eliminate induced voltage or stored energy, before touching them. Where it has been determined that contact with other exposed energized conductors or circuit parts is possible, apply ground connecting devices rated for the available fault duty.

**6.9.** The equipment and/or electrical source is now locked out (tagged out).

**7.0 Restoring the equipment and/or electrical supply to normal condition.**

**7.1.** After the job/task is complete, visually verify that the job/task is complete.

**7.2.** Remove all tools, equipment, and unused materials and perform appropriate housekeeping.

**7.3.** Remove all grounding equipment/conductor/devices.

**7.4.** Notify all personnel involved with the job/task that the lockout (tagout) is complete, that the electrical supply is being restored, and to remain clear of the equipment and electrical supply.

**7.5.** Perform any quality control tests/checks on the repaired/replaced equipment and/or electrical supply.

**7.6.** Remove lockout (tagout) devices by the person who installed them.

**7.7.** Notify the equipment and/or electrical supply owner that the equipment and/or electrical supply is ready to be returned to normal operation.

**7.8.** Return the disconnecting means to their normal condition.

**8.0 Procedure Involving More Than One Person.** For a simple lockout/tagout and where more than one person is involved in the job/task, each person shall install his/her own personal lockout (tagout) device.

**9.0 Procedure Involving More Than One Shift.** When the lockout (tagout) extends for more than one day, the lockout (tagout) shall be verified to be still in place at the beginning of the next day. Where the lockout (tagout) is continued on successive shifts, the lockout (tagout) is considered to be a complex lockout (tagout).

For complex lockout (tagout), the person-in-charge shall identify the method for transfer of the lockout (tagout) and of communication with all employees.

**10.0 Complex Lockout (Tagout).** A complex lockout/tagout plan is required where one or more of the following exist:

(1) Multiple energy sources (more than one)
(2) Multiple crews
(3) Multiple crafts
(4) Multiple locations
(5) Multiple employers
(6) Unique disconnecting means
(7) Complex or particular switching sequences
(8) Continues for more than one shift, that is, new workers

**10.1.** All complex lockout/tagout procedures shall require a written plan of execution. The plan will include the requirements in 1.0 through 3.0, 6.0, 7.0, and 9.0 through 13.0.

**10.2.** A person in charge shall be involved with a complex lockout/tagout procedure. At this location shall be the person in charge.

**10.3.** The person in charge shall develop a written plan of execution and communicate that plan to all persons engaged in the job/task. The person in charge shall be held accountable for safe execution of the complex lockout/tagout plan. The complex lockout/tagout plan must address all the concerns of employees who might be exposed, and they must understand how electrical energy is controlled. The person in charge shall ensure that each person understands the hazards to which they are exposed and the safety-related work practices they are to use.

**10.4.** All complex lockout/tagout plans identify the method to account for all persons who might be exposed to electrical hazards in the course of the lockout/tagout.

Select which of the following methods is to be used:

(1) Each individual will install his or her own personal lockout or tagout device.
(2) The person in charge shall lock his/her key in a "lock box"
(3) The person in charge shall maintain a sign in/out log for all personnel entering the area.
(4) Another equally effective methodology.

**10.5.** The person in charge can install locks/tags, or direct their installation on behalf of other employees.

**10.6.** The person in charge can remove locks/tags or direct their removal on behalf of other employees, only after all personnel are accounted for and ensured to be clear of potential electrical hazards.

**10.7.** Where the complex lockout (tagout) is continued on successive shifts, the person in charge shall identify the method for transfer of the lockout and of communication with all employees.

**11.0 Discipline.**

**11.1.** Knowingly violating this procedure will result in _____ (state disciplinary actions that will be taken).

**11.2.** Knowingly operating a disconnecting means with an installed lockout device (tagout device) will result in _____ (state disciplinary actions to be taken).

**12.0 Equipment.**

**12.1.** Locks shall be _____ (state type and model of selected locks).

**12.2.** Tags shall be _____ (state type and model to be used).

**12.3.** Voltage detecting device(s) to be used shall be _____ (state type and model).

**13.0 Review.** This procedure was last reviewed on ______, and is scheduled to be reviewed again on _____ ______ (not more than one year from the last review).

**14.0 Lockout/Tagout Training.** Recommended training can include, but is not limited to, the following:

(1) Recognizing lockout/tagout devices
(2) Installing lockout/tagout devices
(3) Duty of employer in writing procedures
(4) Duty of employee in executing procedures
(5) Duty of person-in-charge
(6) Authorized and unauthorized removal of locks/tags
(7) Enforcing execution of lockout/tagout procedures
(8) Individual employee control of energy
(9) Simple lockout/tagout
(10) Complex lockout/tagout
(11) Using single line and diagrammatic drawings to identify sources of energy
(12) Use of tags and warning signs
(13) Release of stored energy
(14) Personnel accounting methods
(15) Grounding needs/requirements
(16) Safe use of voltage detecting instruments

# Simplified, Two-Category, Flame-Resistant (FR) Clothing System

ANNEX H

*This annex is not a part of the requirements of this NFPA document but is included for informational purposes only.*

After an employer develops and publishes a procedure that describes how thermal protection is determined, the employer must develop a system that enables the procedure requirements to be administered. Annex H is intended to illustrate one method that enables a personal protective equipment (PPE) program to be administered.

## H.1 Use of Simplified Approach

The use of Table H.1 is suggested as a simplified approach to assure adequate personal protective equipment (PPE) for electrical workers within facilities with large and diverse electrical systems. The clothing listed in Table H.1 fulfills the minimum FR clothing requirements of Table 130.7(C)(9)(a) and Table 130.7(C)(10). The clothing systems listed in this table should be used with the other PPE appropriate for the Hazard/Risk Category. See Table 130.7(C)(10).

*TABLE H.1 Simplified, Two-Category, Flame-Resistant Clothing System*

| Clothing* | Applicable Tasks |
|---|---|
| **Everyday Work Clothing**<br>FR long-sleeve shirt (minimum arc rating of 4) worn over an untreated cotton T-shirt with FR pants (minimum arc rating of 8)<br>*or*<br>FR coveralls (minimum arc rating of 4) worn over an untreated cotton T-shirt (or an untreated natural fiber long-sleeve shirt) with untreated natural fiber pants. | All Hazard/Risk Category 1 and 2 tasks listed in Table 130.7(C)(9)(a).<br>On systems operating at less than 1000 volts, these tasks include work on all equipment *except*<br>• Insertion or removal of low-voltage motor starter "buckets,"<br>• Insertion or removal of power circuit breakers from switchgear cubicles or<br>• Removal of bolted covers from switchgear.<br>On systems operating at 1000 volts or greater, tasks also include the operation of switching devices *with equipment enclosure doors closed.* |
| **Electrical "Switching" Clothing**<br>Multilayer FR flash jacket and FR bib overalls worn over either FR coveralls (minimum arc rating of 4) or FR long-sleeve shirt and FR pants (minimum arc rating of 4), worn over untreated natural fiber long-sleeve shirt and pants, worn over an untreated cotton T-shirt<br>*or*<br>Insulated FR coveralls (with a minimum arc rating of 25, independent of other layers) worn over untreated natural fiber long-sleeve shirt with untreated denim cotton blue jeans ("regular weight," minimum 12 oz/yd2 fabric weight), worn over an untreated cotton T-shirt. | All Hazard/Risk Category 3 and 4 tasks listed in Table 130.7(C)(9)(a)<br>On systems operating at 1000 volts or greater, these tasks include work on exposed live parts of all equipment.<br>On systems of less than 1000 volts, tasks include insertion or removal of low-voltage motor starter MCC "buckets," insertion or removal of plug-in devices into or from busway, insertion or removal of power circuit breakers and removal of bolted covers from switchgear. |

*Note other PPE required for the specific tasks listed in Tables 130.7(C)(9)(a) and 130.7(C)(10), which include arc-rated face shields or flash suit hoods, FR hardhat liners, safety glasses or safety goggles, hard hat, hearing protection, leather gloves, voltage-rated gloves, and voltage-rated tools.

ANNEX I

# Job Briefing and Planning Checklist

*This annex is not a part of the requirements of this NFPA document but is included for informational purposes only.*

Annex I is intended to illustrate the various subjects that should be discussed when a job briefing is held. Other subjects might need to be discussed. The purpose of this checklist is to enable the conversation to occur.

**I.1.** Figure I.1 illustrates considerations for a Job Briefing and Planning Checklist.

**Identify**

- ❑ The hazards
- ❑ The voltage levels involved
- ❑ Skills required
- ❑ Any "foreign" (secondary source) voltage source
- ❑ Any unusual work conditions
- ❑ Number of people needed to do the job
- ❑ The shock protection boundaries
- ❑ The available incident energy
- ❑ Potential for arc flash (Conduct a flash-hazard analysis.)
- ❑ Flash protection boundary

**Ask**

- ❑ Can the equipment be de-energized?
- ❑ Are backfeeds of the circuits to be worked on possible?
- ❑ Is a "standby person" required?

**Check**

- ❑ Job plans
- ❑ Single-line diagrams and vendor prints
- ❑ Status board
- ❑ Information on plant and vendor resources is up to date
- ❑ Safety procedures
- ❑ Vendor information
- ❑ Individuals are familiar with the facility

**Know**

- ❑ What the job is
- ❑ Who else needs to know—Communicate!
- ❑ Who is in charge

**Think**

- ❑ About the unexpected event . . . What if?
- ❑ Lock — Tag — Test — Try
- ❑ Test for voltage — FIRST
- ❑ Use the right tools and equipment, including PPE
- ❑ Install and remove grounds
- ❑ Install barriers and barricades
- ❑ What else . . . ?

**Prepare for an emergency**

- ❑ Is the standby person CPR trained?
- ❑ Is the required emergency equipment available? Where is it?
- ❑ Where is the nearest telephone?
- ❑ Where is the fire alarm?
- ❑ Is confined space rescue available?
- ❑ What is the exact work location?
- ❑ How is the equipment shut off in an emergency?
- ❑ Are the emergency telephone numbers known?
- ❑ Where is the fire extinguisher?
- ❑ Are radio communications available?

***FIGURE I.1*** *Job Briefing and Planning Checklist.*

ANNEX J

# Energized Electrical Work Permit

*This annex is not a part of the requirements of this NFPA document but is included for informational purposes only.*

Annex J illustrates an energized electrical work permit. The entire content of that work permit is not fixed by requirement, although many employers have used this permit successfully. The basic purpose of the permit is to ensure that people in responsible positions are involved in the decision to accept the increased risk associated with working on or near exposed live parts. A side benefit of the work permit is that its review might produce a decision to perform the work deenergized.

**J.1.** Figure J.1 illustrates considerations for an Energized Electrical Work Permit.

## ENERGIZED ELECTRICAL WORK PERMIT

**PART I: TO BE COMPLETED BY THE REQUESTER:**

Job/Work Order Number ____________

(1) Description of circuit/equipment/job location: ____________

(2) Description of work to be done: ____________

(3) Justification of why the circuit/equipment cannot be de-energized or the work deferred until the next scheduled outage: ____________

____________ Requester/Title ____________ Date

**PART II: TO BE COMPLETED BY THE ELECTRICALLY QUALIFIED PERSONS *DOING* THE WORK:**

**Check when Complete**

(1) Detailed job description procedure to be used in performing the above detailed work: ____________ ☐

(2) Description of the Safe Work Practices to be employed: ____________ ☐

(3) Results of the Shock Hazard Analysis: ____________ ☐

(4) Determination of Shock Protection Boundaries: ____________ ☐

(5) Results of the Flash Hazard Analysis: ____________ ☐

(6) Determination of the Flash Protection Boundary: ____________ ☐

(7) Necessary personal protective equipment to safely perform the assigned task: ____________ ☐

(8) Means employed to restrict the access of unqualified persons from the work area: ____________ ☐

(9) Evidence of completion of a Job Briefing including discussion of any job-related hazards: ____________ ☐

(10) Do you agree the above described work can be done safely? ❑ Yes ❑ No (If *no,* return to requester)

____________ Electrically Qualified Person(s) ____________ Date

____________ Electrically Qualified Person(s) ____________ Date

**PART III: APPROVAL(S) TO PERFORM THE WORK WHILE ELECTRICALLY ENERGIZED:**

____________ Manufacturing Manager ____________ Maintenance/Engineering Manager

____________ Safety Manager ____________ Electrically Knowledgeable Person

____________ General Manager ____________ Date

Note: Once the work is complete, forward this form to the site Safety Department for review and retention.

© 2004 National Fire Protection Association NFPA 70E (p. 1 of 1)

***FIGURE J.1*** *Sample Permit for Energized Electrical Work.*

# General Categories of Electrical Hazards

## ANNEX K

*This annex is not a part of the requirements of this NFPA document but is included for informational purposes only.*

Annex K provides an abbreviated discussion of known electrical hazards. The discussion is intended to provide critical information for a worker to use in his or her argument for an improved electrical safety program.

### K.1 General Categories

There are three general categories of electrical hazards: electrical shock, arc-flash, and arc-blast.

### K.2 Electric Shock

Approximately 30,000 nonfatal electrical shock accidents occur each year. The National Safety Council estimates that about 1000 fatalities each year are due to electrocution, more than half of them while servicing energized systems of less than 600 volts. Electrocution is the fourth leading cause of industrial fatalities, after traffic, homicide, and construction accidents. The current required to light a 7½ watt, 120 volt lamp, if passed across the chest, is enough to cause a fatality. The most damaging paths through the body are through the lungs, heart, and brain.

### K.3 Arc-Flash

When an electric current passes through air between ungrounded conductors or between ungrounded conductors and grounded conductors, the temperatures can reach 35,000°F. Exposure to these extreme temperatures both burns the skin directly and causes ignition of clothing, which adds to the burn injury. The majority of hospital admissions due to electrical accidents are from arc-flash burns, not from shocks. Each year more than 2,000 people are admitted to burn centers with severe arc-flash burns. Arc-flashes can and do kill at distances of 10 ft.

### K.4 Arc-Blast

The tremendous temperatures of the arc cause the explosive expansion of both the surrounding air and the metal in the arc path. For example, copper expands by a factor of 67,000 times when it turns from a solid to a vapor. The danger associated with this expansion is one of high pressures, sound, and shrapnel. The high pressures can easily exceed hundreds or even thousands of pounds per square foot, knocking workers off ladders, rupturing eardrums, and collapsing lungs. The sounds associated with these pressures can exceed 160 dB. Finally, material and molten metal is expelled away from the arc at speeds exceeding 700 miles per hour, fast enough for shrapnel to completely penetrate the human body.

# Typical Application of Safeguards in the Cell Line Working Zone

## ANNEX L

*This annex is not a part of the requirements of this NFPA document but is included for informational purposes only.*

### L.1 Application of Safeguards

This section permits a typical application of safeguards in electrolytic areas where hazardous electrical conditions exist. Take, for example, an employee working on an energized cell. The employee uses manual contact to make adjustments and repairs. Consequently, the exposed energized cell and grounded metal floor could present a hazardous electrical condition. Safeguards for this employee can be provided in several ways:

(1) Protective boots can be worn that isolate the employee's feet from the floor and that provide a safeguard from the hazardous electrical condition.
(2) Protective gloves can be worn that isolate the employee's hands from the energized cell and that provide a safeguard.
(3) If the work task causes severe deterioration, wear, or damage to personal protective equipment, the employee might have to wear both protective gloves and boots.
(4) A permanent or temporary insulating surface can be provided for the employee to stand on to provide a safeguard.
(5) The design of the installation can be modified to provide a conductive surface for the employee to stand on. If the conductive surface is bonded to the cell, the hazardous electrical condition will be removed and a safeguard will be provided by voltage equalization.
(6) Safe work practices can provide safeguards. If protective boots are worn, the employee should not make long reaches over energized (or grounded) surfaces such that his or her elbow bypasses the safeguard. If such movements are required, protective sleeves, protective mats, or special tools should be utilized. Training on the nature of hazardous electrical conditions and proper use and condition of safeguards is in itself a safeguard.
(7) The energized cell can be temporarily bonded to ground to remove the hazardous electrical condition.

### L.2 Electrical Power Receptacles

Power supply circuits and receptacles in the cell line area for portable electric equipment should meet the requirements of 430.8(F). However, it is recommended that receptacles for portable electric equipment not be installed in electrolytic cell areas and that only pneumatic powered portable tools and equipment be used.

# Cross-Reference Tables

## ANNEX M

*This annex is not a part of the requirements of this NFPA document but is included for informational purposes only.*

***TABLE M.1*** *Cross Reference from the 2000 edition of NFPA 70E to the 2004 edition of NFPA 70E*

| Topic | 2000 Edition | 2004 Edition |
|---|---|---|
| Introduction | Introduction | Article 90 |
| Scope | Introduction, I-1 | 90.1 |
| Definitions | Introduction, I-2 | Article 100 |
| Installation Safety Requirements | Part I | Chapter 4 |
| General Requirements for Electrical Installations | Part I, Chapter 1 | Article 400 |
| General | Part I, Chapter 1, 1-1 | 400.1 |
| Approval | Part I, Chapter 1, 1-2 | 400.2 |
| Examination, Identification, Installation, and Use of Equipment | Part I, Chapter 1, 1-3 | 400.3 |
| Insulation Integrity | Part I, Chapter 1, 1-3.3 | 400.4 |
| Interrupting Rating | Part I, Chapter 1, 1-3.4 | 400.5 |
| Circuit Impedance and Other Characteristics | Part I, Chapter 1, 1-3.5 | 400.6 |
| Deteriorating Agents | Part I, Chapter 1, 1-3.6 | 400.7 |
| Mechanical Execution of Work | Part I, Chapter 1, 1-3.7 | 400.8 |
| Mounting and Cooling of Equipment | Part I, Chapter 1, 1-3.8 | 400.9 |
| Electrical Connections | Part I, Chapter 1, 1-4 | 400.10 |
| Arcing Parts | Part I, Chapter 1, 1-5 | 400.12 |
| Marking | Part I, Chapter 1, 1-6 | 400.13 |
| Identification of Disconnecting Means | Part I, Chapter 1, 1-7 | 400.14 |
| 600 Volts, Nominal, or Less | Part I, Chapter 1, 1-8 | Article 400, Section B |
| Spaces about Electrical Equipment | Part I, Chapter 1, 1-8.1 | 400.15 |
| Guarding of Live Parts | Part I, Chapter 1, 1-8.2 | 400.16 |
| Over 600 Volts, Nominal | Part I, Chapter 1, 1-9 | Article 400, Section C |
| General | Part I, Chapter 1, 1-9.1 | 400.17 |
| Enclosure for Electrical Installations | Part I, Chapter 1, 1-9.2 | 400.18 |
| Work Space about Equipment | Part I, Chapter 1, 1-9.3 | 400.19 |
| Entrance and Access to Work Space | Part I, Chapter 1, 1-9.4 | 400.20 |
| Table for Minimum Depth of Clear Working Space at Electric Equipment | Part I, Chapter 1, Table 1-9.5.1 | Table 400.21 |
| Wiring Design and Protection | Part I, Chapter 2 | Article 410 |
| Use and Identification of Grounded and Grounding Conductors | Part I, Chapter 2, 2-1 | 410.1 |
| Branch Circuits | Part I, Chapter 2, 2-2 | 410.2 |

*(continues)*

*TABLE M.1 Continued*

| Topic | 2000 Edition | 2004 Edition |
|---|---|---|
| Identification of Ungrounded Conductors | Part I, Chapter 2, 2-2.3 | 410.3 |
| Ground-Fault Circuit-Interrupter Protection for Personnel | Part I, Chapter 2, 2-2.4 | 410.4 |
| Outlet Devices | Part I, Chapter 2, 2-2.5 | 410.5 |
| Cord Connections | Part I, Chapter 2, 2-2.6 | 410.6 |
| Outside Branch Circuit, Feeder, and Service Conductors, 600 Volts, Nominal, or Less | Part I, Chapter 2, 2-3 | 410.7 |
| Services | Part I, Chapter 2, 2-4 | 410.8 |
| Overcurrent Protection | Part I, Chapter 2, 2-5 | 410.9 |
| Grounding | Part I, Chapter 2, 2-6 | 410.10 |
| Wiring Methods, Components, and Equipment for General Use | Part I, Chapter 3 | Article 420 |
| Wiring Methods | Part I, Chapter 3, 3-1 | 420.1 |
| Cabinets, Cutout Boxes, and Meter Socket Enclosures | Part I, Chapter 3, 3-2 | 420.2 |
| Position and Connection of Switches | Part I, Chapter 3, 3-3 | 420.3 |
| Switchboards and Panelboards | Part I, Chapter 3, 3-4 | 420.4 |
| Enclosures for Damp or Wet Locations | Part I, Chapter 3, 3-5 | 420.5 |
| Conductor Identification | Part I, Chapter 3, 3-6 | 420.6 |
| Flexible Cords and Cables, 600 Volts, Nominal, or Less | Part I, Chapter 3, 3-7 | 420.7 |
| Portable Cables Over 600 Volts, Nominal | Part I, Chapter 3, 3-8 | 420.8 |
| Fixture Wires | Part I, Chapter 3, 3-9 | 420.9 |
| Equipment for General Use | Part I, Chapter 3, 3-10 | 420.10 |
| Specific Purpose Equipment and Installations | Part I, Chapter 4 | Article 430 |
| Electric Signs and Outline Lighting | Part I, Chapter 4, 4-1 | 430.1 |
| Cranes and Hoists | Part I, Chapter 4, 4-2 | 430.2 |
| Elevators, Dumbwaiters, Escalators, Moving Walks, Wheelchair Lifts, and Stairway Chair Lifts | Part I, Chapter 4, 4-3 | 430.3 |
| Electric Welders—Disconnecting Means | Part I, Chapter 4, 4-4 | 430.4 |
| Information Technology Equipment—Disconnecting Means | Part I, Chapter 4, 4-5 | 430.5 |
| X-Ray Equipment | Part I, Chapter 4, 4-6 | 430.6 |
| Induction and Dielectric Heating | Part I, Chapter 4, 4-7 | 430.7 |
| Electrolytic Cells | Part I, Chapter 4, 4-8 | 430.8 |
| Electrically Driven or Controlled Irrigation Machines | Part I, Chapter 4, 4-9 | 430.9 |
| Swimming Pools, Fountains, and Similar Installations | Part I, Chapter 4, 4-10 | 430.10 |
| Carnivals, Circuses, Fairs, and Similar Events | Part I, Chapter 4, 4-11 | 430.11 |
| Hazardous (Classified) Locations, Class I, II, and III, Divisions 1 and 2 and Class I, Zones 0, 1, and 2 | Part I, Chapter 5 | Article 440 |
| Scope | Part I, Chapter 5, 5-1 | 440.1 |
| General | Part I, Chapter 5, 5-2 | 440.2 |
| Class I, Zone 0, 1, and 2 Locations | Part I, Chapter 5, 5-4 | 440.3 |
| Special Systems | Part I, Chapter 6 | Article 450 |
| Systems Over 600 Volts, Nominal | Part I, Chapter 6, 6-1 | 450.1 |
| Emergency Systems | Part I, Chapter 6, 6-2 | 450.2 |
| Class 1, Class 2, and Class 3 Remote Control, Signaling, and Power-Limited Circuits | Part I, Chapter 6, 6-3 | 450.3 |
| Fire Alarm Systems | Part I, Chapter 6, 6-4 | 450.4 |
| Communications Systems | Part I, Chapter 6, 6-5 | 450.5 |
| Solar Photovoltaic Systems | Part I, Chapter 6, 6-6 | 450.6 |
| Integrated Electrical Systems | Part I, Chapter 6, 6-7 | 450.7 |

*TABLE M.1 Continued*

| Topic | 2000 Edition | 2004 Edition |
|---|---|---|
| Safety-Related Work Practices | Part II | Chapter 1 |
| General | Part II, Chapter 1 | Article 110 |
| Scope | Part II, Chapter 1, 1-1 | 110.1 |
| Purpose. | Part II, Chapter 1, 1-2 | 110.2 |
| Responsibility | Part II, Chapter 1, 1-3 | 110.3 |
| Organization | Part II, Chapter 1, 1-4 | 110.5 |
| Training Requirements | Part II, Chapter 1, 1-5 | 110.6 |
| General Requirements for Electrical Work Practices | Part II, Chapter 2 | 110.1 |
| Working On or Near Electrical Conductors or Circuit Parts | Part II, Chapter 2, 2-1 | 130.1 |
| General | Part II, Chapter 2, 2-1.1 | 110.1 |
| Working On or Near Deenergized Electrical Conductors or Circuit Parts that Have Lockout/Tagout Devices Applied | Part II, Chapter 2, 2-1.2 | 120.2 |
| Approach Boundaries to Live Parts | Part II, Chapter 2, 2-1.3.4 | 130.2 |
| Table for Approach Boundaries to Live Parts for Shock Protection | Part II, Chapter 2, Table 2-1.3.4 | Table 130.2(B) |
| Work On or Near Uninsulated Overhead Lines | Part II, Chapter 2, 2-2 | 130.5 |
| Electrical Safety Program | Part II, Chapter 2, 2-3 | 110.7 |
| Conductive Articles Being Worn | Part II, Chapter 2, 2-3.5 | 130.6(D) |
| Conductive Materials, Tools, and Equipment Being Handled | Part II, Chapter 2, 2-3.6 | 130.6(E) |
| Insulated Tools and Equipment | Part II, Chapter 2, 2-3.7 | 130.7(D)(1) |
| Personal and Other Protective Equipment | Part II, Chapter 3 | 130.7 |
| Standards for Personal Protective Equipment | Part II, Chapter 3, 3-3.8 | 130.7(C)(8) |
| Table for Standards on Protective Equipment | Part II, Chapter 3, Table 3-3.8 | Table 130.7(C)(8) |
| Selection of Personal Protective Equipment | Part II, Chapter 3, 3-3.9 | 130.7(C)(9) |
| Personal Protective Equipment Required for Various Tasks | Part II, Chapter 3, 3-3.9.1 | 130.7(C)(9)(a) |
| Table for Hazard Risk Category Classifications | Part II, Chapter 3, Table 3-3.9.1 | Table 130.7(C)(9)(a) |
| Protective Clothing and Personal Protective Equipment Matrix | Part II, Chapter 3, 3-3.9.2 | 130.7(C)(10) |
| Table for Protective Clothing and Personal Protective Equipment (PPE) Matrix | Part II, Chapter 3, Table 3-3.9.2 | Table 130.7(C)(10) |
| Protective Clothing Characteristics | Part II, Chapter 3, 3-3.9.3 | 130.7(C)(11) |
| Table for Protective Clothing Characteristics | Part II, Chapter 3, Table 3-3.9.3 | Table 130.7(C)(11) |
| Factors in Selection of Protective Clothing | Part II, Chapter 3, 3-3.9.4 | 130.7(C)(12) |
| Arc Flash Protective Equipment | Part II, Chapter 3, 3-3.9.5 | 130.7(C)(13) |
| Other Protective Equipment | Part II, Chapter 3, 3-4 | 130.7(D) |
| Test Instruments and Equipment | Part II, Chapter 3, 3-4.10 | 110.9(A) |
| Standards for Other Protective Equipment | Part II, Chapter 3, 3-4.11 | 130.7(F) |
| Table for Standards on Other Protective Equipment | Part II, Chapter 3, Table 3-4.11 | Table 130.7(F) |
| Use of Specific Safety-Related Equipment and Work Practices | Part II, Chapter 4 | |
| Test Instruments and Equipment Use | Part II, Chapter 4, 4-1 | 110.9(A) |
| Lockout/Tagout Practices and Devices | Part II, Chapter 5 | 120.3 |
| Appendix for Limits of Approach | Part II, Appendix A | Annex C |
| Appendix for Sample Calculation of Flash Protection Boundary | Part II, Appendix B | Annex D |
| Appendix for Electrical Safety Program | Part II, Appendix C | Annex E |
| Appendix for Hazard/Risk Evaluation Procedure | Part II, Appendix D | Annex F |
| Appendix for Sample Lockout/Tagout Procedure | Part II, Appendix E | Annex G |
| Appendix for Simplified, Two-Category, Flame-Resistant (FR) Clothing System | Part II, Appendix F | Annex H |
| Safety-Related Maintenance Requirements | Part III | Chapter 2 |

*(continues)*

*TABLE M.1 Continued*

| Topic | 2000 Edition | 2004 Edition |
|---|---|---|
| Introduction | Part III, Chapter 1 | Article 200 |
| General Maintenance Requirements | Part III, Chapter 2 | Article 205 |
| Qualified Persons | Part III, Chapter 2, 2-1 | 205.1 |
| Single Line Diagram | Part III, Chapter 2, 2-2 | 205.2 |
| Spaces About Electrical Equipment | Part III, Chapter 2, 2-3 | 205.3 |
| Grounding and Bonding | Part III, Chapter 2, 2-4 | 205.4 |
| Guarding of Live Parts | Part III, Chapter 2, 2-5 | 205.5 |
| Safety Equipment | Part III, Chapter 2, 2-6 | 205.6 |
| Clear Spaces | Part III, Chapter 2, 2-7 | 205.7 |
| Identification of Components | Part III, Chapter 2, 2-8 | 205.8 |
| Warning Signs | Part III, Chapter 2, 2-9 | 205.9 |
| Identification of Circuits | Part III, Chapter 2, 2-10 | 205.10 |
| Single and Multiple Conductors and Cables | Part III, Chapter 2, 2-11 | 205.11 |
| Flexible Cords and Cables | Part III, Chapter 2, 2-12 | 205.12 |
| Substation, Switchgear Assemblies, Switchboards, Panelboards, Motor Control Centers, and Disconnect Switches | Part III, Chapter 3 | Article 210 |
| Enclosures | Part III, Chapter 3, 3-1 | 210.1 |
| Area Enclosures | Part III, Chapter 3, 3-2 | 210.2 |
| Conductors | Part III, Chapter 3, 3-3 | 210.3 |
| Insulation Integrity | Part III, Chapter 3, 3-4 | 210.4 |
| Protective Devices | Part III, Chapter 3, 3-5 | 210.5 |
| Premises Wiring | Part III, Chapter 4 | Article 215 |
| Covers for Wiring System Components | Part III, Chapter 4, 4-1 | 215.1 |
| Open Wiring Protection | Part III, Chapter 4, 4-2 | 215.2 |
| Raceways and Cable Trays | Part III, Chapter 4, 4-3 | 215.3 |
| Controller Equipment | Part III, Chapter 5 | Article 220 |
| Scope | Part III, Chapter 5, 5-1 | 220.1 |
| Protection and Control/Circuitry | Part III, Chapter 5, 5-2 | 220.2 |
| Fuses and Circuit Breakers | Part III, Chapter 6 | Article 225 |
| Fuses | Part III, Chapter 6, 6-1 | 225.1 |
| Molded-Case Circuit Breakers | Part III, Chapter 6, 6-2 | 225.2 |
| Circuit Breaker Testing | Part III, Chapter 6, 6-3 | 225.3 |
| Rotating Equipment | Part III, Chapter 7 | Article 230 |
| Terminal Boxes | Part III, Chapter 7, 7-1 | 230.1 |
| Guards, Barriers, and Access Plates | Part III, Chapter 7, 7-2 | 230.2 |
| Hazardous (Classified) Locations | Part III, Chapter 8 | Article 235 |
| Scope | Part III, Chapter 8, 8-1 | 235.1 |
| Maintenance Requirements for Hazardous (Classified) Locations | Part III, Chapter 8, 8-2 | 235.2 |
| Batteries and Battery Rooms | Part III, Chapter 9 | Article 240 |
| Ventilation | Part III, Chapter 9, 9-1 | 240.1 |
| Eye and Body Wash Apparatus | Part III, Chapter 9, 9-2 | 240.2 |
| Cell Flame Arresters and Cell Ventilation | Part III, Chapter 9, 9-3 | 240.3 |
| Portable Electric Tools and Equipment | Part III, Chapter 10 | Article 245 |
| Maintenance Requirements for Portable Tools and Equipment | Part III, Chapter 10, 10-1 | 245.1 |
| Personal Safety and Protective Equipment | Part III, Chapter 11 | Article 250 |
| Maintenance Requirements for Personal Safety and Protective Equipment | Part III, Chapter 11, 11-1 | 250.1 |

*TABLE M.1 Continued*

| Topic | 2000 Edition | 2004 Edition |
|---|---|---|
| Inspection and Testing of Protective Equipment and Protective Tools | Part III, Chapter 11, 11-2 | 250.2 |
| Safety Grounding Equipment | Part III, Chapter 11, 11-3 | 250.3 |
| Safety Requirements for Special Equipment | Part IV | Chapter 3 |
| Introduction | Part IV, Chapter 1 | Article 300 |
| Scope | Part IV, Chapter 1, 1-1 | 300.1 |
| Responsibility | Part IV, Chapter 1, 1-2 | 300.2 |
| Organization | Part IV, Chapter 1, 1-3 | 300.3 |
| Safety-Related Work Practices for Electrolytic Cells | Part IV, Chapter 2 | Article 310 |
| Scope | Part IV, Chapter 2, 2-1 | 310.1 |
| Definitions | Part IV, Chapter 2, 2-2 | 310.2 |
| Safety Training | Part IV, Chapter 2, 2-3 | 310.3 |
| Employee Training | Part IV, Chapter 2, 2-4 | 310.4 |
| Safeguarding of Employees in the Cell Line Working Zone | Part IV, Chapter 2, 2-5 | 310.5 |
| Portable Tools and Equipment | Part IV, Chapter 2, 2-6 | 310.6 |
| Appendix for Typical Application of Safeguards in the Cell Line Working Zone | Part IV, Chapter 2, Appendix A | Annex L |
| Safety Requirements Related to Batteries and Battery Rooms | Part IV, Chapter 3 | Article 320 |
| Scope | Part IV, Chapter 3, 3-1 | 320.1 |
| Definitions | Part IV, Chapter 3, 3-2 | 320.2 |
| Battery Connections | Part IV, Chapter 3, 3-3 | 320.3 |
| Installations of Batteries | Part IV, Chapter 3, 3-4 | 320.4 |
| Battery Room Requirements | Part IV, Chapter 3, 3-5 | 320.5 |
| Battery Enclosure Requirements | Part IV, Chapter 3, 3-6 | 320.6 |
| Protection | Part IV, Chapter 3, 3-7 | 320.7 |
| Personnel Protective Equipment | Part IV, Chapter 3, 3-8 | 320.8 |
| Tools and Equipment | Part IV, Chapter 3, 3-9 | 320.9 |
| Safety Related Work Practices for Use of Lasers | Part IV, Chapter 4 | Article 330 |
| Scope | Part IV, Chapter 4, 4-1 | 330.1 |
| Definitions | Part IV, Chapter 4, 4-2 | 330.2 |
| Safety Training | Part IV, Chapter 4, 4-3 | 330.3 |
| Safeguarding of Employees in the Laser Operating Area | Part IV, Chapter 4, 4-4 | 330.4 |
| Employee Responsibility | Part IV, Chapter 4, 4-5 | 330.5 |
| Safety Related Work Practices: Power Electronic Equipment | Part IV, Chapter 5 | Article 340 |
| Scope | Part IV, Chapter 5, 5-1 | 340.1 |
| Application | Part IV, Chapter 5, 5-2 | 340.3 |
| Reference Standards | Part IV, Chapter 5, 5-3 | 340.4 |
| Definition | Part IV, Chapter 5, 5-4 | 340.2 |
| Hazards Associated with Power Electronic Equipment | Part IV, Chapter 5, 5-5 | 340.5 |
| Hazards Associated with Power Electronic Equipment | Part IV, Chapter 5, 5-6 | 340.6 |
| Specific Measures for Personnel Safety | Part IV, Chapter 5, 5-7 | 340.7 |
| Appendix for Tables, Notes, and Charts [Reserved] | Appendix A | Annex A |
| Appendix for Referenced Publications | Appendix B | Annex B |

*TABLE M.2 Cross Reference from the 2004 edition of NFPA 70E to the 2000 edition of NFPA 70E*

| Topic | 2004 Edition | 2000 Edition |
|---|---|---|
| Introduction | | |
| Scope | 90.1 | I-1 |
| Standard Arrangement | 90.2 | I-1.3 |
| Safety-Related Work Practices | Chapter 1 | Part II |
| Definitions | Article 100 | I-2 |
| General Requirements for Electrical Safety-Related Work Practices | Article 110 | Part II, Chapter 1 |
| Scope | 110.1 | Part II, 1-1 |
| Purpose | 110.2 | Part II, 1-2 |
| Responsibility | 110.3 | Part II, 1-3 |
| Multiemployer Relationship | 110.4 | |
| Organization | 110.5 | Part II, 1-4 |
| Training Requirements | 110.6 | Part II, 1-5 |
| Electrical Safety Program | 110.7 | Part II, 2-3 |
| Working On or Near Electrical Conductors or Circuit Parts | 110.8 | Part II, 2-1 |
| Use of Equipment | 110.9 | Part II, 1-8 |
| Establishing an Electrically Safe Work Condition | Article 120 | Part II, Chapter 1 |
| Process of Achieving an Electrically Safe Work Condition | 120.1 | Part II, 2-1 |
| Working On or Near Deenergized Electrical Conductors or Circuit Parts That Have Lockout/Tagout Devices Applied | 120.2 | Part II, 2-2 |
| Lockout/Tagout Practices and Devices | 120.3 | Part II, 2-3 |
| Temporary Protective Grounding Equipment | 120.4 | Part II, 3-4.2 |
| Working On or Near Live Parts | Article 130 | Part II, Chapter 2 |
| Justification for Work | 130.1 | Part II, 2-1.1.1 |
| Energized Electrical Work Permit | 130.1(A) | New |
| Approach Boundaries to Live Parts | 130.2 | Part II, 2-1.3.4 |
| Approach to Live Parts | 130.2(B) | Part II, 2-1.3.3.2 |
| Table for Approach Boundaries to Live Parts for Shock Protection | Table 130.2(B) | Part II, Table 2-1.3.4 |
| Flash Hazard Analysis | 130.3 | Part II, 2-1.3.3 |
| Test Instruments and Equipment Use | 130.4 | Part II, 4-1 |
| Work On or Near Uninsulated Overhead Lines | 130.5 | Part II, 2-2 |
| Other Precautions for Personnel Activities | 130.6 | New |
| Personal and Other Protective Equipment | 130.7 | Part II, Chapter 3 |
| Standards for Personal Protective Equipment | 130.7(C)(8) | Part II, 3-3.8 |
| Table for Standards on Protective Equipment | Table 130.7(C)(8) | Part II, Table 3-3.8 |
| Selection of Personal Protective Equipment | 130.7(C)(9) | Part II, 3-3.9 |
| Personal Protective Equipment Required for Various Tasks | 130.7(C)(9)(a) | Part II, 3-3.9.1 |
| Table for Hazard/Risk Category Classifications | Table 130.7(C)(9)(a) | Part II, Table 3-3.9.1 |
| Protective Clothing and Personal Protective Equipment Matrix | 130.7(C)(10) | Part II, 3-3.9.2 |
| Table for Protective Clothing and Personal Protective Equipment (PPE) Matrix | Table 130.7(C)(10) | Part II, Table 3-3.9.2 |
| Protective Clothing Characteristics | 130.7(C)(11) | Part II, 3-3.9.3 |
| Protective Clothing Characteristics | Table 130.7(C)(12) | Part II, Table 3-3.9.3 |
| Factors in Selection of Protective Clothing | 130.7(C)(12) | Part II, 3-3.9.4 |
| Other Protective Equipment | 130.7(D) | Part II, 3-4 |
| Insulated Tools and Equipment | 130.7(D)(1) | Part II, 2-3.7 |
| Table for Standards on Other Protective Equipment | Table 130.7(F) | Part II, Table 3-4.11 |
| Safety-Related Maintenance Requirements | Chapter 2 | Part III |
| Introduction | Article 200 | Part III, Chapter 1 |
| General | 200.1 | Part III, 1-1 |
| General Maintenance Requirements | Article 205 | Part III, Chapter 2 |

*TABLE M.2 Continued*

| Topic | 2004 Edition | 2000 Edition |
|---|---|---|
| Qualified Persons | 205.1 | Part III, 2-1 |
| Single Line Diagram | 205.2 | Part III, 2-2 |
| Spaces About Electrical Equipment | 205.3 | Part III, 2-3 |
| Grounding and Bonding | 205.4 | Part III, 2-4 |
| Guarding of Live Parts | 205.5 | Part III, 2-5 |
| Safety Equipment | 205.6 | Part III, 2-6 |
| Clear Spaces | 205.7 | Part III, 2-7 |
| Identification of Components | 205.8 | Part III, 2-8 |
| Warning Signs | 205.9 | Part III, 2-9 |
| Identification of Circuits | 205.10 | Part III, 2-10 |
| Single and Multiple Conductors and Cables | 205.11 | Part III, 2-11 |
| Flexible Cords and Cables | 205.12 | Part III, 2-12 |
| Substations, Switchgear Assemblies, Switchboards, Panelboards, Motor Control Centers, and Disconnect Switches | Article 210 | Part III, Chapter 3 |
| Enclosures | 210.1 | Part III, 3-1 |
| Area Enclosures | 210.2 | Part III, 3-2 |
| Conductors | 210.3 | Part III, 3-3 |
| Insulation Integrity | 210.4 | Part III, 3-4 |
| Protective Devices | 210.5 | Part III, 3-5 |
| Premises Wiring | Article 215 | Part III, Chapter 4 |
| Covers for Wiring System Components | 215.1 | Part III, 4-1 |
| Open Wiring Protection | 215.2 | Part III, 4-2 |
| Raceways and Cable Trays | 215.3 | Part III, 4-3 |
| Controller Equipment | Article 220 | Part III, Chapter 5 |
| Scope | 220.1 | Part III, 5-1 |
| Protection and Control Circuitry | 220.2 | Part III, 5-2 |
| Fuses and Circuit Breakers | Article 225 | Part III, Chapter 6 |
| Fuses | 225.1 | Part III, 6-1 |
| Molded-Case Circuit Breakers | 225.2 | Part III, 6-2 |
| Circuit Breaker Testing | 225.3 | Part III, 6-3 |
| Rotating Equipment | Article 230 | Part III, Chapter 7 |
| Terminal Boxes | 230.1 | Part III, 7-1 |
| Guards, Barriers, and Access Plates | 230.2 | Part III, 7-2 |
| Hazardous (Classified) Locations | Article 235 | Part III, Chapter 8 |
| Scope | 235.1 | Part III, 8-1 |
| Maintenance Requirements for Hazardous (Classified) Locations | 235.2 | Part III, 8-2 |
| Batteries and Battery Rooms | Article 240 | Part III, Chapter 9 |
| Ventilation | 240.1 | Part III, 9-1 |
| Eye and Body Wash Apparatus | 240.2 | Part III, 9-2 |
| Cell Flame Arresters and Cell Ventilation | 240.3 | Part III, 9-3 |
| Portable Electric Tools and Equipment | Article 245 | Part III, Chapter 10 |
| Maintenance Requirements for Portable Tools and Equipment | 245.1 | Part III, 10-1 |
| Personal Safety and Protective Equipment | Article 250 | Part III, Chapter 11 |
| Maintenance Requirements for Personal Safety and Protective Equipment | 250.1 | Part III, 11-1 |
| Inspection and Testing of Protective Equipment and Protective Tools | 250.2 | Part III, 11-2 |
| Safety Grounding Equipment | 250.3 | Part III, 11-3 |
| Safety Requirements for Special Equipment | Chapter 3 | Part IV |
| Introduction | Article 300 | Part IV, Chapter 1 |
| Scope | 300.1 | Part IV, 1-1 |
| Responsibility | 300.2 | Part IV, 1-2 |

*(continues)*

*TABLE M.2 Continued*

| Topic | 2004 Edition | 2000 Edition |
|---|---|---|
| Organization | 300.3 | Part IV, 1-3 |
| Safety-Related Work Practices for Electrolytic Cells | Article 310 | Part IV, Chapter 2 |
| Scope | 310.1 | Part IV, 2-1 |
| Definitions | 310.2 | Part IV, 2-2 |
| Safety Training | 310.3 | Part IV, 2-3 |
| Employee Training | 310.4 | Part IV, 2-4 |
| Safeguarding of Employees in the Cell Line Working Zone | 310.5 | Part IV, 2-5 |
| Portable Tools and Equipment | 310.6 | Part IV, 2-6 |
| Safety Requirements Related to Batteries and Battery Rooms | Article 320 | Part IV, Chapter 3 |
| Scope | 320.1 | Part IV, 3-1 |
| Definitions | 320.2 | Part IV, 3-2 |
| Battery Connections | 320.3 | Part IV, 3-3 |
| Installations of Batteries | 320.4 | Part IV, 3-4 |
| Battery Room Requirements | 320.5 | Part IV, 3-5 |
| Battery Enclosure Requirements | 320.6 | Part IV, 3-6 |
| Protection | 320.7 | Part IV, 3-7 |
| Personnel Protective Equipment | 320.8 | Part IV, 3-8 |
| Tools and Equipment | 320.9 | Part IV, 3-9 |
| Safety Related Work Practices for Use of Lasers | Article 330 | Part IV, Chapter 4 |
| Scope | 330.1 | Part IV, 4-1 |
| Definitions | 330.2 | Part IV, 4-2 |
| Safety Training | 330.3 | Part IV, 4-3 |
| Safeguarding of Employees in the Laser Operating Area | 330.4 | Part IV, 4-4 |
| Employee Responsibility | 330.5 | Part IV, 4-5 |
| Safety Related Work Practices: Power Electronic Equipment | Article 340 | Part IV, Chapter 5 |
| Scope | 340.1 | Part IV, 5-1 |
| Definition | 340.2 | Part IV, 5-4 |
| Application | 340.3 | Part IV, 5-2 |
| Reference Standards | 340.4 | Part IV, 5-3 |
| Hazards Associated with Power Electronic Equipment | 340.5 | Part IV, 5-5 |
| Hazards Associated with Power Electronic Equipment | 340.6 | Part IV, 5-6 |
| Specific Measures for Personnel Safety | 340.7 | Part IV, 5-7 |
| Installation Safety Requirements | Chapter 4 | Part I |
| General Requirements for Electrical Installations | Article 400 | Part I |
| Scope | 400.1 | Part I, 1-1 |
| Approval | 400.2 | Part I, 1-2 |
| Examination, Identification, Installation, and Use of Equipment | 400.3 | Part I, 1-3 |
| Insulation Integrity | 400.4 | Part I, 1-3.3 |
| Interrupting Rating | 400.5 | Part I, 1-3.4 |
| Circuit Impedance and Other Characteristics | 400.6 | Part I, 1-3.5 |
| Deteriorating Agents | 400.7 | Part I, 1-3.6 |
| Mechanical Execution of Work | 400.8 | Part I. 1-3.7 |
| Mounting and Cooling of Equipment | 400.9 | Part I, 1-3.8 |
| Electrical Connections | 400.10 | Part I, 1-4 |
| Flash Protection | 400.11 | New |
| Arcing Parts | 400.12 | Part I, 1-5 |
| Marking | 400.13 | Part I, 1-6 |
| Identification of Disconnecting Means | 400.14 | Part I, 1-7 |
| 600 Volts, Nominal, or Less | Section B | Part I, 1-8 |
| Spaces about Electrical Equipment | 400.15 | Part I, 1-8.1 |
| Guarding of Live Parts | 400.16 | Part I, 1-8.2 |

*TABLE M.2 Continued*

| Topic | 2004 Edition | 2000 Edition |
|---|---|---|
| Over 600 Volts, Nominal | Section C | Part I, 1-9 |
| General | 400.17 | Part I, 1-9.1 |
| Enclosure for Electrical Installations | 400.18 | Part I, 1-9.2 |
| Table for Minimum Distance from Fence to Live Parts | Table 400.18 | Part I, Table 1-9.2 |
| Work Space about Equipment | 400.19 | Part I, 1-9.3 |
| Entrance and Access to Work Space | 400.20 | Part I, 1-9.4 |
| Work Space and Guarding | 400.21 | Part I, 1-9.5 |
| Wiring Design and Protection | Article 410 | Part I, Chapter 2 |
| Use and Identification of Grounded and Grounding Conductors | 410.1 | Part I, 2-1 |
| Branch Circuits | 410.2 | Part I, 2-2 |
| Identification of Ungrounded Conductors | 410.3 | Part I, 2-2.3 |
| Ground-Fault Circuit-Interrupter Protection for Personnel | 410.4 | Part I, 2-2.4 |
| Outlet Devices | 410.5 | Part I, 2-2.5 |
| Cord Connections | 410.6 | Part I, 2-2.6 |
| Outside Branch Circuit, Feeder, and Service Conductors, 600 Volts, Nominal, or Less | 410.7 | Part I, 2-3 |
| Services | 410.8 | Part I, 2-4 |
| Overcurrent Protection | 410.9 | Part I, 2-5 |
| Grounding | 410.10 | Part I, 2-6 |
| Wiring Methods, Components, and Equipment for General Use | Article 420 | Part I, Chapter 3 |
| Wiring Methods | 420.1 | Part I, 3-1 |
| Cabinets, Cutout Boxes, and Meter Socket Enclosures | 420.2 | Part I, 3-2 |
| Position and Connection of Switches | 420.3 | Part I, 3-3 |
| Switchboards and Panelboards | 420.4 | Part I, 3-4 |
| Enclosures for Damp or Wet Locations | 420.5 | Part I, 3-5 |
| Conductor Identification | 420.6 | Part I, 3-6 |
| Flexible Cords and Cables | 420.7 | Part I, 3-7 |
| Portable Cables Over 600 Volts, Nominal | 420.8 | Part I, 3-8 |
| Fixture Wires | 420.9 | Part I, 3-9 |
| Equipment for General Use | 420.10 | Part I, 3-10 |
| Specific Purpose Equipment and Installations | Article 430 | Part I, Chapter 4 |
| Electric Signs and Outline Lighting | 430.1 | Part I, 4-1 |
| Cranes and Hoists | 430.2 | Part I, 4-2 |
| Elevators, Dumbwaiters, Escalators, Moving Walks, Wheelchair Lifts, and Stairway Chair Lifts | 430.3 | Part I, 4-3 |
| Electric Welders–Disconnecting Means | 430.4 | Part I, 4-4 |
| Information Technology Equipment–Disconnecting Means | 430.5 | Part I, 4-5 |
| X-Ray Equipment | 430.6 | Part I, 4-6 |
| Induction and Dielectric Heating Equipment | 430.7 | Part I, 4-7 |
| Electrolytic Cells | 430.8 | Part I, 4-8 |
| Electrically Driven or Controlled Irrigation Machines | 430.9 | Part I, 4-9 |
| Swimming Pools, Fountains, and Similar Installations | 430.10 | Part I, 4-10 |
| Carnivals, Circuses, Fairs, and Similar Events | 430.11 | Part I, 4-11 |
| Hazardous (Classified) Locations, Class I, II, and III, Divisions 1 and 2 and Class I, Zones 0, 1, and 2 | Article 440 | Part I, Chapter 5 |
| Scope | 440.1 | Part I, 5-1 |
| General | 440.2 | Part I, 5-2 |
| Class I, Zone 0, 1, and 2 Locations | 440.3 | Part I, 5-4 |
| Special Systems | Article 450 | Part I, Chapter 6 |
| Systems Over 600 Volts, Nominal | 450.1 | Part I, 6-1 |
| Emergency Systems | 450.2 | Part I, 6-2 |

*(continues)*

*TABLE M.2 Continued*

| Topic | 2004 Edition | 2000 Edition |
| --- | --- | --- |
| Class 1, Class 2, and Class 3 Remote Control, Signaling, and Power-Limited Circuits | 450.3 | Part I, 6-3 |
| Fire Alarm Systems | 450.4 | Part I, 6-4 |
| Communications Systems | 450.5 | Part I, 6-5 |
| Solar Photovoltaic Systems | 450.6 | Part I, 6-6 |
| Integrated Electrical Systems | 450.7 | Part I, 6-7 |
| Tables, Notes, and Charts | Annex A | Appendix A |
| Referenced Publications | Annex B | Appendix A |
| Limits of Approach | Annex C | Part II, Appendix A |
| Sample Calculation of Flash Protection Boundary | Annex D | Part II, Appendix B |
| Electrical Safety Program | Annex E | Part II, Appendix C |
| Hazard/Risk Evaluation Procedure | Annex F | Part II, Appendix D |
| Annex G, Sample Lockout/Tagout Procedure | Annex G | Part II, Appendix D |
| Simplified, Two-Category, Flame-Resistant (FR) Clothing System | Annex H | Part II, Appendix F |
| Job Briefing and Planning Checklist | Annex I | New |
| Annex J, Energized Electrical Work Permit | Annex J | New |
| Annex K, General Categories of Electrical Hazards | Annex K | New |
| Appendix L, Typical Application of Safeguards in the Cell Line Working Zone | Annex L | Part IV, Chapter 2, Appendix A |
| Annex M, Cross-Reference Tables | Annex M | New |

# Index

-C-

**-D-**

-H-

-O-



-P-

**-Q-**



**-R-**